INTERNATIONAL SERIES OF MONOGRAPHS ON

NUCLEAR ENERGY

GENERAL EDITOR: J. V. DUNWORTH

Division XI: REACTOR OPERATIONAL PROBLEMS

Volume 1

REACTOR SAFEGUARDS

OTHER TITLES IN THE SERIES ON NUCLEAR ENERGY

Division I. ECONOMICS AND LAW

Vol. 1. ALLARDICE (Ed.)–*Atomic Power: An Appraisal*

Division II. NUCLEAR PHYSICS

Vol. 1. HUGHES–*Neutron Cross Sections*
Vol. 2. BRADLEY (Trans.)–*Physics of Nuclear Fission*
Vol. 3. *Soviet Reviews of Nuclear Science–Translation of "Atomnaya Energiya", Vol.* 3, *no.* 11 (1957).
Vol. 4. YIFTAH, OKRENT and MOLDAUER–*Fast Reactor Cross Sections*
Vol. 5. *Isobaric Nuclei Nos.* 1–6

Division IV. ISOTOPES AND RADIATION

Vol. 1. GROSHEV, LUTSENKO, DEMIDOV and PELEKHOV–*Atlas of Gamma-ray Spectra from Radiative Capture of Thermal Neutrons*

Division V. HEALTH PHYSICS

Vol. 1. HANDLOSER–*Health Physics Instrumentation*
Vol. 2. AMPHLET–*Radioactive Wastes: Treatment and Disposal*

Division VI. MEDICINE

Vol. 1. MEAD and HOWTON–*Radioisotope Studies of Fatty Acid Metabolism*

Division VII. REACTOR ENGINEERING

Vol. 1. KOMAROVSKII–*Shielding Materials for Nuclear Reactors*

Division X. REACTOR DESIGN PHYSICS

Vol. 1. LITTLER and RAFFLE–*An Introduction to Reactor Physics*
Vol. 2. PRICE, HORTON and SPINNEY–*Radiation Shielding*
Vol. 3. GALANIN–*Thermal Reactor Theory*
Vol. 4. DRESNER–*Resonance Absorption in Nuclear Reactors*
Vol. 5. THIE–*Heavy Water Exponential Experiments using* ThO_2 *and* UO_2

Division XIV. PLASMA PHYSICS AND THERMONUCLEAR RESEARCH

Vol. 1. SIMON–*An Introduction to Thermonuclear Research*

REACTOR SAFEGUARDS

by

CHARLES R. RUSSELL
Ph.D., P.E.

PERGAMON PRESS
OXFORD · LONDON · NEW YORK · PARIS
1962

PERGAMON PRESS LTD.
Headington Hill Hall, Oxford
4 & 5 Fitzroy Square, London W.1

PERGAMON PRESS INC.
122 East 55th Street, New York 22, N.Y.

GAUTHIER-VILLARS
55 Quai des Grands-Augustins, Paris 6

PERGAMON PRESS G.m.b.H.
Kaiserstrasse 75, Frankfurt am Main

Library of Congress Card Number 62-8704

Set in Times New Roman 10 on 12 pt. by Santype Limited, Salisbury and printed in Great Britain by Cox & Wyman Ltd., London, Fakenham and Reading

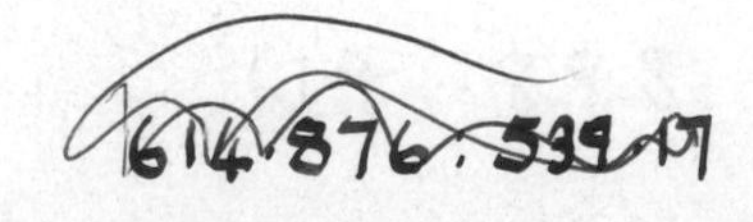

CONTENTS

PREFACE

THIS book is intended to provide information for all who are interested in the subject of reactor safeguards, and therefore much of the material is descriptive although some sections are written for the engineer or physicist directly concerned with hazards analysis or site selection problems. Such detailed sections can be omitted without losing an essential understanding about reactor safeguards from the other sections of this book.

The author wishes to acknowledge the cooperation by the Industrial Information Branch of the United States Atomic Energy Commission in granting approval to use material from unclassified AEC reports and in providing a copy of the unpublished manuscript "AEC Safety Monograph" prepared under an AEC contract by Walter Brooks and Harry Soodak of the Nuclear Development Corporation of America. Extensive material particularly for Chapters III, IV, and VII was drawn from this source. Permission was received to use material from certain documents published by other organizations including the Publications Board of the United Nations, and Atomic Energy of Canada, Limited. The Air Force Institute of Technology, Argonne National Laboratory, U.S. Department of Commerce, Hanford Atomic Products Operation, and other organizations have all been generous in providing material.

However, any opinions which may be expressed or implied in the material presented in this publication are those of the author and are not necessarily the opinions of any other organization. In addition, whereas the author has endeavored to make all content as up-to-date and as factual as possible, the author neither is responsible nor accepts responsibility for the safety of personnel following procedures described herein or for losses and damages which may arise as a result of errors and omissions in the presented material. Neither does the author represent that the performance of any work or effort in accordance with the material or techniques referred to will produce the results herein described. A conscientious effort has been made to provide adequate references to original documents. These original works should be consulted whenever their content is involved.

CHARLES R. RUSSELL

FOREWORD

WHILE serving as Secretary of the Reactor Safeguards Committee and its successor, the Advisory Committee on Reactor Safeguards, it was the author's privilege to observe from an especially advantageous position the considerations given unprecedented problems of reactor safeguards which came as the development of atomic energy progressed from production reactors at remote sites to privately owned nuclear power stations near centers of population. Many situations, which have long since become commonplace, were originally controversial and recommendations on adequate safeguards were arrived at with much less information than is now available for hazards evaluation. Fortunately the dedicated individuals who served on the committees had the ability to penetrate the confusion which sometimes surrounded these new problems and always to arrive at recommendations which ensured the public safety and yet permitted progress in reactor development.

The rapid expansion of the atomic energy programs with several dozens of reactors in operation has led to the development of well-organized procedures for hazards evaluation, licensing and inspection of nuclear reactors as required by regulations. Extensive experimental programs are in progress to obtain needed information on reactor behaviour and to develop more safeguards. To date experience with reactor operations has been that they are remarkably free from major accidents which might cause exposure of the public to radiation hazards. It is hoped that this compilation of information which has been accumulated over several years and used for the preparation of lectures in reactor safeguards at the North Carolina State College, may contribute to the continuation of this record for safe reactor operations.

CHAPTER 1

INTRODUCTION

THE development of nuclear reactors has established a remarkable record for safe operation despite the unprecedented magnitude and nature of the hazards and the war-time urgency with which much of the work was done. The talented individuals who were responsible for the atomic energy program clearly foresaw the hazards presented by highly radioactive materials accumulating in nuclear fuel elements in quantities many orders of magnitude greater than the equivalent of all the radium produced up to that time, and under conditions which could result in temperatures ranging beyond the boiling point of any material within time intervals shorter than human reaction times. Using the information available in the 1942 period on the fission process and reactor kinetics, large installations were designed and sites suitable for nuclear work of all types, ranging from laboratories to large manufacturing operations, were selected with a judgment which experience since that time has proved to be excellent. These early studies and discussions form the background for our present policies on reactor safeguards although information is being continually augmented by experience and extensive experimentation and studies. These programs are now leading to international standards for the design and safe operation of nuclear installations.

RADIATION HAZARDS

THE discovery and development of sources of ionizing radiation has not always been attended by the record for human safety that has been possible in reactor development. Soon after the discovery of X-rays in Germany in 1895, it was found that biological damage can be done by radiations which cannot be seen or otherwise sensed.[1] Following the separation of radium from uranium ores in 1898, there were reports of skin ulcers from radiation burns. The first death due to exposure to ionizing radiation occurred in 1901 and the following year there was a diagnosis of a radiation cancer. Soon this was followed by cases of leukemia and anemia. It has been estimated that there were more than 100 fatalities due to exposure to ionizing radiation among early experimenters with X-rays and radium. The insidiousness of radioactive materials ingested and deposited in the body was learned from the experience in the radium dial industry, particularly during the First World War, when there were few precautions taken in working with radium paints. After a period of 10 to 20 years many unfortunate people developed cancer or anemia or suffered progressive degeneration of the bone structure due to ingested radium. Much is now known about the effects of external exposure to ionizing radiation and internal exposure to radioactive materials as a result of the great effort which has gone into the atomic energy program and of all environmental hazards to which man is exposed, there is probably the most information available on radiation hazards. However, as information and experience accumulate, the estimates of the amounts of radiation exposure considered safe are being regularly decreased as indicated in the following table:

Table I

Permissible Occupational Exposure Levels[2]

	Level (rem/year)
1956	5
1950	15
1934	60
Prior to 1934	100

The early exposure levels were based on evidence of damage to body cells whereas the recent reductions in exposure levels result from evidence of genetic effects distributed over future generations up to perhaps 50 in number. It is well known that man is continually subjected to ionizing radiation in the form of cosmic rays from unknown sources in space and from natural and man-made radioactive materials around us in the food we eat, the water we drink and in building materials. For example, a brick or stone building subjects the inhabitants to five times more radiation exposure from natural radioactive materials than a wood structure. Also there is a great variation between places in the amount of natural radioactivity and in some areas the established exposure limits are exceeded by natural radioactivity without apparent effects on the inhabitants. Man lives under conditions of continuous exposure to background radiation. The amount of additional exposure to which he should be routinely subjected is a matter of judgment in weighing the benefits of atomic energy against possible damages in this and successive generations. However, it is even more clearly understood today than when the first sustaining chain reaction was attained, that the human damages, that could result from the release of radioactive materials from a reactor into the environs of a populated area, are so great that every reasonable precaution must be taken to prevent such an event.

CHICAGO PILES 1, 2, AND 3

The first nuclear reactor was constructed during November, 1942, in a squash court under the West Stands of Stagg Field on the University of

Chicago campus. This was a thermal reactor in which graphite was used to reduce the energy of neutrons causing other atoms of uranium fuel to fission. Layers of graphite bricks were piled on a timber framework with lumps of uranium metal or oxide at alternate corners of the squares to form a cubic lattice and the assembly was to have been spherical. Three slots near the center of the structure were provided for control and safety rods of neutron absorbing materials, which were strips of cadmium or boron steel. Cadmium

FIG. 1. View of west end of Stagg Field, University of Chicago location of CP–1, the world's first atomic pile. (Argonne National Laboratory photograph.)

strips were used for manual control of the reactor and, in addition, there were two safety rods and later an automatic control. Boron trifluoride neutron detectors and gamma-ray ionization chambers were placed in and around the reactor. Foils (indium) were used for comparison of induced activity with the neutron counters.

The kinetic behavior of the reactor was predicted with reasonable accuracy. Measurements had been made and verified of the yield and time delay of the more important delayed neutrons from the fissioning of uranium-235 and it did appear, if the excess of neutrons produced in each generation was kept well below the fractional yield of delayed neutrons, that the reactor could be easily controlled manually. Long before reaching the critical layer, cadmium strips were inserted in the core during the addition of material and these strips were carefully removed once each day for a measurement of neutron multiplication. The design of the core was necessarily conservative, and

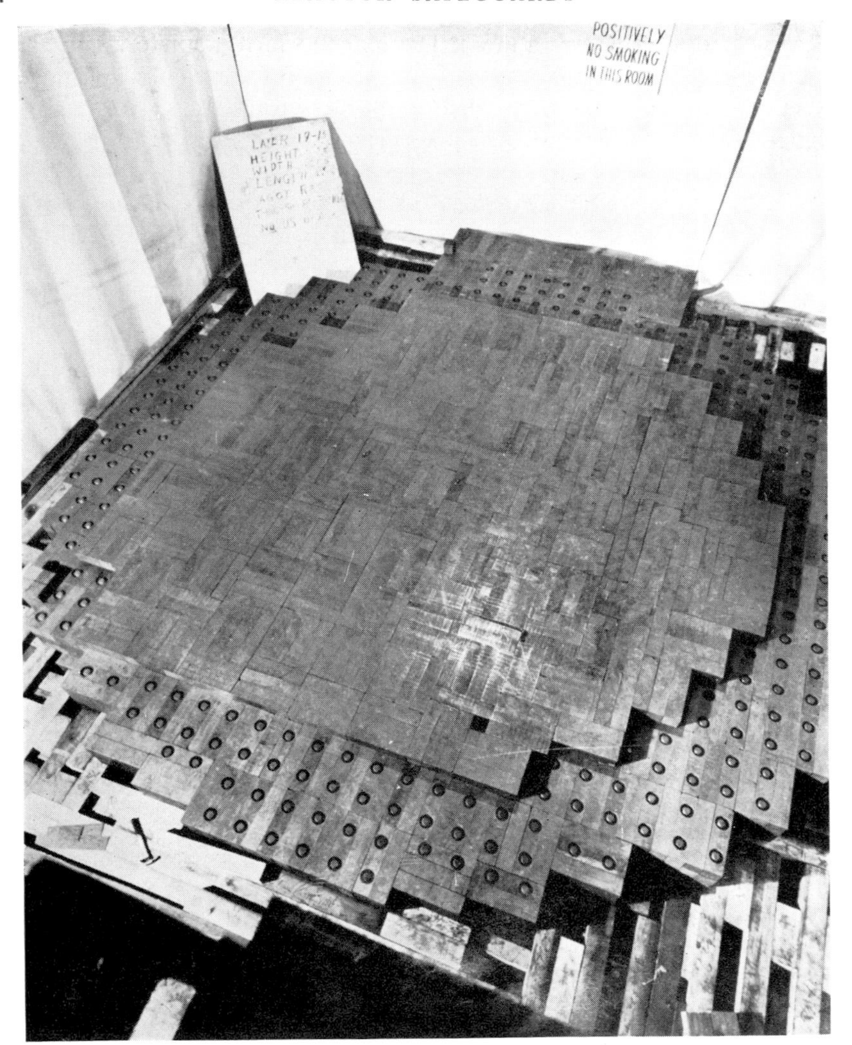

FIG. 2. CP–1 graphite lattice. (Argonne National Laboratory photograph.)

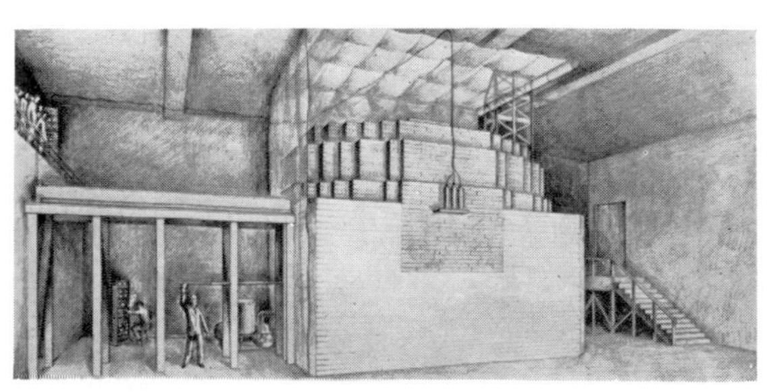

FIG. 3. Artist's sketch of CP–1. (Argonne National Laboratory photograph.)

therefore, criticality was reached before a completely spherical shape had been attained in the process of adding layers and measuring neutron multiplication.

A self-sustaining chain reaction was achieved in this reactor on 2 December, 1942, and it was operated initially at a power level of $\frac{1}{2}$ W. The operation of the reactor was quite sluggish and for a movement of a control rod of 1 cm out from the critical position, some four hours were required for the power of the reactor to double. The power level was increased only to about 200 W. and it was not considered safe to operate above this level because of radiation to personnel in the vicinity.

FIG. 4. CP–2 Reactor, Rebuilt West Stands Uranium–Graphite Reactor. (Argonne National Laboratory photograph.)

Although the operation of this reactor was innocuous except for some direct radiation hazards due to a lack of adequate shielding, it was decided later that month that CP-1 should be dismantled and rebuilt in the Palos Hills Forest Preserve at the location known as Site A. This was some 25 miles southwest of downtown Chicago and 17 miles northwest of Joliet, Illinois, near the present site of the Argonne National Laboratory. Site A was originally selected for some of the chemical engineering pilot plants for the Manhattan District program. However, the location was later considered to be too near major centers of population for large scale operations and the use of the site was limited to laboratory work including the reconstruction of the graphite reactor with adequate shielding and later the heavy water moderated and cooled research reactor, CP-3.

OAK RIDGE

A site was recommended in July, 1942, for all of the Manhattan District production plants in the sparsely settled eastern Tennessee countryside some 25 miles west of Knoxville, where there was isolation from large centers of population with a good supply of water and in an area easily accessible by rail and motor transportation. The 58,800 acres originally acquired were

FIG. 5. CP–3 heavy water reactor. (Argonne National Laboratory photograph.)

large enough to accommodate several huge plants in flat areas separated by wooded ridges, although it was later decided that this area was not adequate for the large plutonium production reactors and associated chemical processing plants. Additional acreage has been purchased since 1942, while other land has been sold or relinquished, and the present site with a greatest length of about 17 miles and a greatest width of about 9 miles containing 54,000 acres is bounded on the east, southeast and southwest by the Clinch River which serves as a boundary.

The intermediate sized air-cooled graphite reactor with pilot plants for the plutonium separation process were constructed, starting in February, 1943, at the X-10 site some 10 miles from the town of Oak Ridge. This reactor was originally designed to operate at 500 kW of heat but was quickly raised to several times that power level. The first significant quantities of plutonium for research, leading to the building of the Hanford production plant, were made in the X-10 reactor, which has served long and usefully in research and the production of isotopes. Several other reactors have been built and operated at the X-10 site and in the adjoining valley including the new Oak Ridge Research Reactor, LITR, the Bulk Shielding Reactor, the Tower Shielding Reactor and Homogeneous Reactor Experiments.

FIG. 6. Top of CP–3 showing fuel rods, central thimble and connections. (Argonne National Laboratory photograph.)

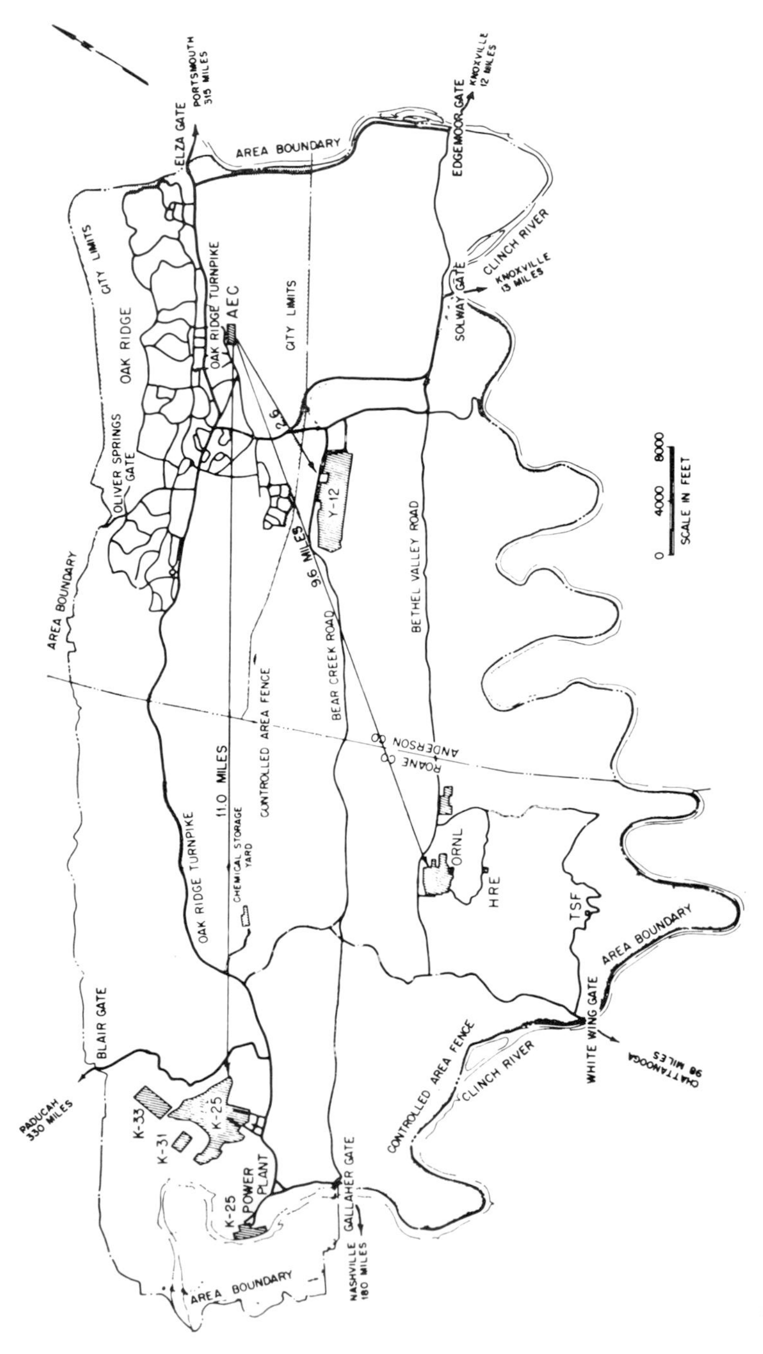

FIG. 7. Oak Ridge site.

HANFORD

The criteria for a site for the plant to manufacture plutonium were established at a meeting on 14 December, 1942, based upon the preliminary work that had been done on the production and separation of plutonium up to that time. The possibility of explosions of catastrophic proportions, and the possibility of releasing into the atmosphere intensely radioactive gases, dictated the selection of a site of sufficient area to permit the several manufacturing areas to be separated by distances of miles. The size of the manufacturing site required was based upon the tentative decision to construct six

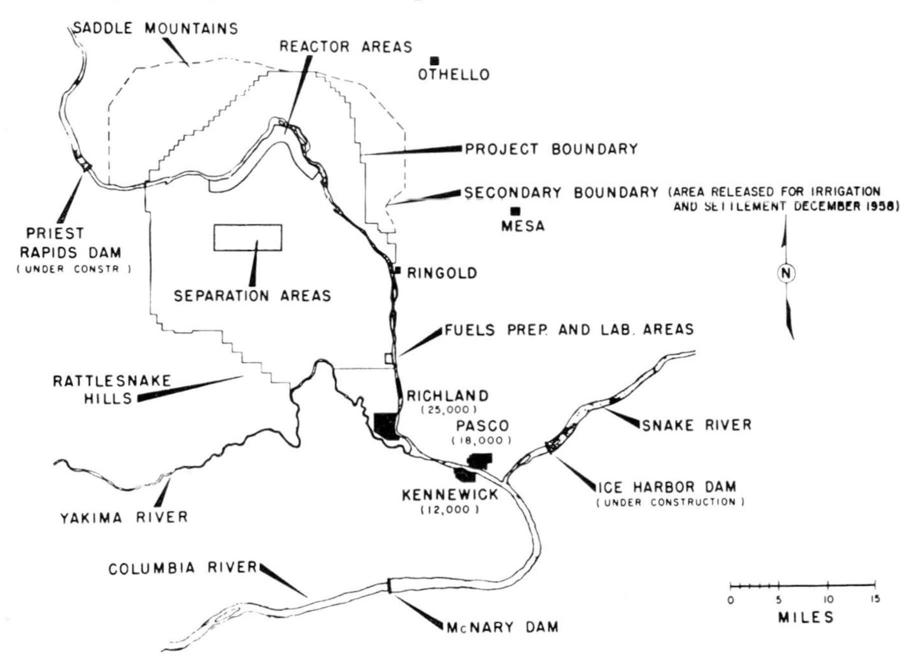

FIG. 8. Hanford plant environs.

primary manufacturing areas separated from each other and from the nearest primary manufacturing areas by not less than 4 miles. Moreover, it was decided that a boundary outside the manufacturing areas should be established not less than 6 miles distance from any area. It was agreed that complete control would have to be exercised over not only the manufacturing site of 12 miles by 16 miles, but also the 6 mile strip surrounding the area, with all persons other than plant employees excluded, and that residential occupancy should be prohibited within a larger area of about 44 by 48 miles, centred on the manufacturing site.

The military importance of the project and the potential hazards were considered to require the selection of an area of small existing population isolated to the maximum extent from any centers of population. It was agreed, therefore, that no large town or city should be less than 20 miles distance from the nearest manufacturing area and that housing for workers should be not less than 10 miles, preferably not less than 20 miles, from the nearest manufacturing area. Although access to railroad and main highway facilities was an essential requirement, it was determined, because of the possible hazards involved and the necessity of maintaining security, that no main line railroad or public highway should be less than 10 miles from the manufacturing areas.

A large and dependable supply of relatively pure water of reasonably low temperature was necessary throughout the year in order to dissipate the enormous quantities of heat released in the reactors. It was estimated that the minimum requirement would be approximately 25,000 gallons per minute. The distance the water would have to be pumped and the pumping head involved were also considerations.

A dependable source of electric power capable of supplying at least 100,000 kilowatts was necessary for driving the pumps for circulating the large quantities of cooling water through the reactors.

FIG. 9. Hanford production area. (Reproduced by permission of United States Atomic Energy Commission.)

A party of three men left Washington, D.C., on 16 December, 1942, to make the field investigations which led to the selection of the region near Hanford and White Bluffs, in southeastern Washington as most nearly meeting the site requirements. The site was acquired in the early months of 1943, and ground was broken for the construction camp in April 1943. The primary area of 429 square miles is bounded on the north and east by the Columbia River and on the south and west by a perimeter fence, with a supplementary primary area of 138 square miles. Some 164 square miles on the north side of the river, known as the Wahluke Slope, was held as a secondary area until some of this area was released for irrigation and settlement in December, 1958. At that time it was considered that the continuing improvements in the design and operation of the reactors had substantially reduced the probability of a serious accident, thereby making possible the release of an area of 105,500 acres. The present site is approximately a rectangle 30 miles north and south and 25 miles east and west.

Eight plutonium-producing reactors are located along the river in the northern part of the primary area and the various chemical separation plants are on the interior plateau about 10 miles from the river. In the southeast

FIG. 10. Hanford production area. (Reproduced by permission United States Atomic Energy Commission.)

corner of the project are the major research laboratories and fuel preparation facilities. A few miles south of the reserved area is the city of Richland. The nearest other community of any considerable size is Yakima, about 40 miles to the west, with a population of 30,000 and the larger cities of Seattle, Tacoma, Portland, and Spokane lie well outside a 100 mile radius.

REACTOR SAFEGUARDS COMMITTEE

Although considerable effort had been put into the evaluation of the possible hazards from the Hanford production reactors, based on the best information available at the time, and remarkably good judgment was shown in the selection of the location and size of the site suitable for such operations, shortly after the war the need for a careful re-evaluation of the situation at Hanford was recognized. There were questions about the safety of continuing the operations of the water-cooled graphite reactors and also there were influential groups seeking the release of some of the government controlled area, particularly the land which could be irrigated as part of the Coulee Dam irrigation project. There were other problems which at the time were controversial such as the proposed location of the Brookhaven Research Reactor on Long Island and the Knolls Intermediate Breeder Reactor. Soon after the transfer of the Atomic Energy Program from the Army's Manhattan Engineering District to the Atomic Energy Commission, the several requirements for expert advice and consultation on reactor safety problems led to plans for a committee of experts in the various areas of science and engineering related to reactor hazards evaluation. In 1947, invitations were extended to a group of distinguished contributors to the war-time atomic energy project to assist in this evaluation and in early 1948, the first meeting of the Reactor Safeguards Committee was held. The initial membership included Richard P. Feynman, Joseph W. Kennedy, Edward Teller (Chairman) and John A. Wheeler. George L. Weil served as the Atomic Energy Commission representative. Other members added after the first meeting included Manson Benedict, Benjamin G. Holzman and Abel Wolman.

EXCLUSION DISTANCE FORMULA

The initial studies of the possible hazards to off-site personnel led to the derivation of a simplified formula relating reactor power to distance at which a given exposure would be received from fission products released into the atmosphere in the event of a reactor catastrophe.[6] This formula is of interest because it illustrates the nature of the problem and the assumptions involved in making such exposure estimates. Also this formula has been widely used as an approximate "rule of thumb" in the past for estimating hazards and in making some important decisions on site requirements.

It was assumed that the reactor had operated for a sufficiently long time so that fission products had accumulated to approximately the equilibrium value.

This assumption is conservative by about 30 per cent for the fission products in a reactor that has operated for 100 days as compared with the infinite limit and variations of this magnitude are not significant when compared to the possible changes in atmospheric diffusion and wind conditions. The rate of release of ionizing radiation from these fission products decreases with time according to the negative one-fifth power of time. In an operating reactor the fission products of long half-life accumulate and, therefore, the composition and rate of decay differ from fresh fission products formed in a power burst such as an exploding atomic weapon or a power excursion in a criticality experiment. For such fission products the rate of decay of activity is much more rapid, being proportional to the negative six-fifths power of time.

It has been observed that a cloud of material released into the atmosphere will disperse by diffusion through the air so that the cloud diameter is approximately proportional to one-seventh the distance traveled from the source. This rate of spread can vary greatly with meteorological conditions; however, this value of one-seventh is a typical value. The cloud may increase in height with distance traveled or may be limited in height by an overhead layer of air called an inversion layer. This inversion condition may prevail some seasons of the year and limit vertical mixing.

Now the total quantity of fission products in the reactor is proportional for long time operation to the power of the reactor. It was assumed that 50 per cent of the radioactive material escaped into the atmosphere as the result of overheating of the fuel element with possibly a fire in the core contributing to this dispersal. By using the relations for fission product activity decay, the amount of ionizing radiation as a function of time after release was calculated. Since the dimensions of the cloud were known from the distance traveled down wind (one-seventh the travel distance horizontally by 1500 ft vertically) the concentration of radioactive material and the radiation released per unit volume of air were determined.

The range of beta radiation in the air is of the order of 5 ft, whereas the range of gamma radiation from fission products is some 1,000 ft before being greatly attenuated. After the cloud has traveled several miles, its dimensions are much greater than the range of beta and gamma radiation and, therefore, near the center of the cloud the rate of energy absorption in a unit volume of air is about equal to the rate of energy release in this same volume. Since the rate of exposure is determined by the rate of absorption of ionizing radiation per unit volume, the rate of exposure was calculated.

The exposure time was determined by the time required for the cloud to pass a point. Since the formula was derived as a means for estimating exclusion distances for the public outside the controlled area around the reactor, an estimate of the time required to notify people outside the reactor site and evacuate these people from the area was needed. It appeared that 3 hr was a reasonable value based on experience in evacuating people from the areas in floods and similar situations. Using this time for the travel of the cloud from the reactor to the site boundary, the wind velocity corresponding to this cloud travel time was calculated. It can be shown that the exposure decreases with increasing wind velocity. Now having the wind speed and hence the time for a cloud of known dimension to pass a given point, the total exposure from the overhead cloud was calculated.

For any selected value of direct exposure to this cloud of fission products moving across the land, it was then possible to calculate a minimum distance corresponding to this exposure. At an exposure of 300 roentgen it is expected that there will be acute sickness in all cases and some fatalities; thus it is clear that this level of exposure is very serious, although it would be expected that the majority of the people thus exposed would survive but with some effects on their life-span. Taking 300 r as the exposure level at the boundary of the exclusion area, the exclusion distance formula became:

$$R \text{ (exclusion distance in miles from reactor)} = 0.01\sqrt{\text{(kilowatts reactor heat power)}}$$

$$R = 0.01\sqrt{P}$$

The derivation of this formula is presented in more detail in Appendix B, along with estimates by the Reactor Safeguards Committee of the deposition and exposure from fission products on the ground (Appendix C) and of the energy developed in an exploding reactor (Appendix D).

The exposure from inhalation of radioactive material from the passing cloud was considered but for acute exposure the distances involved were of the same order as for direct exposure and, therefore, the inhalation problem was not included in the above derivation, although inhalation was recognized as an important problem in reactor hazards evaluation as is also the case for ground deposition. In some situations these may be the controlling factors. Also it should be noted from the above derivation that any people just outside the exclusion area would not be safe from a significant exposure unless evacuated and there is, therefore, a secondary zone outside the exclusion area where there could be a serious exposure problem.

In reviewing the problems presented to the Reactor Safeguards Committee, the members developed a general philosophy on reactor hazards from which has grown, as additional information became available, the current policies. The considerations given various aspects of the problems by the Committee are frequently referred to and it is of interest to review some of the analyses that were made.

Only in certain unusual types of reactors did the Reactor Safeguards Committee find evidence for the possibility of a nuclear explosion and then of a much smaller magnitude than of an atomic bomb. Such an explosion in a reactor might approach 1 ton of TNT equivalent as compared to the 20,000 tons of TNT equivalent in the Hiroshima bomb. Since considerable ingenuity was required to devise the systems required for detonation of a bomb, it would not be expected that the necessary conditions would exist for such an occurrence. However, there may be conditions caused by earthquakes, bombing, or more probably sabotage, which could expose the reactor to the danger of an uncontrolled chain reaction through dislocation of reactor parts, loss of power for auxiliaries, or loss of coolant. In every case the safety system, which is incorporated into the reactor, is expected to shut down the reactor in time to prevent a serious casualty. It was in the failure of the safety measures themselves that the Committee considered the real danger to life, since a combination of material failures could conceivably leave the reactor with no effective check on neutron multiplication so that the danger of an explosion or at least the destruction of the reactor could exist.

The magnitude of the blast that could be expected from such a casualty would depend on the time rate of increase of the reactor power that resulted and the time that elapsed before the reactor shut itself off. The reactor would eventually shut itself off because of the increase in size caused by the internal pressures that could be built up. In this process of reactor power increase, the time required for the number of neutrons and therefore the power output to increase is of significance. This time depends on two factors: (1) the average life of a neutron in the reactor, and (2) the rate of reproduction of the neutrons.

The average lifetime of the neutrons in a reactor depends, in part, on the energy of the neutrons causing the majority of the fissions in the reactor. In a fast reactor, the neutrons are still at high energy when they multiply, whereas in intermediate or thermal reactors, the majority of the neutrons must be slowed down to lower energies before the nuclear reaction takes place. In these latter reactors, the dangers from an accidental increase in the reactivity are less because of the necessary delay in the multiplication process; however, the delayed neutrons are as effective in a fast reactor as in intermediate and thermal reactors in limiting the rate of power increase for the conditions that are controlled by delayed neutrons.

When the rate of neutron reproduction is just equal to the rate of consumption, the reactivity is said to be unity and since the neutrons would not increase in number the average power of the reactor remains constant. If the reactivity is greater than unity, the number of neutrons and the power increase exponentially with time. The Committee, therefore, carefully investigated for each reactor the maximum reactivity that might be obtained under all conditions. The reactivity may also be a function of temperature and, if the neutron reproduction rate increases with temperature, and instability exists which might lead to an uncontrolled reaction. The question of temperature stability is further complicated by the possible selective heating of the fissionable materials. In fact, in a sudden power burst, the fuel elements may be heated to a considerably higher temperature than the surrounding coolant and moderator unless they are intimately mixed with the fissionable materials.

The physical dimensions of a reactor were also considered to have a bearing on accidental changes in reactivity of the assembly and, therefore, on safety. If the reactor parts are arranged in a small volume, there will be a greater change in reactivity corresponding to a given displacement of the parts than if the reactor were larger. Such displacements from mechanical reasons or thermal expansion, and through the action of earthquakes or bombing were studied.

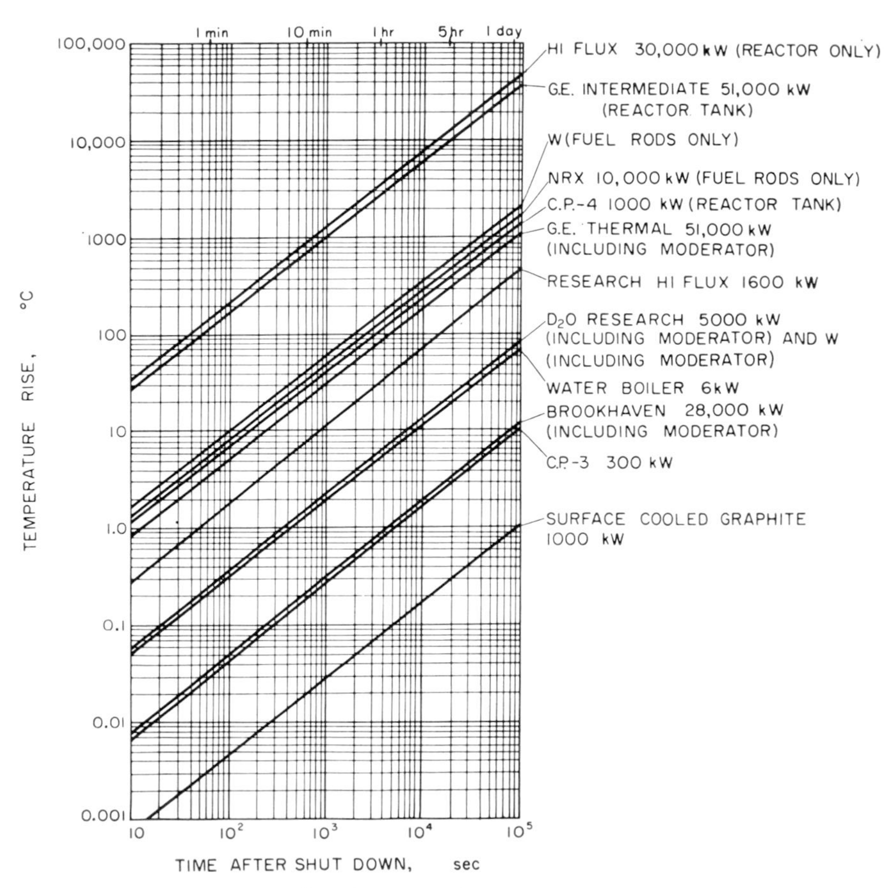

FIG. 11. Temperature rise of reactors after shut-down following loss of coolant. (Reproduced from Report Wash–3 (Rev.) by permission of the United States Atomic Energy Commission.)

Careful attention was given to the design of the control system and particularly to the limits on the speed at which the controls might be withdrawn by any means whatsoever since it was recognized that a well-informed saboteur might be able to cause large changes in reactivity, if special precautions were not taken. Also attention was given to the possibility of accidental or deliberate blocking of the control rod entrance holes or passages.

Another factor considered important in determining the safety of the reactor is the physical size, as it determines the heat capacity of the reactor parts. In a large reactor, the time after an accident before the inner pressure increase will scatter the reactor assembly will be greater and a greater heat release will result. Also the parts which are scattered will have acquired a greater velocity in a large reactor because they will have moved further from their original position before the reactor stops and the total impulse will be greater.

The behaviour of a reactor when the flow of coolant stopped was also studied since this is a casualty that is difficult to rule out of consideration completely. If the coolant is subject to boiling, as for instance water, the entire coolant is liable to be expelled from the reactor. This may either increase or decrease reactivity depending on the reactor design. Even if the safety system operates in the case of loss of coolant flow and the reactor is shut down, there remains a danger of damage to the reactor from the heat evolved from the fission products and the action of delayed neutrons after shut-down. The temperature rises of a number of reactors were calculated as a function of time after shut-down assuming there was no coolant flow and the curves, Fig. 11, indicate that reactors of large volume and heat capacity, such as Hanford and Brookhaven, which operate at relatively great power and, therefore, contain very large quantities of radioactivity, are nevertheless, from this point of view, quite safe, since the large mass of graphite moderator and natural uranium fuel provide a large heat capacity. This statement assumed the timely operation of some emergency cooling systems. The high flux reactors with fuel elements of low heat capacity such as the Materials Testing Reactor were found to have temperature rises of the order of 1000°C in the first 10 min after shut-down with no coolant and such reactors are the most dangerous in this respect by a considerable margin.

The effects of fire on a reactor were also considered. An ordinary fire which might involve say the graphite moderator of a reactor, would also release radioactive gases or smoke. An even greater danger would arise if the fire were sufficient to vaporize or melt the main body of the reactor with the release of all its fission products. Sodium and sodium–potassium alloy are also inflammable, particularly in the presence of water, and since sodium becomes radioactive under neutron irradiation, the results of such a fire would be a cloud of smoke which might be dangerously radioactive.

The potential hazards associated with the chemical plants and radioactive waste storage areas that might be part of a reactor installation were also considered by the Reactor Safeguards Committee. Explosions might occur within the chemical plant due to maloperation, or damage to both plant and storage areas might be caused by sabotage, bombing, or earthquakes. Some of these accidents could result in the release of radioactive material into the air. Since the quantities of active material could equal or exceed that stored in a reactor which has been operating for any length of time, this hazard was considered to be of considerable magnitude.

Each reactor usually has special design features that might effect hazards and, therefore, it was considered by the Committee that each reactor must be considered as a separate problem in hazards evaluation. Since the accumulation of environmental and design data and its analysis with regard to reactor safety requires extensive studies in several fields of science and engineering, a formal procedure was developed for preparing and presenting a Hazards Summary Report for each reactor by the organization planning to build a new reactor or significantly modify an existing one. This report for each reactor contains a description of the facility and of the processes to be performed including a statement of the quantities of radioactive effluent that might be released. The procedures for routine and non-routine operations and plans for emergency conditions are presented in detail. Information on the location of the reactor including meteorological, hydrological, geological and seismological data are included and evaluated with regard to routine operation and the maximum credible accidents involving release of radioactive material that might occur. These Hazards Summary Reports were reviewed by the Reactor Safeguards Committee to determine completeness and accuracy and the recommendations of the Committee including possible suggestions for further studies were sent to the Atomic Energy Commission for their final decision as to whether the facility should be constructed and operated.

This general procedure for reactor hazards evaluation remained in effect for several years until the licensing of reactors was initiated and the Hazards Summary Report then became a part of the application for a license. The volume of work involved in reviewing the hazards problems for the many existing and planned reactors made it necessary to establish a permanent staff within the Commission to handle more routine problems. A statutory Advisory Committee on Reactor Safeguards was established in 1957 to review safety studies and facility license applications referred to the Committee by the AEC. The Committee's reports on applications for facility licenses are a part of the record of the application and are available to the public, except for security material.

NATIONAL REACTOR TESTING STATION

With the establishment of reactor developments as an important objective of the Atomic Energy Commission and the formulation of plans for a series of experimental and test reactors, the need for a reactor test site was recognized. The requirements for this site included an area sufficiently large and isolated, so that many new types of reactors could be operated and components even tested to destruction without public hazards. After an engineering survey of many possible sites, the Naval Proving Ground in Idaho, previously used for testing naval ordnance, was selected. This is in a barren desert section of Idaho some 40 miles from Idaho Falls, a city of 20,000 population. Additional land has been acquired particularly to the northeast where irrigated areas were being expanded in the direction of the National Reactor Testing Station. The aircraft reactor test facilities are located in the north end of the property and the tests conducted at these facilities are considered to require a large exclusion area. The Atomic Energy Commission only recently announced the withdrawal of about 124,000 acres of public land to be used as a "safety buffer zone" near the aircraft reactor experimental area. The land newly set aside brings the total area of the National Reactor Testing Station to approximately 570,000 acres or about 890 square miles which is nearly the area (1,214 square miles) of the state of Rhode Island.

Areas have been developed within the National Reactor Testing Station with separation between each area of several miles. These include the facilities for fast reactors EBR I & II near the south end of the site, central facilities where general services are provided, the Army Reactors test site along the entrance road where up to five small power reactors may be located, the area for transient testing of reactors, the Material Testing Reactor site where this and the Engineering Test Reactor are located, the Naval Reactors site and the Aircraft Reactors site. In addition, there is the Chemical Processing Plant which is a major facility with the safeguards problems associated with the separation processes for spent fuel elements.

There are several reasons for locating more than one reactor at most of the areas with separation distances of the order of miles between areas. The original site plan was made on the basis of adequate separations between such nuclear facilities so that not more than one unit could be lost in event of major contamination by release of radioactive materials. As the reactor program developed there were requirements for additional reactors at the National Reactor Testing Station and it was found that major economies resulted from locating similar reactors at the same site so that common facilities could serve more than one unit. There has also been sufficient favorable experience with the operation of the existing reactors to indicate that the probability of losing any of the facilities due to an accident is very low indeed. Therefore, with the savings in construction and operating cost that can be realized and the simplified administration problems associated with more than one reactor at a site, the concept of multiple reactor installations has developed.

There are obvious inconveniences and expenses associated with the 40 mile separations between the National Reactor Testing Station and Idaho Falls. Experience has shown that many of the reactors could have been tested in a more populated area; however, there have been other reactor tests for which the isolation at the National Reactor Testing Station was needed.

For those groups that have been operating reactors at the Idaho site and built up competent staffs at the various areas, the inconveniences associated with this remote location are much less than might be predicted.

CONTAINMENT

The concept of containment for reactors to limit the release of radioactive materials was developed for the Knolls Intermediate Breeder Reactor since a location within some 30 miles of Schenectady, New York, was desired in order to permit participation by the personnel at the laboratories in the area in this reactor project. The construction of this reactor at a site some 10 miles from any heavy populated area and location of the reactor within a gas enclosing building was proposed to the Reactor Safeguards Committee at their first meeting in November 1947, in Schenectady, New York. This concept eventually led to the construction near West Milton, New York, in 1953, of a steel sphere with a diameter of 225 ft. designed to contain an internal pressure of 23.8 psig. The Submarine Intermediate Reactor prototype for the *Seawolf* was operated in this facility. The maximum pressure for design of the structure was determined by considering a disruptive core explosion from nuclear energy release, followed by sodium–water and air reactions, but without detonation of a hydrogen–air mixture since adequate sources for ignition of a combustible mixture were assumed to be present at least before any large accumulation of such a mixture could be formed. Missile protection was provided by explosion mats and steel plates. This attack on the problem of containment and resulting large spherical shell design, construction and testing set a pattern for most future work. It is of interest to note that the prototype Submarine Thermal Reactor for the submarine *Nautilus* was built and tested at the National Reactor Testing Station where the isolation provided by this site was considered to make containment unnecessary.

Another interesting example of containment is provided by the gas tight concrete building housing the Argonne Research Reactor, CP-5. When design work was started in 1948, it was evident that the proximity of the Argonne National Laboratory site to large population centers made it necessary that this reactor be extremely safe under all conditions. A heavy-water-moderated reactor with natural-uranium fuel was selected because of the long neutron lifetime, prompt negative temperature coefficient, and other safety features. In addition, it was considered appropriate to locate the reactor within a reinforced concrete cylindrical building specially constructed to be gas tight with air locks at the entrances and a connected external expansion chamber for the air in the building.

UNIVERSITY RESEARCH REACTORS

A university research reactor to be of maximum usefulness for research and instruction purposes needs to be located close to the campus where there

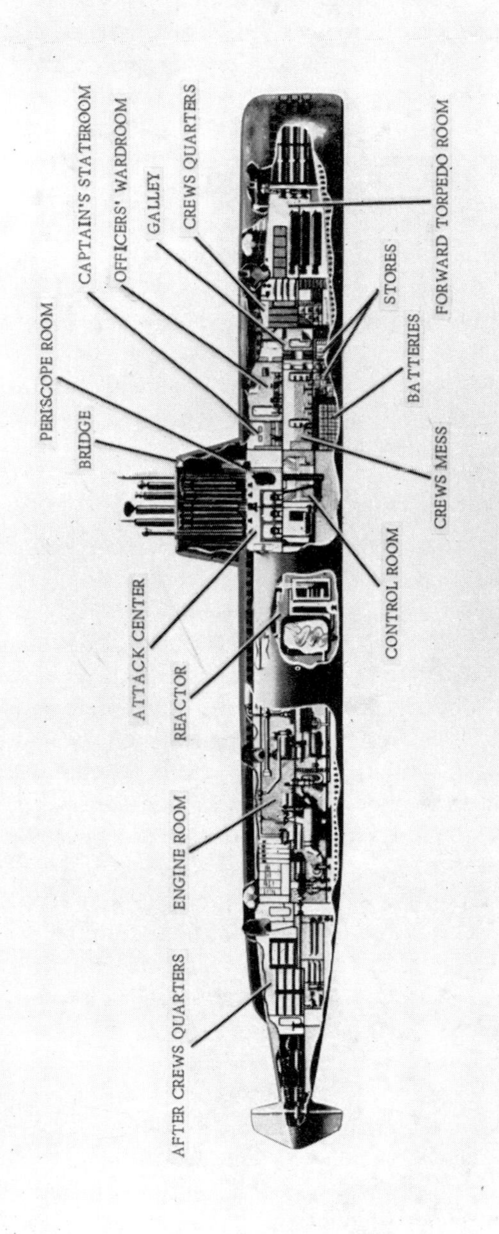

FIG. 13. Nuclear submarine. (Reproduced by permission of United States Atomic Energy Commission.)

may be large concentrations of people. The North Carolina State College Research Reactor which was placed in operation on the Raleigh campus in 1953, established a precedence but not without considerable study and discussion of the hazard problems involved. This small homogeneous reactor was of a type with strong inherent safety features which had been well established from several years' operating experience with this type of reactor. The reactor operated at low power and the location of the beam holes below grade reduce the problem of open beam holes. Members of the staff who were building the reactor and who became responsible for its operation were well experienced in reactor technology from work at atomic energy installations. After carefully considering these factors and the importance of having reactors available at universities, the location of this reactor on the Raleigh campus for operation without containment was approved by the Atomic Energy Commission and this project was soon followed by proposals for many other university research reactors, some of which operate at sufficiently high power levels to require somewhat more isolated sites and in some cases provisions for limited containment.

POWER REACTORS

The operation of the nuclear reactor in the submarine *Nautilus* in the vicinity of Groton, Connecticut, also established a precedence for containment of nuclear power reactors to be operated near centers of population. The safety characteristics of the *Nautilus* reactor and containment provided by the submarine hull and internal structures under all conceivable conditions were studied in great detail including the problems presented by loading and unloading the core of this reactor. The general policies and detailed design features and operating regulations for the safe operation of nuclear submarines in confined waters were developed over a considerable period of time by the Navy and the Atomic Energy Commission and their contractors before the launching of the *Nautilus.*

From the civilian power reactor program came requirements for locating very large power reactors close to centers of population in order to avoid the capital costs and energy losses associated with long transmission line. Also isolation areas corresponding to the power levels of civilian power reactors are usually not available, and would cost an exorbitant amount in areas where civilian power reactors are being built. For these reasons the concept of containment without a large exclusion area was considered essential to the economics of civilian power reactors.

The Pressurized Water Reactor was built near Shippingport, Pennsylvania, after several sites, as proposed by the utility companies interested in this program, had been evaluated on a preliminary basis. The reactor is located on a tract of just over 400 acres on the bank of the Ohio River in western Pennsylvania about 25 miles west of Pittsburgh. The immediate area surrounding the plant is thinly populated and there is only one major industrial plant in the vicinity. A railroad traverses the property; however, the traffic is limited. The Ohio River provides a natural barrier to future population growth on the sides bounded by water and contributes to the exclusion area. A high bluff along the river provides similar protection. A considerable number of safety features are included in the design. The reactor and primary loops are located within an underground containment system consisting of four large steel pressure vessels connected by large tubular ducts. The reactor vessel is located in a central spherical container. This system is conservatively sized to contain the pressure created by a rupture of the primary coolant system including the effect of the stored energy in the water and the metal as well as any non-explosive energy release due to a zirconium–water reaction. Very careful attention was given to every detail of the design and construction of this system. The earth and concrete around the containment vessel provide shielding against radioactive materials in the systems or that which might be released from the reactor. Since the reactor is owned by the government, although operated by the Duquesne Light Company, the insurance and liability problems were not major considerations for this plant.

There were soon four privately owned large power reactor proposals from utility companies under consideration by the U.S. Atomic Energy Commission. All of these power reactors were to have steel containment vessels and three of them were in the general vicinity of large cities (Commonwealth Edison Company's Dresden Boiling Water Reactor Plant, 35 miles southwest of Chicago, Illinois; Consolidated Edison Company of New York's Indian Point Pressurized Water Reactor Plant, 24 miles north of New York City boundary; and Power Reactor Development Company's Enrico Fermi Fast Breeder Reactor Plant, 25 miles south of Detroit). The Yankee Atomic Electric Company's Pressurized Water Plant in western Massachusetts is in a region of relatively low population density and is remote from any large cities. Since the containment vessels for these reactors are above ground, at least in part, the direct radiation problem from radioactive material released into the containment vessel by an accident had to be considered. Also, since these reactors are privately owned the insurance and liability problems were major considerations. Sufficiently encouraging answers to the reactor safeguards and other problems were found to permit these projects to proceed and there are now many additional power reactor projects underway.

The British development of a large gas-cooled reactor for power generation, which went into operation in October, 1956, preceding the Shippingport reactor, established the precedent in the United Kingdom for reactors of this type without the usual containment. In 1946, the British selected the air-cooled graphite reactor for the production of plutonium. It was originally intended that these production reactors would be graphite-moderated and water-cooled of the Hanford type; however, this type of reactor cannot be made inherently stable, if operated solely on a natural uranium fuel. Since larger exclusion areas were considered necessary for the safe operation of that type of reactor than were available in the thickly populated United Kingdom it was decided to select a reactor that is inherently stable and which, therefore, could be safely operated without great isolation. It was for this reason that the graphite-moderated gas-cooled natural uranium fuel reactor was selected for British development. Two small air-cooled reactors were built in 1947 and 1948, for research and the two Windscale production reactors were placed in operation in the 1950–51 period. Although it was recognized that operation at pressure would reduce the blower power required in cooling, the decision was made to operate at atmospheric pressure and cool with air in order to avoid the time required to solve the problem of a very large pressure vessel. Cooling air was supplied by a number of large centrifugal blowers

FIG. 14. Model of the N.S. *Savannah*. (Maritime Commission photograph.)

and the inlet air is drawn through both wet and dry filters. The exhaust air was discharged through a 400 ft stack with a filter gallery at the top of the stack. The filters were removed by remote control since they could be radioactive. The Windscale plutonium factory and later the Calder Hall power reactors are located on the Calder River close to the sea on the west coast of England in open farm country.

Success with the Windscale production reactors led the British to decide in 1953, to construct the gas-cooled graphite Calder Hall reactors. These are cooled by carbon dioxide under pressure and the pressure vessel around the reactor is made of 2 in. welded steel plate and is about 40 ft in diameter and approximately 60 ft high. The hot carbon dioxide from the reactor generates steam in a large heat exchanger for power generation. These highly successful nuclear power reactors have provided the basic design from which have developed the several more advanced gas-cooled reactors for electric power generation and plutonium production.

The nuclear powered merchant ship presents combinations of problems from the nuclear reactor with those of a ship subject to collision and other damages. Such accidents could occur in any port of call, which might be a large city. Also, an accident involving release of radioactivity from the reactor might occur at sea under conditions that present a hazard to the passengers and crew. There have been no nuclear powered merchant ships in operation up to this time and the favorable experience with submarines does not provide an adequate precedence since the problems with a merchant ship are certainly

FIG. 15. Construction of Experimental Breeder Reactor–11 at the National Reactor Testing Station. (Argonne National Laboratory photograph.)

more varied and difficult. Therefore, very careful study is being given the design criteria and operating regulations required for safe operation in domestic and foreign confined waters.

A nuclear powered aircraft would combine the possible problems of a high power nuclear reactor of small heat capacity and, therefore, short melt-down time in event of loss of cooling, with the normal hazards of aircraft. Since the deviations in distance possible for a high-speed aircraft from a scheduled flight path could be large in the event of a malfunction of equipment, nuclear powered flight in the vicinity of centers of population has been a controversial

FIG. 16. Model of the first experimental nuclear rocket shell. (Los Alamos Scientific Laboratory.)

FIG. 17. Artist's conception of Test Area No. 1, Jackass Flats, near Las Vegas, Nevada, for testing nuclear rocket propulsion (Los Alamos Scientific Laboratory.)

subject. It was planned that the first nuclear powered aircraft would be based at the National Reactor Testing Station; but, later considerations in the Atomic Energy Commission of the value and importance of the other installations at this site have led to a decision that any nuclear powered aircraft flight testing be done elsewhere. However sufficient information is becoming available from experimental programs to reduce the uncertainties in nuclear flight testing. The release of substantial quantities of fission products from a nuclear ramjet or nuclear rocket probably can not be avoided at the extremely high operating temperatures required for the fuel elements in such advanced reactors. The facilities for testing experimental reactors in the Pluto and Rover programs are located in an isolated area in Nevada with provisions for moving reactors from the test stand to the disassembly area without exposure of personnel.

REFERENCES

1. *Selected Materials on Employee Radiation Hazards and Workman's Compensation*, Join Committee on Atomic Energy, February, 1956.
2. *The Control of Radiation Hazards in Industry*, 1957, State of New York, Department of Labor.
3. *Selected Materials on Radiation Protection Criteria and Standards: Their Basis and Use*, Joint Committee on Atomic Energy, May, 1960.
4. SMYTH, H. D., *Atomic Energy for Military Purposes*, Princeton University Press, 1948.
5. STEVENSON, C. G., G. E. Hanford Atomic Products Operation, Private Communication.
6. AEC Report Wash-3 (Rev.), "Summary Report of Reactor Safeguard Committee," March, 1950, Decl. March 1957.

CHAPTER 2

RADIOACTIVE MATERIALS

THE primary hazards associated with nuclear reactors result from the radioactive materials formed in the fission process or by interaction of neutrons with reactor materials; and therefore a more detailed consideration of the production and properties of fission products and other radioactive materials is required, as a basis for hazards evaluations.

Following the discovery of the natural radioactivity of uranium minerals, the properties of the natural radioactive elements and their radiations have been intensively studied, and the ideas and techniques developed in these studies have been applied to artificial or induced radioactivity. The laws of radioactive change were found to be valid for artificial radionuclides including those resulting from the fission process and the properties of fission products can be described quantitatively by these elementary laws. The quantity of radiation from a sample of radioactive material decreases with time in an exponential manner although the energies of the radiations that are emitted remain constant. It was found that the number of atoms disintegrating in a time interval is directly proportional to the number of atoms of the active species present at that time. If the number of atoms present is N and the number which disintegrate in a time interval Δt is ΔN atoms then the rate of disintegration is $-\Delta N/\Delta t$ and the ratio of this rate to N is the constant of proportionality λ

$$\frac{-\Delta N}{N\Delta t} = \lambda$$

or in terms of simple calculus

$$\frac{-\mathrm{d}N}{\mathrm{d}t} = \lambda N$$

This disintegration constant, λ, is a characteristic of each radioactive species, and the value of the constant cannot be changed by ordinary physical conditions such as temperature or chemical combinations. The measurement of the decay constant is a frequently used method for identifying a radioelement.

By integration of the equation for radioactive decay over the interval of $t = 0$ to $t = t$ it is found that

$$\ln\left(\frac{N_t}{N_0}\right) = -\lambda t$$

where N_0 is the number of atoms of the species present at $t = 0$. This expression can be put into the exponential form

$$N_t = N_0 e^{-\lambda t}$$

or expressed in common logarithms

$$\log N_t = \log N_0 - 0.434 t\lambda$$

Another constant frequently used is the time required for half of the atoms in a given sample to decay or

$$\ln \tfrac{1}{2} = -\lambda t_{\frac{1}{2}}$$

Hence the half-life of the nuclide, $t_{\frac{1}{2}}$, is

$$t_{\frac{1}{2}} \quad \frac{0.693}{\lambda}$$

The value of the half-life may vary from a small fraction of a second (3×10^{-7} sec for polonium-212 to billions of years (4.5×10^9 yr for uranium-238). The rate at which a given quantity of polonium-212 decays and emits ionizing radiation is, therefore, the order of 10^{16} times the rate for the same quantity of uranium-238. In fact, the disintegration constant λ for uranium-238 is so small that pure uranium-238 and its compounds do not present any significant radiation hazards and can be, of course, handled without shielding with only the precautions necessary because of its chemical toxicity. This is also the case for uranium-235 with a half-life of 7.1×10^8 yr, except that added precautions appropriate for a fissionable material are also required.

The curie is a unit used to describe the rate of radioactive decay and is usually defined as 3.7×10^{10} disintegrations per sec. A quantity of radioactive material has one curie of activity when disintegrations occur at this rate in the sample. One gram of radium has one curie of activity and this was the original basis for the unit. The quantity of uranium-238 required for a curie of activity can be calculated from the half-life and atomic weight to be 3.5 tons, whereas the quantity of polonium-212 corresponding to a curie would be many orders of magnitude smaller than the sensitivity of a fine chemical balance. Before the discovery of nuclear fissioning, a curie of radioactivity was considered a very large quantity and up to the time of the construction of the first nuclear reactors, the total production of radium was less than 2000 curies. However, the quantities of radioactive materials formed in a large nuclear reactor can be expressed in units of megacuries (millions of curies) and many new radionuclides are formed in the fission process and these are still being studied even though a vast amount of work has already been done.

In the early studies of natural radioactive materials it was found that three series of radionuclides can be identified. One series called the uranium series starts with uranium-238; another series starts with uranium-235 and is called the actinium series; and the thorium series starts with thorium-232. In these series the parent nuclide decays with the emission of ionizing radiation and the daughter nuclide also decays in turn until eventually a stable nuclide is formed ending the series. In the uranium series some of the transformations and emissions are as follows:

Radionuclide	Emission	Half-life
Uranium-238	α 4.2 MeV	4.5×10^9 yr
Thorium-234	β 0.2	24 d
Protactinium-234	β 2.3	1.1 min
Uranium-234	α 4.8	2.5×10^5 yr
Thorium-230	α 4.7	8×10^4 yr
Radium-226	α 4.8	1620 yr
↓		
·		
↓		
·		
↓		
·		
↓		
·		
↓		
Lead-206	stable	

Through the ages the intermediate nuclides in a series have come to an equilibrium composition at which each species is formed from a parent at the same rate at which decomposition takes place. For this special equilibrium situation the decay law leads to the following simple relations between composition and decay constants:

$$N_{\mathrm{Th}^{234}}\lambda_{\mathrm{Th}^{234}} = N_{\mathrm{Pa}^{234}}\lambda_{\mathrm{Pa}^{234}} = N_{\mathrm{U}^{234}}\lambda_{\mathrm{U}^{234}}$$

A sample of uranium or thorium ore can be separated chemically from the other elements in the series and the radioactivity of the purified uranium or thorium is greatly reduced; however, on standing, the radioactivity will slowly increase as daughter products are formed. This increase in activity is of some significance for thorium and there has been at least one incident of release of a quantity of radioactive material into a sewage system following the treatment of thorium which records showed had been highly purified and of low activity but after storage for several years had increased greatly in activity.

In the fission process radionuclides are formed which decay to daughters which are also usually radioactive and on an average, there are three decay steps before a stable nuclide is formed. This decay chain can be indicated as follows:

$$A \xrightarrow{\lambda_A} B \xrightarrow{\lambda_B} C \xrightarrow{\lambda_C} D_{\text{stable}}$$

If the reactor operates at a constant rate and produces species A at a constant rate S, then

$$\frac{dN}{dt} A = S - \lambda_A N_A$$

$$\frac{dN}{dt} B = \lambda_A N_A - \lambda_B N_B$$

$$\frac{dN}{dt} C = \lambda_B N_B - \lambda_C N_C$$

For the case of a reactor that has operated for some time and accumulated an inventory of radionuclides and then shut down, so that the primary source becomes zero, the amount present at shut down will be indicated by $N_A(0)$, $N_B(0)$, and $N_C(0)$, and then the above relations will have the following solutions:

$$N_A(t) = N_A(0)\, e^{-\lambda_A t}$$

$$N_B(t) = N_B(0)\, e^{-\lambda_B t} + \lambda_A N_A(0)\left(\frac{e^{-\lambda_A t}}{\lambda_B - \lambda_A} + \frac{e^{-\lambda_B t}}{\lambda_A - \lambda_B}\right)$$

$$N_C(t) = N_C(0)\, e^{-\lambda_C t} + \lambda_B N_B(0)\left(\frac{e^{-\lambda_B t}}{\lambda_C - \lambda_B} + \frac{e^{-\lambda_C t}}{\lambda_B - \lambda_C}\right)$$

$$+ \lambda_A \lambda_B N_A(0)\left(\frac{e^{-\lambda_A t}}{(\lambda_B - \lambda_A)(\lambda_C - \lambda_A)} + \frac{e^{-\lambda_B t}}{(\lambda_A - \lambda_B)(\lambda_C - \lambda_B)} + \frac{e^{-\lambda_C t}}{(\lambda_A - \lambda_C)(\lambda_B - \lambda_C)}\right).$$

Solutions for other cases are available in the literature.

Three types of ionizing radiation have been identified from studies of the rays from the natural radioactive materials, although all types of radiation are not emitted simultaneously by all radioactive substances. These radiations have been identified as α-rays having a mass of four units corresponding to the helium atom and a positive charge of two units; β-rays having the negative charge and the mass of an electron; and γ-rays which have been shown to consist of electromagnetic waves and are similar to X-rays. Some elements emit α-rays and others emit β-rays and either type of radiation may or may not be accompanied by γ-rays depending on the particular radionuclide.

The β-particles emitted by most radionuclides have energies that are continuously distributed over a range up to a maximum value characteristic of each species. The α-particles are, however, emitted at specific energy levels and these have been measured and studied in some detail. When γ-rays are associated with β or α-rays, there may be one or several energy levels for the γ-radiation depending on the radionuclide.

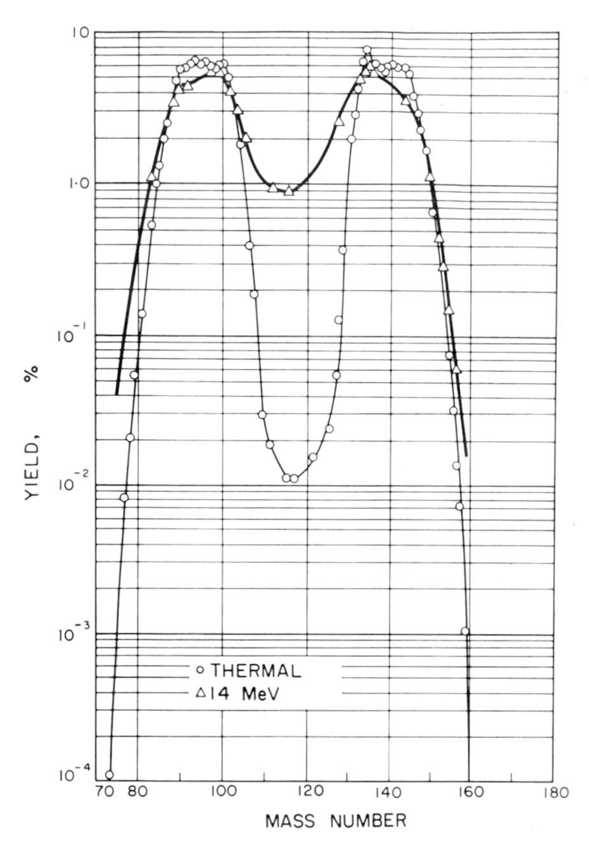

FIG. 1. Yield vs. mass number for U–235 fissions as a function of incident neutron energy. (Reproduced from Report ANL–5800, by permission of the United States Atomic Energy Commission.)

The ranges of radiations in air depend on their energy and the density of the air. Also, there may be a slight difference in range between different particles; however, an α-particle with an energy of 6 MeV, would have a mean range in air at 760 mm pressure and 15°C of 1.8 in. and at 10.5 MeV, a mean range of 4.6 in. Beta particles are much more penetrating and the determination of their range is more complicated because of the continuous energy distribution and the more complex absorption processes; however, an average range can be measured for the most energetic particles. For many radionuclides that emit β-particles, the maximum energy is less than 3 MeV and at this energy the maximum range in air is of the order of 40 ft, and for a more typical energy of 1 MeV, the β-ray range for particles of maximum energy is 12 ft although most of the particles are absorbed in 5 ft. The ionization produced per unit length in air by an α-particle is much greater than for a β-particle of the same energy. Since γ-radiations are electromagnetic waves, the processes for absorption are quite different than for the electrically charged beta and alpha particles, and gamma rays are much more penetrating. The intensity of γ-radiation decreases exponentially as it passes through material and, therefore, in theory an infinite distance would be required to completely absorb this type of radiation. Energetic gamma rays may travel through more than 1000 ft of air before being greatly attenuated.

Apparently an atom of fissionable material can break in many different ways in the fission process, although two primary fission product nuclides are always formed in addition to neutron, gamma, beta and neutrino radiations, and energy in the form of kinetic energy of the fission fragments. The weight of two fission fragments plus the neutrons equals the weight of the fissioning atom with a very small allowance for the energy released. However, fission products with atomic weights over the range from 72 (zinc) to 160 (gadolinium) have been identified. Most (97 per cent) of the species fall into a light group with masses from 85 to 104 and a heavy group with atomic weights from 130 to 149. It can be seen, therefore, that a fission yield can be defined as the proportion of nuclear fissions yielding a particular species. Since there are two fission products per fission the sum of all fission yields is 200 per cent. A plot of fission yield versus atomic weight is shown in Fig. 1 and it is seen that symmetrical curves result with a left and right hand limb which are mirror images.

Since an atom of uranium-235 can apparently fission in some thirty different ways, thereby producing sixty primary fission products each of which is unstable and may undergo up to three stages of radioactive decay producing a new radionuclide in each stage, it is seen that the fission products are a complicated radiochemical system and much work remains to be done in determining the characteristics of all of the species involved. The fission product chain starting with the nuclide produced directly in fission are shown schematically in Table I. The total chain yield from thermal neutron fissions in uranium-235 may be less than the cumulative yields listed in Table II since some species are formed both directly by fission and later by the decay of a parent nuclide. The cumulative fission product yield from fission spectrum (fast) neutron-induced fissions in plutonium-239, uranium-238 and thorium-232 are listed in Table III. It will be noted by comparing values from these tables that there may be a considerable difference in the fission product yield from different fissionable species and between thermal and fast fission.

The neutrons in a reactor are absorbed by the various atoms present and in this process of absorption, produce new nuclides in addition to causing

Table II.

Cumulative Yield of Fission Products from Thermal Fission

Mass No.	Fission Product	U^{233} % Yield	U^{235} % Yield	Pu^{239} % Yield	Mass No.	Fission Product	U^{233} % Yield	U^{235} % Yield	Pu^{239} % Yield
72	Zn^{72} (49 h)			0.00012	127	Sb^{127} (91 h)			0.39
77	Ge^{77} (12 h)	0.010	0.0031[a]		127	Te^{127*} (105 d)		0.035	
77	As^{77} (38.7 h)	0.019			129	Te^{129*} (37 d)		0.35	
78	Ge^{78} (86 m)		0.020[a]		131	Sb^{131} (23 m)		2.6	
81	Se^{81*} (57 m)		0.0084		131	Te^{131*} (30 h)		0.44	
83	Se^{83} (25 m)		0.22		131	I^{131} (8.05 d)	2.7	3.1[a]	3.8
83	Br^{83} (2.4 h)	0.79	0.51	0.085	131	Xe^{131} (stable)	3.74		2.87
83	Kr^{83} (stable)	1.14			132	Te^{132} (77 h)		4.7	5.2
84	Br^{84} (31.8 m)		0.90		132	Xe^{132} (stable)	5.10		4.02
84	Kr^{84} (stable)	1.90			133	Sb^{133} (4.1 m)		4.0	
85	Kr^{85} (10.3 y)	0.56	0.293[a]		133	Te^{133*} (63 m)		4.9	
86	Kr^{86} (stable)	3.18			133	I^{133} (21 h)		6.9	5.3
87	Br^{87} (55 s)		3.1		133	Xe^{133} (5.27 d)		6.62	5.27
89	Sr^{89} (51 d)	6.5		1.9	133	Ca^{133} (stable)	6.18		5.27
90	Sr^{90} (28 y)		5.8[a]		134	Te^{134} (44 m)		6.9[a]	
91	Sr^{91} (9.7 h)		5.8[a]	2.4	134	Xe^{134} (stable)	6.54		5.69
91	Sr^{91} (2.7 h)		5.3		135	I^{135} (6.7 h)	5.1	6.1[a]	5.8
91	Y^{91} (58 d)		5.35	3.0	135	Ca^{135} (2.6×10^6 y)	>4.9		5.53
91	Zr^{91} (stable)	6.53			136	I^{136} (86 s)	1.7	3.1	2.1
92	Zr^{92} (stable)	6.70			136	Xe^{136} (stable)	<8.9		5.06
93	Zr^{93} (1.1×10^6 y)	7.10			137	Ca^{137} (29 y)	7.16	6.15	5.24
94	Y^{94} (16.5 m)		5.4		138	Ca^{138} (32 m)		5.74	

94	Zr^{94}	(stable)	6.82			139	Ba^{139}	(84 m)			5.7
95	Zr^{95}	(65 d)	5.9	6.2	5.9	140	Ba^{140}	(12.8 d)	6.0	6.44*	5.68
95	Mo^{95}	(stable)	6.10			140	Ce^{140}	(stable)	5.6		5.68
96	Zr^{96}	(stable)	5.60			141	Ce^{141}	(33 d)			5.2
97	Zr^{97}	(17 h)		5.9[a]	5.7	142	Ce^{142}	(stable)	5.6		6.69
97	Mo^{97}	(stable)	5.35			143	Ce^{143}	(33 h)		5.7	5.4
98	Mo^{98}	(stable)	5.18			143	Nd^{143}	(stable)	5.2		6.31
99	Mo^{99}	(66 h)	4.8		5.9	144	Ce^{144}	(285 d)	4.1	6.0	5.28
100	Mo^{100}	(stable)	4.40			144	Nd^{144}	(stable)	4.0		5.29
101	Mo^{101}	(14.6 m)		~5.6		145	Nd^{145}	(stable)	3.0		4.24
101	Ru^{101}	(stable)	3.00			146	Nd^{146}	(stable)	2.3		3.53
102	Mo^{102}	(11.5 m)		4.3		147	Nd^{147}	(11 d)		2.7	
102	Ru^{102}	(stable)	2.37			147	Sm^{147}	(stable)		1.71	2.92
103	Ru^{103}	(39.7 d)	1.6		5.8	148	Nd^{148}	(stable)	1.15		2.28
104	Ru^{104}	(stable)	0.96			149	Pm^{149}	(5.6 h)		1.4	
105	Rh^{105}	(35.3 h)			3.9	149	Sm^{149}	(stable)	0.61		1.89
106	Ru^{106}	(1.01 y)	0.28		5.0	150	Nd^{150}	(stable)	0.48		1.38
109	Pd^{109}	(13.4 h)	0.040	0.030	1.5	151	Sm^{151}	(80 y)	0.27		1.17
111	Ag^{111}	(7.6 d)	0.025		0.27	152	Sm^{152}	(stable)	0.17		0.83
112	Pd^{112}	(21 h)	0.016		0.10	153	Sm^{153}	(47 h)	0.095		0.41
115	Ag^{115}	(21 m)		0.0077		154	Sm^{154}	(stable)		0.037	0.32
115	Cd^{115*}	(43 d)	0.001	0.0007	0.003	155	Sm^{155}	(24 m)			0.22
115	Cd^{115}	(53 h)	0.019	0.0097	0.038	155	Eu^{155}	(1.9 y)		0.03	
121	Sn^{121}	(27.5 h)	0.018		0.044	156	Sm^{156}	(10 h)		0.013	
123	Sn^{123}	(136 d)		0.0013		156	Eu^{156}	(15.4 h)			0.12
125	Sn^{125}	(9.6 d)	0.050	0.013	0.072						

s = second m = minute h = hour d = day y = year * = metastable

(Reproduced from Report ANL-5800 by permission of the U.S. Atomic Energy Commission.)

Table III.

Cumulative Percentage Yields from Fission Spectrum Neutron Fissions in Pu^{239}, U^{238}, *and* Th^{232}

Mass No.	Fission Product	Pu^{239}	U^{238}	Th^{232}	Mass No.	Fission Product	Pu^{239}	U^{238}	Th^{232}
72	Zn^{72} (49 h)			0.00033	111	Ag^{111} (7.6 d)		0.073	0.052
73	Ga^{73} (5.0 h)			0.00045	112	Pd^{112} (21 h)	0.14	0.046	0.057
77	Ge^{77} (12 h)			0.009	115	Cd^{115*} (43 d)		0.003	0.003
77	As^{77} (39 h)		0.0038	0.020	115	Cd^{115} (53 h)	0.069	0.037	0.072
83	Br^{83} (2.4 h)			1.9	127	Sb^{127} (93 h)		0.12	
83	Kr^{83} (stable)		0.40	1.99	131	I^{131} (8.05 d)			1.2
84	Kr^{84} (stable)		0.85	3.65	131	Xe^{131} (stable)		3.2[d]	1.62
85	Kr^{85} (10.3 y)		0.153	0.87	132	Te^{132} (77 h)		4.7	2.4
86	Kr^{86} (stable)		1.38	6.0	132	Xe^{132} (stable)		4.7[d]	2.87
89	Sr^{89} (51 d)		2.9	6.7	133	Cs^{133} (stable)		5.5 (8.08)[e]	
90	Sr^{90} (28 y)	2.2	3.2	6.8	134	Xe^{134} (stable)		6.6[d]	5.38
91	Sr^{91} (9.7)			7.2	135	Cs^{135} (2.6×10^6 y)		6.0[e]	
95	Zr^{95} (65 d)		5.7		136	Xe^{136} (stable)		5.9[d]	5.65
97	Zr^{97} (17 h)	5.2		5.2	137	Cs^{137} (29 y)	6.6	6.2 (7.11)[e]	6.3
99	Mo^{99} (67 h)	5.9	6.3	2.7	140	Ba^{140} (12.8 d)	5.0	5.7	6.2
103	Ru^{103} (40 d)		6.6	0.16	141	Ce^{141} (33 d)			9.0
105	Rh^{105} (35 h)			0.07	144	Ce^{144} (290 d)		4.9	7.1
106	Ru^{106} (1.0 y)		2.7	0.042	153	Sm^{153} (47 h)	0.48		
109	Pd^{109} (13.4 h)	1.9	0.32	0.055	156	Eu^{156} (15.4 d)		0.066	

s = second m = minute h = hour d = day y = year * = metastable

(Reproduced from Report ANL-5800 by permission of the U.S. Atomic Energy Commission.)

fissioning in atoms of uranium-235 or other fissionable material. Many of the nuclides formed by neutron absorption are radioactive and even the fission products absorb neutrons to form other radionuclides. It can be seen that the calculation of the quantity of a fission product in a reactor should include neutron reactions and is, therefore, more involved than just the consideration of fission yields.

The probability that a neutron will be absorbed by an atom of a material is expressed in terms of the absorption cross section σ_a, in units of 10^{-24} cm^2 per nucleus which might be considered as a target area. If there are N target nuclei per cubic centimeter, the product $N\sigma_a$ is called the macroscopic cross section, Σ_a, for a neutron absorption and then

$$N\sigma_a = \Sigma_a.$$

The neutron density, N, neutrons per cm^3, multiplied by the average neutron velocity, v, gives the flux, ϕ, as neutrons per cm^2 sec. The rate of formation of the nuclide, N_2, per unit volume by absorption of neutrons by atoms of species, N_1, is, therefore,

$$\frac{dN_{2f}}{dt} = \phi\Sigma_{1a}.$$

However, the species, N_2, will usually be radioactive and decay with a decay constant λ_2 and the net rate of formation of this species is then

$$\frac{dN_2}{dt} = \phi\Sigma_{1a} - \lambda N_2 - \phi\Sigma_{2a}.$$

The systems of differential equations for the formation of radioisotopes in nuclear fuel by the fission process followed by decay systems and reactions of the materials with neutrons have been arranged for machine computations and the results of this work are reported by BLOMEKE and TODD in ORNL 2127. Here are given tables and plots as a function of neutron flux, irradiation time and time after shutdown of total fission product activity, gamma photons per sec for different energy groups from less than 0.25 MeV to those with energies greater than 1.70 MeV, and radioisotope yields. Since these graphs are useful in many calculations of fission product activity, they are reproduced as Figs. 2–7. The results of computation for individual fission products are also given in Report 2127 and in Reference 4.

Although the total beta and gamma power as a function of time after shutdown represents the summation of many decay schemes for the individual radioisotopes, the data can be approximated by reasonably simple analytical expressions which are sufficiently accurate for some computations. It is noted that the total beta and gamma powers are approximately equal for both the cases of a burst and infinite irradiation. The slopes of the plots indicate that

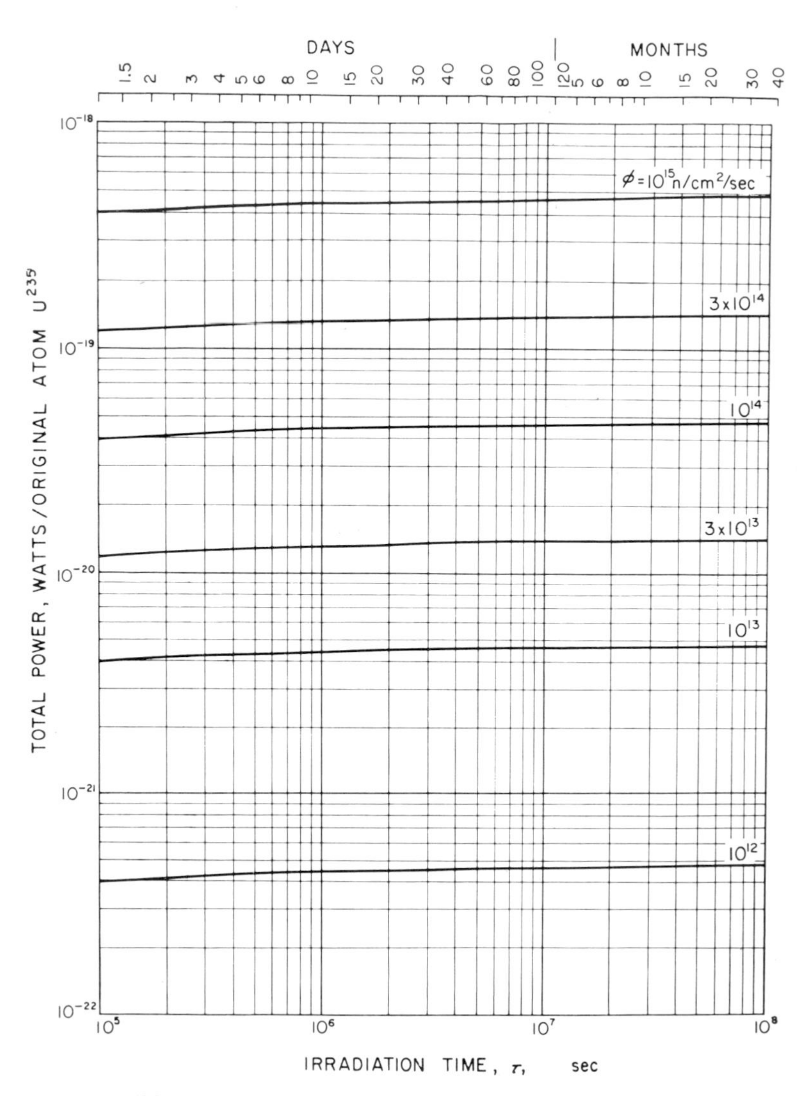

FIG. 2. Build up of total fission product activity with irradiation time and thermal neutron flux. (Reproduced from Report ORNL–2127 by permission of United States Atomic Energy Commission.)

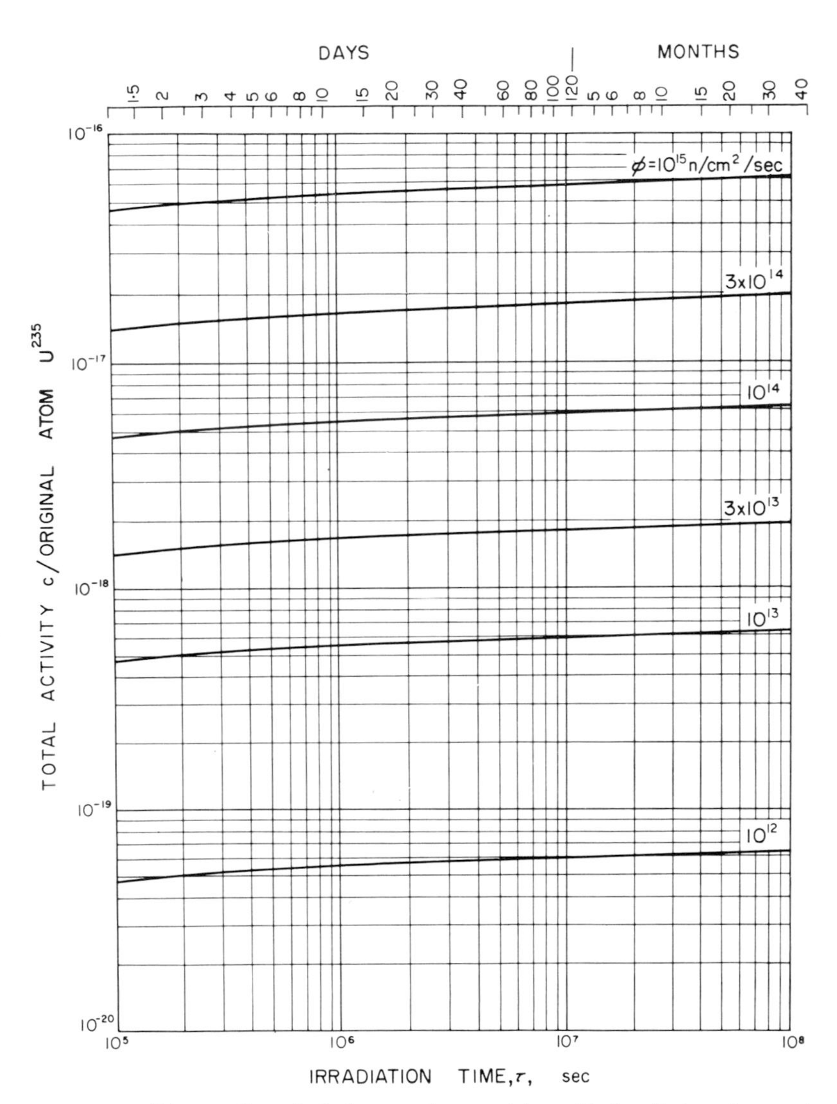

FIG. 3. Build up of total fission product activity with irradiation time and thermal neutron flux. (Reproduced from Report ORNL–2127 by permission of the United States Atomic Energy Commission.)

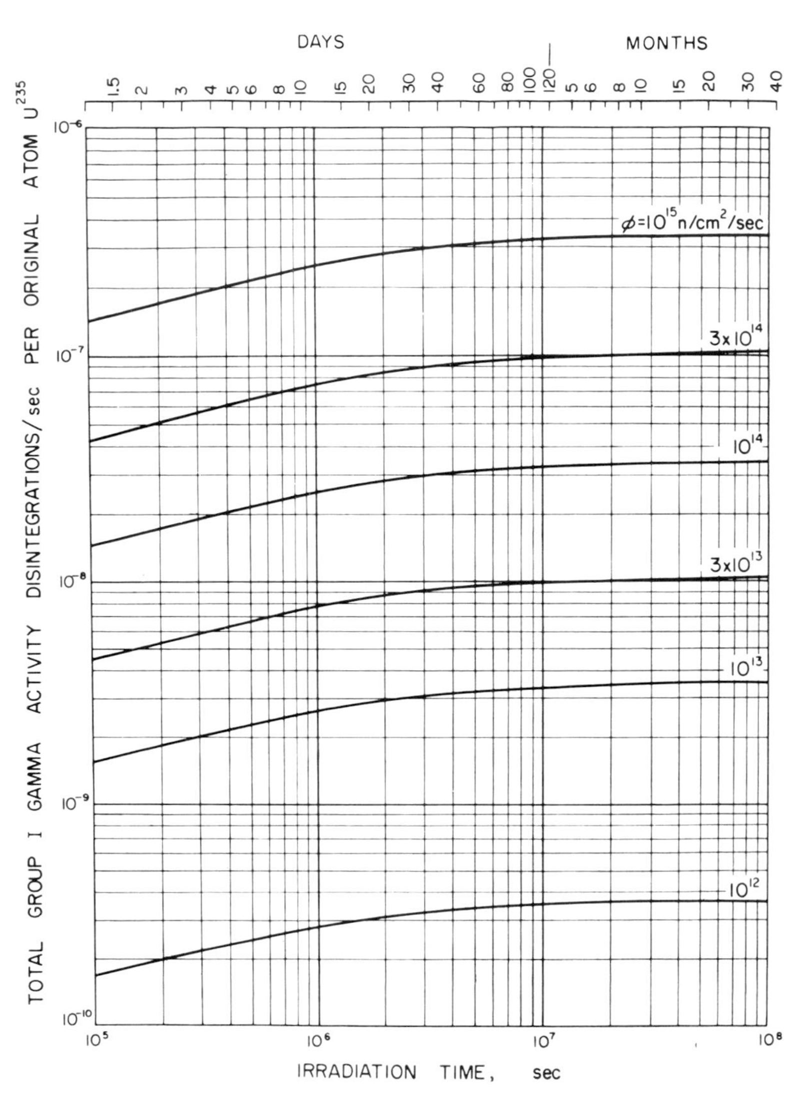

FIG. 4. Build up of Group I (0–0.25 MeV), gamma activity with irradiation time and thermal neutron flux. (Reproduced from Report ORNL–2127 by permission of the United States Atomic Energy Commission.)

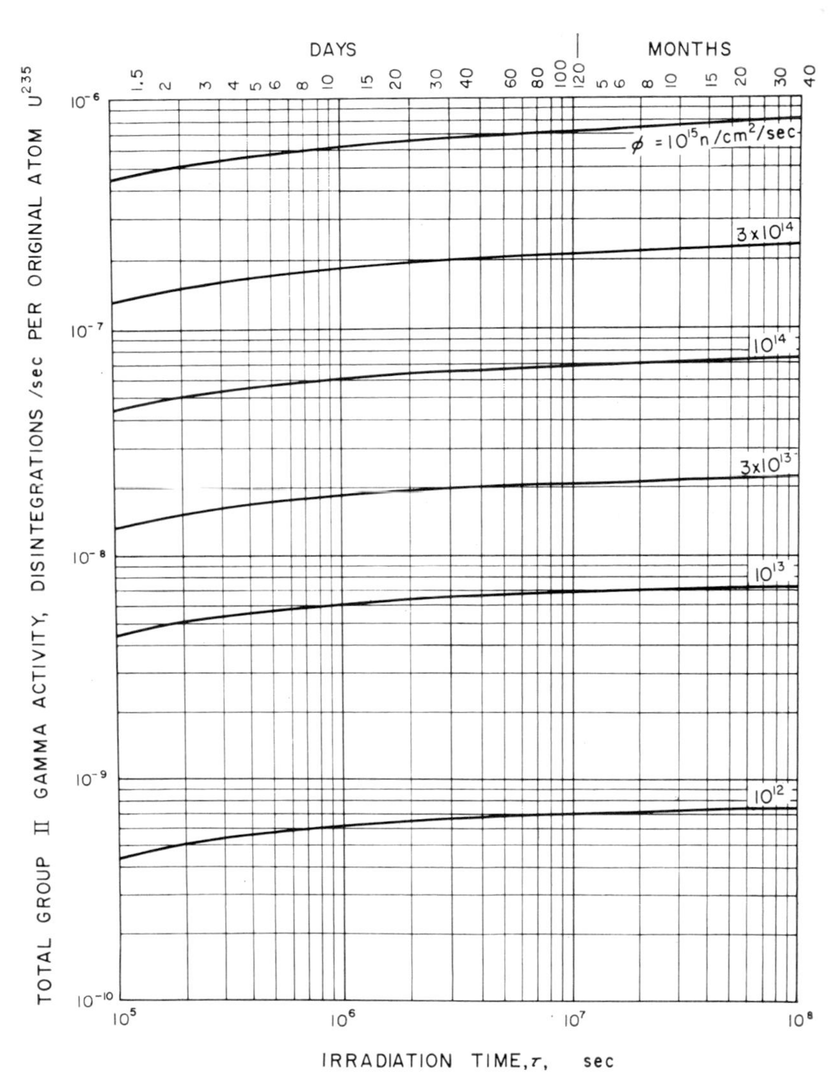

FIG. 5. Build up of Group II (0.26–1.00 MeV) gamma activity with irradiation time and thermal neutron flux. (Reproduced from Report ORNL–2127, by permission of the United States. Atomic Energy Commission.)

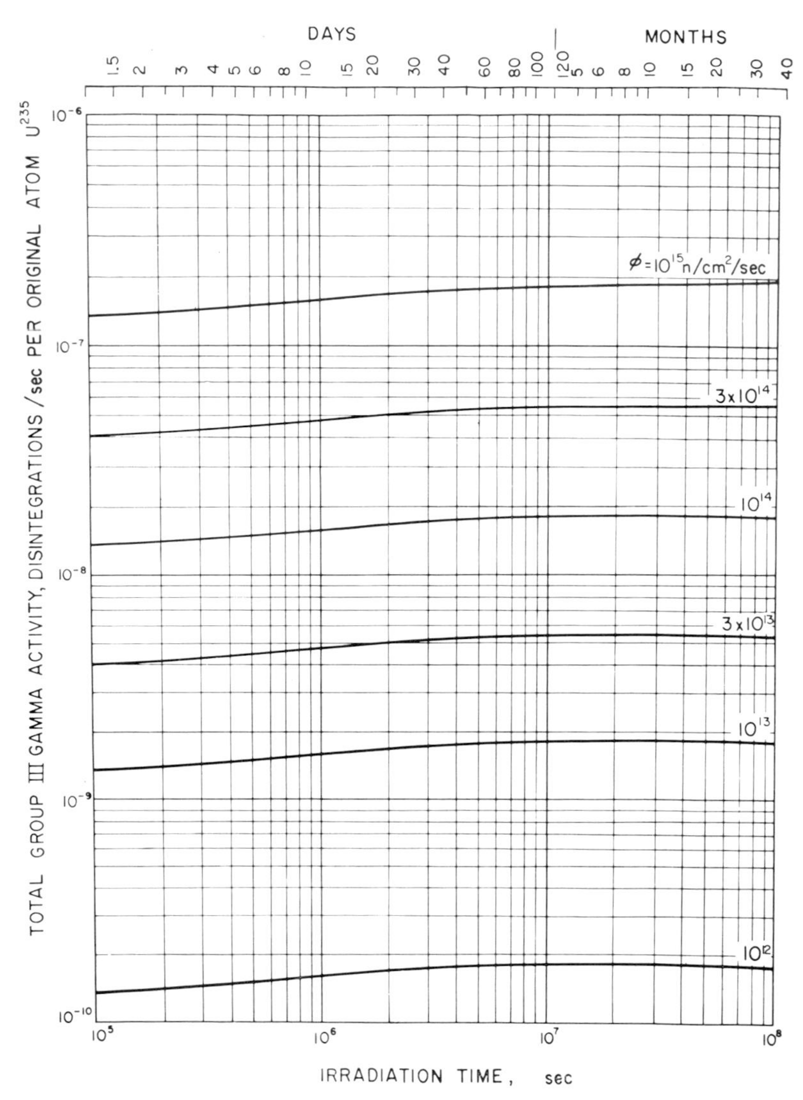

FIG. 6. Build up of Group III (1.01–1.10 MeV) gamma activity with irradiation time and thermal neutron flux. (Reproduced from Report ORNL–2127 by permission of the United States Atomic Energy Commission.)

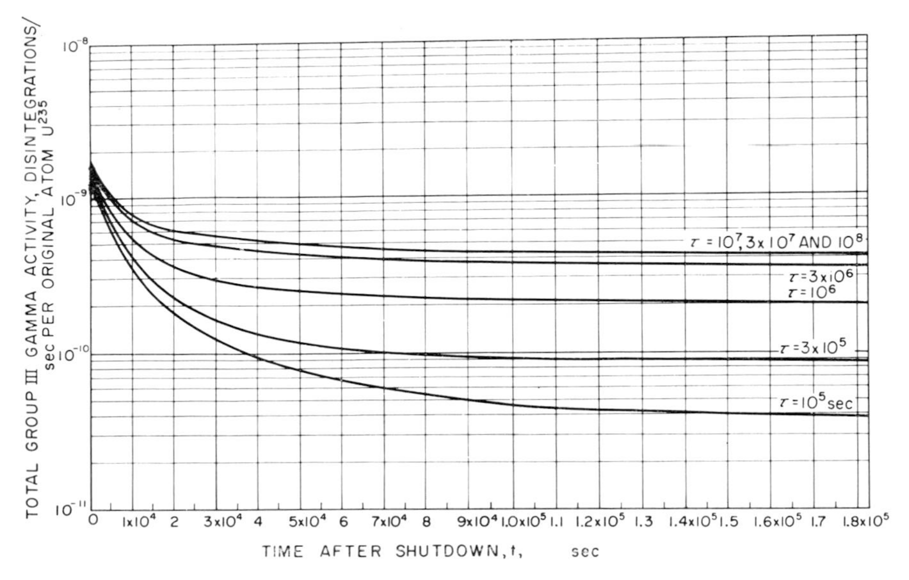

FIG. 7. Total Group III (1.01–1.70 MeV) gamma activity as a function of irradiation time and shutdown time for $\phi = 10^{13}$n/cm² sec. (Reproduced from Report ORNL–2127 by permission of the United States Atomic Energy Commission.)

for the burst the fission product power decays approximately as the negative 1.2 power of time, and for infinite irradiation approximately as the negative 0.2 power of time. The total beta and gamma energy emitted by the fission products from the fissioning of one atom of fuel is 14 MeV or 7 per cent of the total energy released in the fission process. According to Reference 4 the decay power for a time interval can be approximated by

$$P = C \times t^{-1.2} \text{ MeV/sec per fission}$$

where t is time in seconds following the burst and the constant, C, varies with the decay interval between values of 4.1 for short decay periods of 10 sec to 3 hr to 2.46 for long periods of decay (1 day to 100 days). This can be integrated over the total time interval, t_t, from start of reactor operation and including the period of shutdown, t_s, with the fission product power, P_{fp}, expressed in the same units as the steady state of power of the reactor, P,

$$P_{fp} = 0.07P(t_s^{-0.2} - t_t^{-0.2}).$$

In this relation the units of time are seconds. It can be observed from the graphs showing fission product power as a function of decay time that the power remains more nearly constant during the first second following shutdown, and, therefore, this relation and the one following are applicable only for times greater than one second. It can be seen that for very long periods of

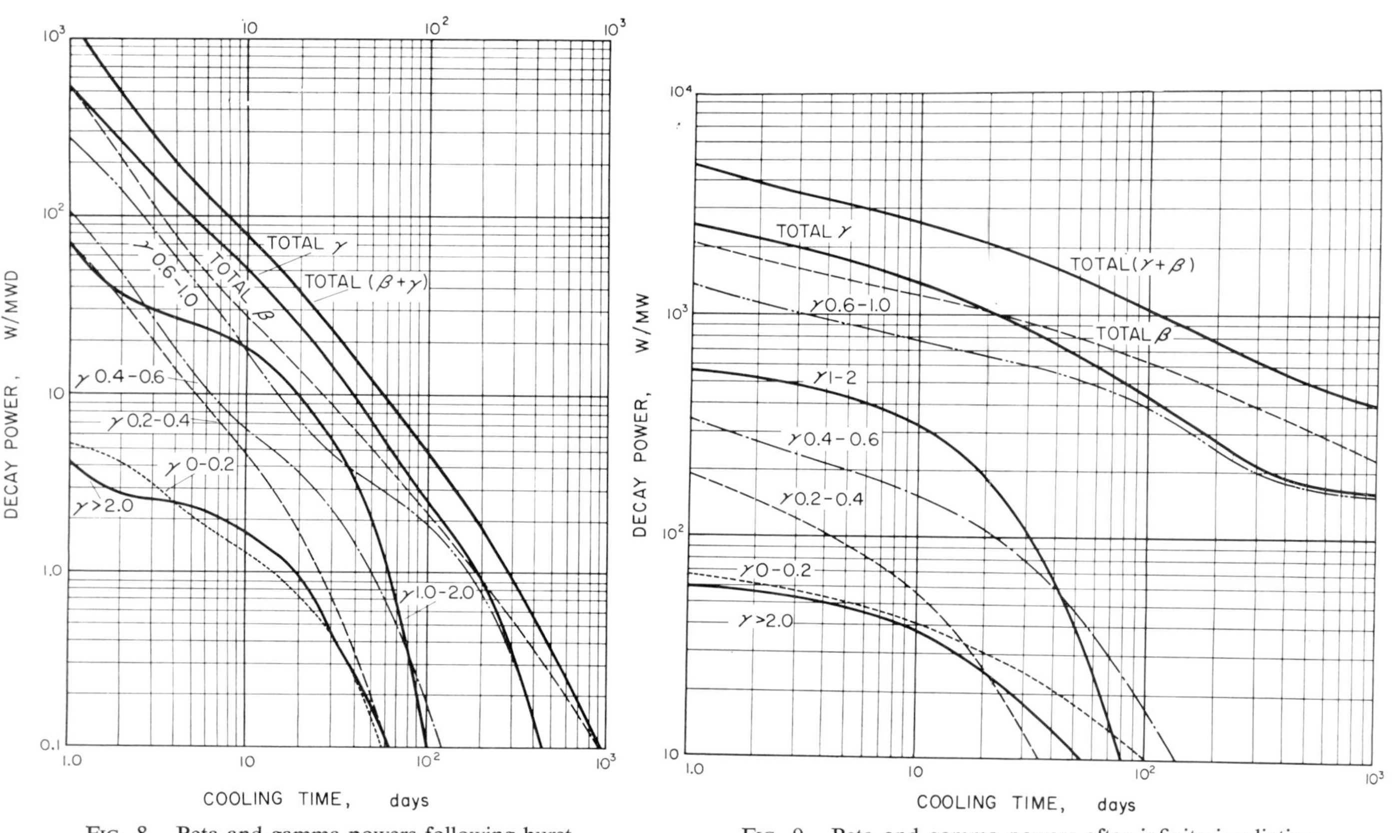

FIG. 8. Beta and gamma powers following burst.

FIG. 9. Beta and gamma powers after infinite irradiation.

(Reproduced from "Fission Product Radioactivity and Heat Generation," by J. R. Stehn and E. F. Clancy, Vol. 13 *Proc. 2nd United Nations Int. Conf. on the Peaceful Uses of Atomic Energy*, United Nations, New York.)

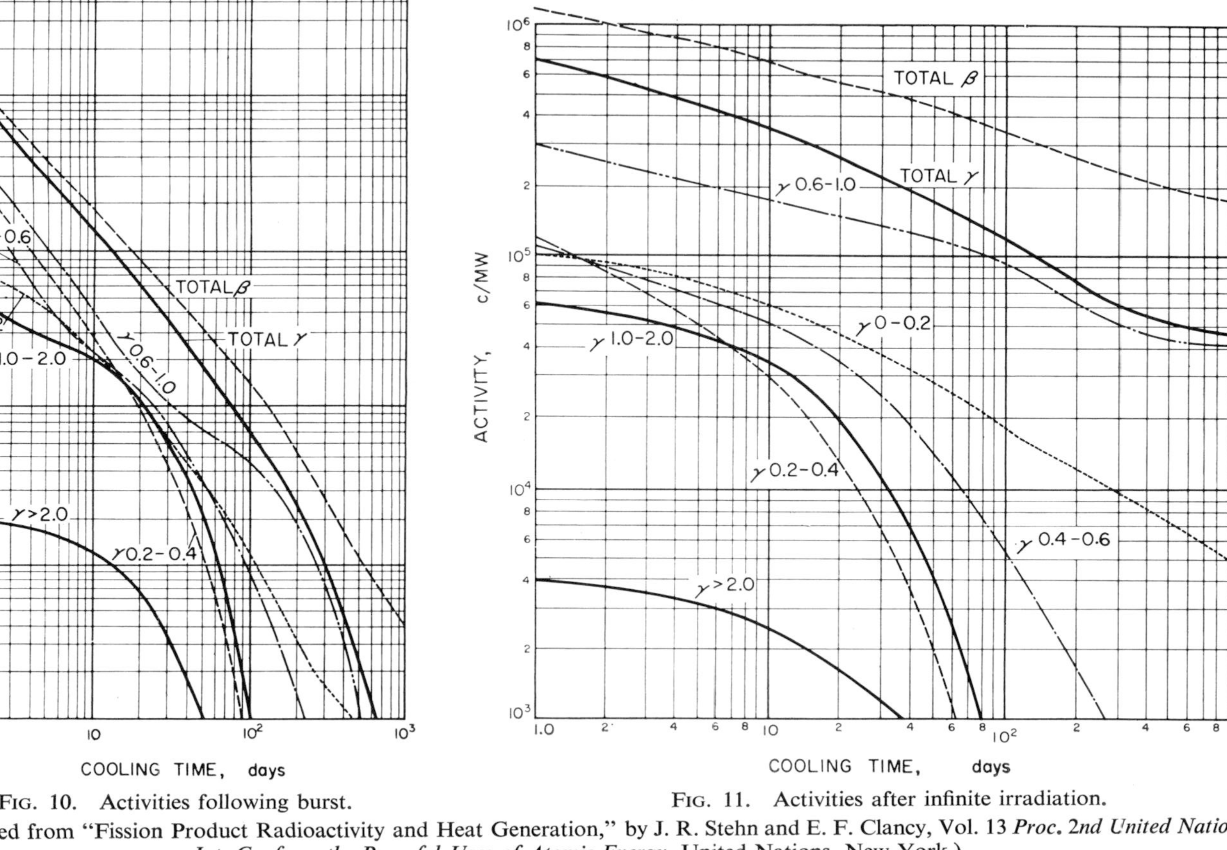

FIG. 10. Activities following burst.

FIG. 11. Activities after infinite irradiation.

(Reproduced from "Fission Product Radioactivity and Heat Generation," by J. R. Stehn and E. F. Clancy, Vol. 13 *Proc. 2nd United Nations Int. Conf. on the Peaceful Uses of Atomic Energy*, United Nations, New York.)

reactor operation, the value of $t_t^{-0.2}$ will approach zero and for this case the relation will reduce to

$$P_{fp} = \frac{0.07P}{t_s^{0.2}}.$$

The value of either the beta or gamma power would be half the value of the total rate of release of fission product energy.

Certain groups of fission products are of particular interest in reactor hazards evaluation such as the volatile radionuclides, since they would be expected to escape more readily from a fuel element.

The material of construction of reactors and the coolant all become radioactive by neutron absorption and the quantity of radioactive material formed in this way may be significant in hazards evaluation. The fertile materials, thorium-232 and uranium-238 are converted respectively to uranium-233 and plutonium-239 for which the maximum permissible concentrations in the body are very small, particularly when expressed in weight units. The oxygen

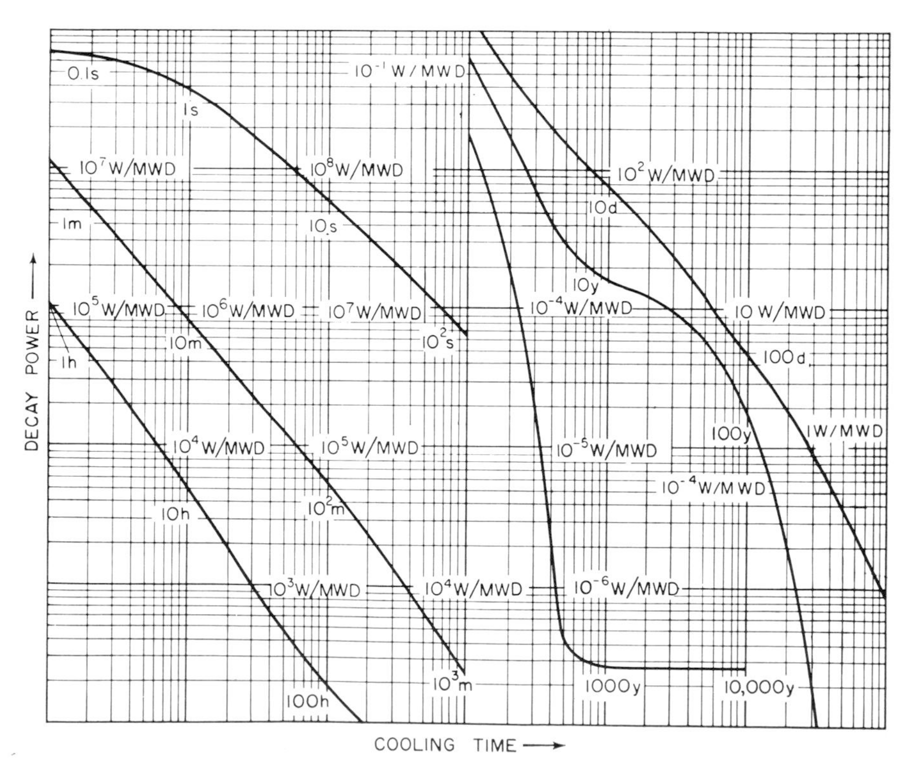

FIG. 12. Decay power following burst. (Reproduced from "Fission Product Radioactivity and Heat Generation," by J. R. Stehn and E. F. Clancy, Vol. 13, *Proc. 2nd United Nations Int. Conf. on the Peaceful Uses of Atomic Energy*, United Nations, New York.)

in water or air used to cool a reactor becomes intensely radioactive by reacting with neutrons to form nitrogen-16 which emits very high energy gamma rays of 6 MeV and has a half-life of 7.3 sec. This reaction to form nitrogen-16 from oxygen-16 only takes place with high energy neutrons (threshold about 10 MeV) whereas most of the neutron activation reactions result from the absorption of thermal neutrons. Other sources of activity in water include the activation products of water impurities including sodium, magnesium, manganese and aluminum. The argon and carbon in carbon dioxide, form radioactive products that may present radiation hazards from air-cooled reactors. Sodium becomes intensely radioactive when used as a

Table IV

Constants for Calculation of Neutron Activation of Reactor Materials for Uranium-235 Fueled Thermal Reactor

	Original Isotope (and A_i)	f_i Fractional Isotopic Abundance	σa_i Activation Cross Section	Radio‡ Nuclide Formed		λ_i Decay Constant 1/seconds	F_i $\frac{10^{10}\sigma a_i f_i}{A_i}$
1	Zr^{92}	0.171	0.2	Zr^{93}	βS	2.45×10^{-14}	3.72×10^{6}
2	Ni^{58}	0.68	4.3	Ni^{59}	KS	2.77×10^{-13}	5.04×10^{8}
3	C^{13}	0.0111	0.0009	C^{14}	βS	3.94×10^{-12}	4.68×10^{3}
4	A^{38}	0.00063	0.8	A^{39}	βS	8.45×10^{-11}	1.33×10^{5}
5	Ni^{62}	0.037	15	Ni^{63}	βS	2.77×10^{-10}	8.95×10^{7}
6	H^{2}	0.00015	0.00057	H^{3}	βS	1.79×10^{-9}	4.28×10^{2}
7*	Co^{59}	0.514	19	Co^{60}	βS	4.22×10^{-9}	1.66×10^{9}
8	Fe^{54}	0.059	2.2	Fe^{55}	KS	7.56×10^{-9}	2.40×10^{7}
9	Zr^{94}	0.174	0.1	Zr^{95}†	Nb^{95}	1.23×10^{-7}	1.85×10^{6}
10	Fe^{58}	0.0033	0.9	Fe^{59}	βS	1.73×10^{-7}	5.12×10^{5}
11	A^{36}	0.00337	6	A^{37}	KS	2.29×10^{-7}	5.62×10^{6}
12	Zr^{96}	0.028	0.1	Zr^{97}	Nb^{97}	1.13×10^{-6}	2.92×10^{6}
13	Na^{23}	1.000	0.53	Na^{24}	βS	1.28×10^{-5}	2.30×10^{8}
14	K^{41}	0.068	1.1	K^{42}	βS	1.54×10^{-5}	1.83×10^{7}
15	Ni^{64}	0.010	2	Ni^{65}	βS	7.48×10^{-5}	3.12×10^{6}
16	A^{40}	0.996	0.53	A^{41}	βS	1.06×10^{-4}	1.32×10^{8}
17*	Co^{59}	0.486	18	Co^{60}	βS	1.10×10^{-3}	1.48×10^{4}
18	Al^{27}	1.000	0.23	Al^{28}	βS	5.02×10^{-3}	8.52×10^{7}
19	O^{18}	0.00204	0.00022	O^{19}	βS	2.39×10^{-2}	2.49×10^{2}
20	N^{15}	0.0037	0.00002	N^{16}	βS	9.35×10^{-2}	4.93×10^{1}
21	B^{11}	0.812	0.05	B^{12}	βS	27.7	3.69×10^{7}
22	Bi^{209}	1.000	0.19	RaE	Po^{210}	1.6×10^{-7}	9.10×10^{6}
			0.15	Bi^{210}	Po^{210}	8.5×10^{-15}	7.18×10^{6}

* Isomers
† Daughter beta or alpha-decay to stable granddaughters
‡ Indicates beta-decay, S indicates stable nucleus formed directly

(Reproduced from Report ANL-5948 by permission of the U.S. Atomic Energy Commission.)

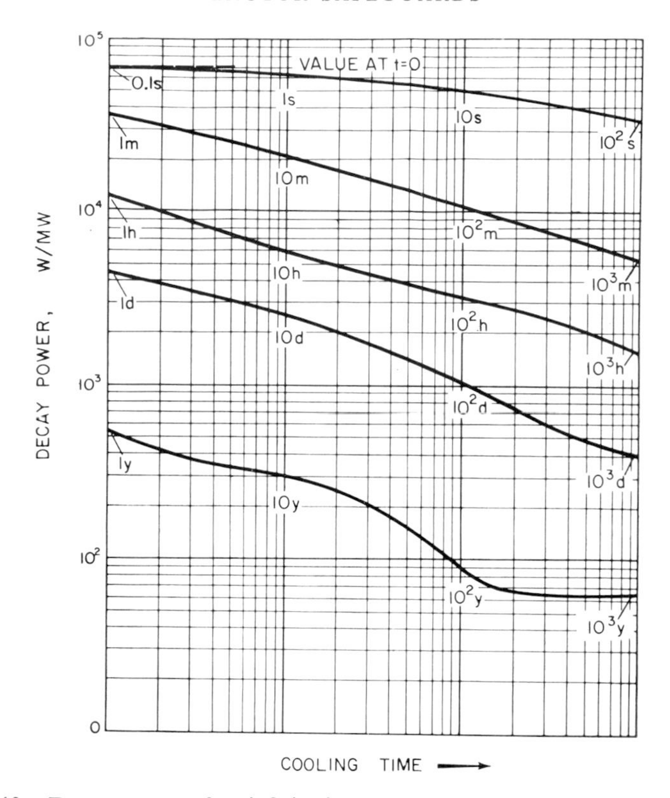

FIG. 13. Decay power after infinite irradiation. (Reproduced from "Fission Product Radioactivity and Heat Generation," by J. R. Stehn and E. F. Clancy, Vol. 13, *Proc. 2nd United Nations Int. Conf. on the Peaceful Uses of Atomic Energy*, United Nations, New York.)

reactor coolant by the formation of sodium-24 which emits strong gamma rays. The primary coolant loops on both water-cooled and sodium-cooled power reactors require special shielding because of the gamma radiations which result from these activation reactions. Although the quantities of neutron activated reactor material can be calculated by the relations given above, a convenient method has been developed by Brittan in terms of the reactor operating power as follows:

$$C_{Ti} = F_i \frac{P}{K_f} \frac{(1 - e^{\lambda_i T})}{\lambda_i}$$

where

C_{Ti} is nuclei per gram of original natural element

$$F_i = 10^{10} \frac{(\sigma_{ai} f_i)}{A_i}$$

σ_{ai} = activation cross section of material irradiated

A_i = atomic weight of isotope
f_i = isotopic abundance fraction
P = reactor operating power, kilowatts
K_f = kilograms of uranium-235 in reactor
T = irradiation time
λ_i = decay constant of the isotope in consistent time units

If the material being irradiated is not exposed to the same neutron flux as the fuel, the above relation should be corrected by the ratio of the actual flux where the material is exposed to the flux at the fuel. Following exposure, the number of activated nuclei per gram of original natural element will decay according to the exponential decay law. The values of λ_i and F_i are given in Table IV for structural materials and coolants. It should be noted that this table does not include values for the formation of nitrogen-16 since only fast neutrons are involved. Also the nitrogen-16 has a very short half-life of 7 sec and, therefore, the activity decays so rapidly that this source of radiation is usually not of major importance in considering reactor accidents although the radiation requires careful consideration in designing shields for routine operation.

REFERENCES

1. Reactor Handbook, Vol. I, *Physics*, AECD–3645.
2. KAPLAN, I., *Nuclear Physics*, Cambridge, Mass., Addison–Wesley, 1955.
3. BLOMEKE, J. O. and TODD, MARY F., Uranium-235 fission-product production as a function of thermal neutron flux, irradiation time and decay time, ORNL 2127, June 10, 1958.
4. STEHN, J. R., and CLANCY, E. F., Fission product radioactivity and heat generation, Vol. 13, *Proc. United Nations Int. Conf. on the Peaceful Uses of Atomic Energy*, United Nations, New York.
5. BRITTAN, R. O., Reactor containment, Report ANL–5948.

CHAPTER 3

REACTOR KINETICS

INTRODUCTION

THE stability of a nuclear reactor under all possible conditions and the possible rates of change of power are important considerations in evaluating the safety of a reactor. At the time the first nuclear reactor was built in 1942, the theory of the fission process in thermal reactors was sufficiently developed and there were data on neutron processes including the delayed neutrons to permit the design of this machine with some assurance that it could be controlled. Since that time much additional information has been accumulated and experimental and theoretical studies related to reactor safety now are a major part of the reactor development program of the Atomic Energy Commission[(1)]. One of the important advances in this area has come somewhat indirectly from the development of methods of machine computations which have made possible much more rigorous analysis of the complex processes involved in the neutron cycle. The processes can be approximated only within limited areas by reasonably simple equations. However such equations are extensively used in the literature and are very helpful in understanding the behaviour of a reactor even though their applications have limitations.

The neutrons released in the fission process have an energy spectrum ranging from 10 eV to 10 MeV with about 2 MeV being the most probable energy as shown in Fig. 1. At least one of the neutrons produced in the fissioning of an atom of fuel must survive the various processes which are involved in the life history of the neutron and then cause the fissioning of another atom of fuel in order for there to be a self-sustaining chain reaction.

The number of prompt neutrons released per fission depends on the type of fuel and the neutron energy as shown in Table I.

Table I[(2)]

Average Number of Prompt Neutrons per Fission

Isotope	Neutron energy	Neutrons
U^{235}	Thermal	2.47
	1.25 MeV	2.65
Pu^{239}	Thermal	2.91
	80 KeV	3.05
U^{233}	Thermal	2.51
	Fission spectrum	2.70

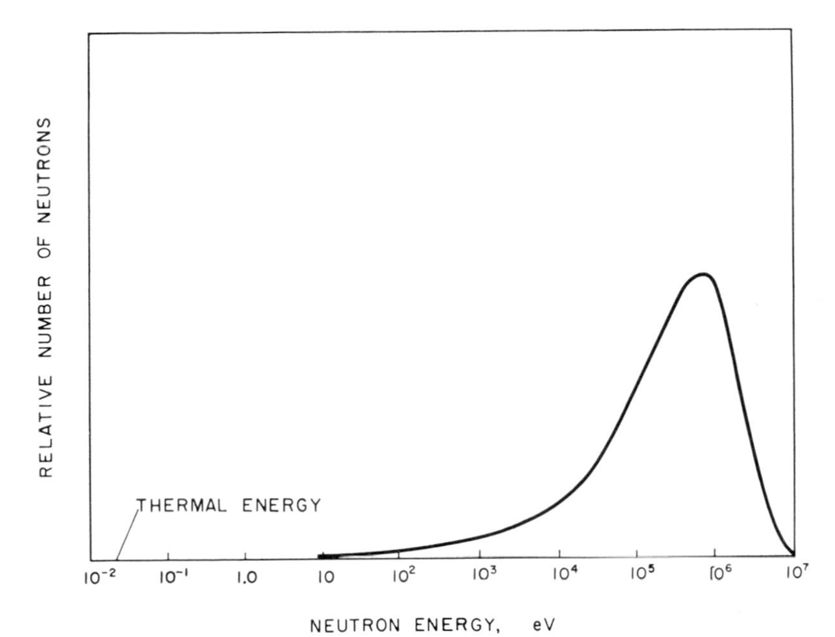

FIG. 1. Energy spectrum of neutrons from U^{235} Fission. (Reproduced from ORNL Report K–1380 by permission of the United States Atomic Energy Commission.)

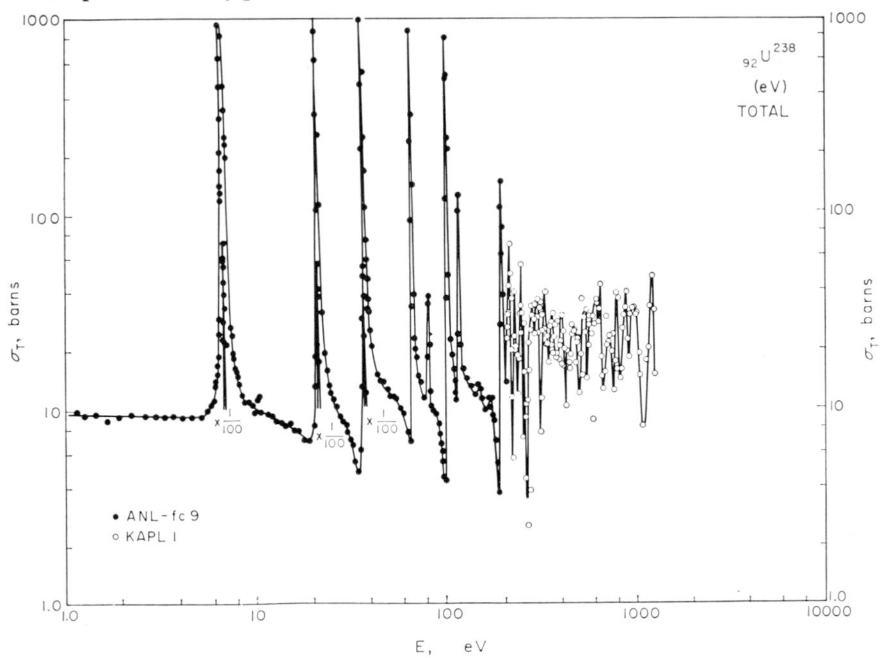

FIG. 2. Cross section of U^{238}. (Reproduced from *Neutron Cross Sections*, by D. J. Hughes and R. B. Schwartz, Report BNL–325, Supplement Number 1[(3)].)

Since only one neutron per fission is required to sustain the process, the remaining neutrons must be lost by absorption in reactor material and by leakage.

The neutron absorption cross sections for many materials decrease linearly with neutron velocity (or with the square root of the neutron energy) although some materials have strong absorption peaks at certain neutron energies as shown in Fig. 2 for uranium-238 and resonance absorption can have an effect on reactor stability since the resonances are broadened by temperature and also the energy spectrum of neutrons in a reactor changes with reactor temperature. The absorption cross sections for fast neutrons are much less than for thermal neutrons as illustrated in Fig. 3 for boron and cadmium, and it can be seen that some materials which are very effective for absorbing thermal neutrons may be only weak absorbers of high energy neutrons. The probability that a neutron will cause fissioning also depends on the neutron energy as shown in Fig. 4 for the fission cross section of uranium-235 as a function of neutron energy. The reason for the great influence of materials of low atomic weight such as those containing hydrogen, beryllium or carbon, which are effective in removing energy from neutrons with minimum absorption, can be seen from this figure, since the fission cross section for a thermal neutron is much greater than for a high energy fission neutron. The rate of interaction of neutrons with the moderator is proportional to the scattering cross section and the number of moderator atoms per unit volume.

The energy of thermal neutrons is some 0.025 eV which is many orders of magnitude less than the average energy of fission neutrons. Since the neutron cross section and other properties of reactor materials may change greatly with neutron energy, it is difficult to use average values and, therefore, this wide energy range is usually divided for computations into several steps sufficiently small so that average values of properties will apply. It is convenient to use logarithmic steps and a term, lethargy, is used to express these energy steps and is defined by the relation

$$u = \ln \frac{E_0}{E}$$

where u is the lethargy, E_0 is some reference energy usually taken as 10 MeV, and E is the neutron energy for a lethargy of u.

Reactor calculations are based on a series of equations accounting for the neutrons in a small unit volume. They are generated or diffused into the unit volume and lost by absorption and diffusion. The difference between these terms is the time rate of change of neutron density. Also, this same principle of conservation of neutrons is applied to each of the energy or lethargy groups into which the energy range may have been divided. A neutron can enter an energy range by starting at a higher energy and reacting with moderator atoms, to lose energy, and also in the higher energy range, by fission. A neutron will leave the energy group by further reactions with reactor materials.

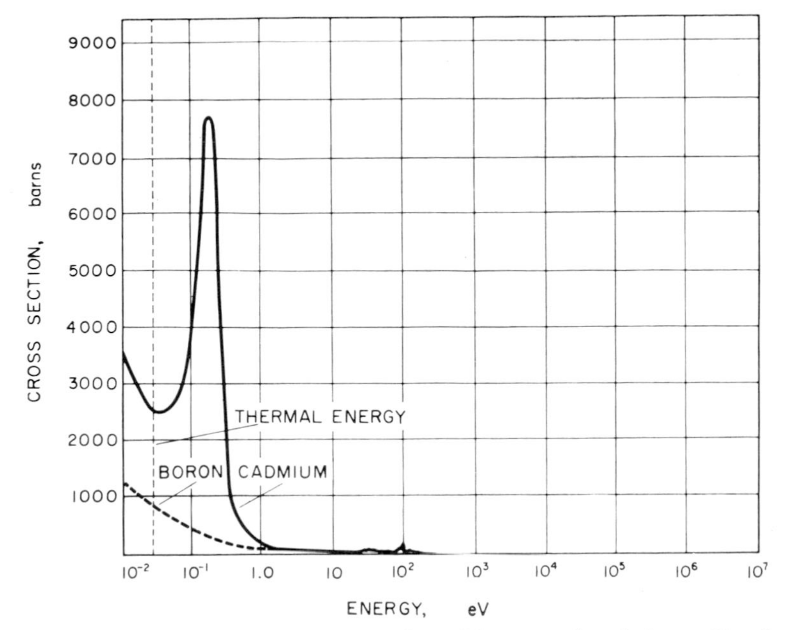

FIG. 3. The neutron capture cross section of boron and cadmium. (Reprint from ORNL Report K–1380 by permission of the United States Atomic Energy Commission.)

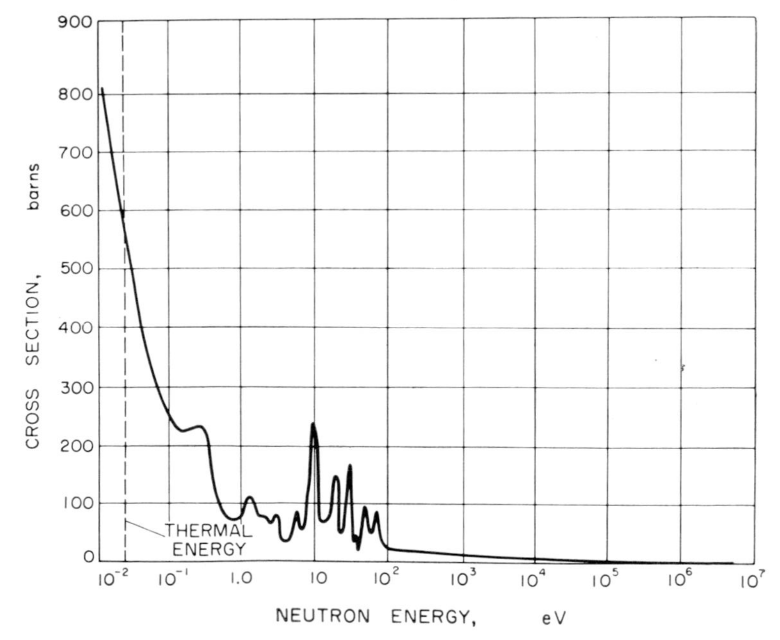

FIG. 4. The fission cross section of U^{235}. (Reprinted from ORNL Report K–1380 by permission of the United States Atomic Energy Commission.)

The number of energy groups used in a calculation may vary from one group calculations considering only thermal neutrons to 20 or more group calculations requiring a digital computer. Procedures for machine computations are highly developed and are usually only limited by the accuracy of the data on neutron processes. It can be seen that the multigroup-multiregion calculations over the core volume in two or three dimensions involve a vast number of computations and only certain special assemblies of the largest computers will have the capacity required for this problem, although most reactor calculations can be made quite adequately using a limited system suitable for the computers that are usually available for this purpose.

The neutron leakage from a reactor is determined by the geometry and arrangement of materials. From simple diffusion theories it can be seen that the leakage of neutrons is proportional to the surface area of the core and for a spherical core the surface area per unit of volume is a minimum. Therefore, for a spherical core the fraction of neutrons that are lost by leakage is less than for any other geometry, and a spherical core will have maximum reactivity for a given core volume. Also, distortion of a core of some other geometry into a shape that is more nearly spherical may decrease neutron leakage and increase reactivity. The average neutron leakage per unit volume of a spherical core is proportional to R^2/R^3 or $1/R$ where R is the radius and, therefore, the leakage per unit volume vanishes for an infinitely large reactor. It is convenient to refer many calculations to an infinite core where by definition leakage is zero. Also, the fraction of high energy neutrons lost by leakage would be expected to be greater than that of thermal neutrons for a given core design. Neutron leakage can be decreased and reactivity increased by placing materials of high scattering and low absorption cross section outside the core. Water (light or heavy) beryllium and graphite are used for this purpose. Placing such materials and particularly any material containing hydrogen near an unreflected core may greatly increase reactivity.

The neutrons not lost by leakage nor absorbed in non-fission processes must react with atoms of fuel to produce fissioning. The number of neutrons thus remaining from those produced in the fissioning of an atom of fuel and available for continuing the reaction is termed the multiplication factor, k. The neutron population will increase by the factor $k-1$ each neutron generation since one neutron is required to produce fissioning. If l is the average time interval between successive generations of neutrons the rate of change of neutron density is therefore

$$\frac{dN}{dt} = N\frac{(k-1)}{l}$$

and this can be integrated to give neutron density as an exponential function of time

$$N = N_0\, e^{(k-1)t/l}$$

where N_0 is the neutron density at the start of the time interval t. The average time interval between successive generations is the sum of the time required for the fission process which is very short indeed and of the order of 10^{-14} sec and the average time after release for the neutron to interact with the moderator and diffuse as a thermal neutron before absorption. The neutron lifetime or generation time, l, varies greatly according to the type of reactor and for thermal reactors, order of magnitude estimates for orientation purposes can often be made by taking for l the slowing down time plus the diffusion time. These times are given for several moderators in the table below for thermal neutrons.

Table II[4]

Slowing Down and Diffusion Time for Thermal Neutrons

Moderator	Slowing down time, sec	Diffusion time, sec
Water	10^{-5}	2×10^{-4}
Heavy water	5×10^{-5}	0.15
Beryllium	7×10^{-5}	4×10^{-3}
Graphite	1.5×10^{-4}	10^{-2}

It can be seen that the neutron lifetime in heavy water is several orders of magnitude greater than in light water. As a result the kinetics of a heavy water reactor are much slower and a sudden change in conditions in a light water moderated reactor which might result in the release of energy with explosive violence may only result in a slower non-violent power increase in a heavy water moderated reactor.

For an effective multiplication factor of 1.001 and a neutron lifetime of 10^{-4} sec the above relation for neutron density as an exponential function of time would indicate that the neutron density would increase by some 22,000 in one sec. Since the power of a reactor is proportional to the neutron density, it can be seen that a device with such a rapid response could not be manually controllable and certainly would be extremely dangerous. Actually, it was known at the time the first reactor was built that some neutrons are emitted by short half-life fission products and are, therefore, appreciably delayed after the prompt neutrons released at the instant of fissioning. If the effective delay time of the delayed fission products is 0.1 sec. and the excess of the multiplication factor over unity does not exceed the fraction of the neutrons that are delayed by this effective time, (each generation of neutrons must wait for these delayed neutrons) the value of l becomes 0.1 sec. From the above relation the increase in neutron density or reactor power in a sec is only 0.14 and, therefore, the reactor would be easily controled within human reaction times. Although the above relations apply to neutrons of only one energy group and some gross approximations have been made about the delayed neutrons, their importance to reactor dynamics is usefully illustrated.

Delayed neutrons are actually released by 6 fission product chains that have been identified with half-lives ranging from 55 sec to 0.277 sec. The yields of delayed neutrons seem to be independent of neutron energy but vary greatly between different fissionable isotopes as shown in Table III.

Table III

Delayed Neutron Yield from Thermal Fission

Isotope	Half-life T_i, sec	Decay constant λ	Delayed neutron fraction
U^{233}	55.00	0.0126	0.00023
	20.57	0.0337	0.00078
	5.00	0.139	0.00066
	2.13	0.325	0.00073
	0.615	1.13	0.00014
	0.277	2.50	0.00009
			0.0026
U^{235}	55.72	0.0124	0.00021
	22.72	0.0305	0.00140
	6.22	0.111	0.00125
	2.30	0.301	0.00252
	0.61	1.13	0.00074
	0.23	3.00	0.00027
			0.0064
Pu^{239}	54.28	0.0128	0.00007
	23.04	0.0301	0.00063
	5.60	0.124	0.00044
	2.13	0.325	0.00068
	0.618	1.12	0.00018
	0.257	2.69	0.00009
			0.0021

The yields of delayed neutrons do not change greatly with the energy of the neutron causing fission, and are therefore nearly the same in a fast and a thermal reactor.

The significant differences between the yields of delayed neutrons by the reactor fuels may be of some importance to reactor safety since a change in reactivity which would be less than the fraction of delayed neutrons in a reactor fueled with uranium-235 may be much greater than the fraction of delayed neutrons for a plutonium fueled reactor. The kinetics of a plutonium fuel reactor may therefore be much different from one containing uranium-235 fuel.

More complete equations for the time behavior of a reactor can now be developed including the delayed neutrons but still limited to one energy group of neutrons and one position in the reactor[5]

$$\frac{dN}{dt} = (k_p - 1)\frac{N}{l_0} + \Sigma\lambda_i C_i + S$$

$$\frac{dC_i}{dt} = -\lambda_i C_i + kB_i \frac{N}{l_0}$$

where N is the effective number of neutrons in the reactor, C_i is the effective number of delayed neutron emitters with disintegration constant λ_i and effective delayed neutron fraction β_i, S is the neutron source strength as neutrons per sec, l_0 is the neutron lifetime, k_p is the prompt multiplication factor given by

$$k_p = k(1 - \beta)$$

where k is the multiplication factor, and β is the total effective delayed neutron fraction

$$\beta = \Sigma\beta_i.$$

A frequently used quantity in reactor kinetics calculations is the reactivity ρ also called the excess reactivity and defined by the relation

$$\rho = \frac{k - 1}{k} = \frac{\Delta k}{k}$$

The dollar is a unit of reactivity defined as that amount of reactivity required to make the reactor critical on prompt neutrons alone, and then

$$1 = \beta$$

where β is the effective delayed neutron fraction. The quantity ρ/β is then the excess reactivity in dollars, and a cent of reactivity is, of course, one hundredth of this value.

An exponential equation for the rate of change of neutron density or reactor power can be written in the form

$$N = N_0 \, e^{t/T}$$

where T is called the reactor period or e-folding time since it is the time required for the reactor power to change by the factor of e or 2.7183.

When the multiplication factor k is unity the reactivity ρ is zero and the neutron density or power of a reactor does not change with time. When k is less than unity, the power of a reactor decreases exponentially to some value depending on the sources of neutrons in the reactor which may include the reactions of gamma radiation with any beryllium or heavy water in the reactor to produce neutrons or a separate polonium–beryllium neutron source.

These neutron sources produce a background neutron flux which also causes some fissioning of uranium and a multiplication of the source strength as k approaches unity. The amount of this multiplication of a source of S neutrons per sec in a reactor, since each such neutron will produce k additional neutrons in each generation, is the summation of those produced in successive generations of the series[6]

$$1 + k + k^2 + k^3 + k^4 \cdots$$

which for $k < 1$ can be expressed as $1/(1 - k)$ and, therefore,

$$N = \frac{S}{1 - k}.$$

The source strength is, therefore, multiplied in a sub-critical reactor by the factor $1/(1 - k)$. For this reason it is very difficult in starting up a reactor to determine just when criticality is reached since a substantial neutron flux may be produced by this multiplication of the source strength when just below criticality. Also, it is possible with an extremely weak source of neutrons such as in a new core or a criticality experiment without a separate neutron source to pass quickly through the critical point without a significant neutron flux and unknowingly get into a dangerous condition of large excess reactivity. The only real test of the condition of criticality is the change of neutron flux with time. Therefore, in starting up a reactor which has an appreciable amount of excess reactivity, it may be considered appropriate to withdraw the control rods in intervals with several minutes of waiting between the intervals to determine whether the neutron flux is coming to an equilibrium value and it is safe to withdraw the rods further or whether the flux is increasing exponentially, thereby indicating a super-critical condition.

Another significant property of a chain reacting system is that a reactor can operate at any flux or power as far as the nuclear processes are concerned, since the condition of criticality only determines the rate of change of flux or power. At criticality the reactor may operate at a few watts or at several megawatts power. The maximum power at which the reactor can be safely operated is determined by the cooling system and it is, therefore, necessary to provide means for accurately determining and regulating power levels to prevent excessive temperatures.

Solutions to the reactor kinetics equations are available in the literature for various conditions such as the prompt jump approximation which is the condition of an instantaneous change in reactivity which then remains constant for a time interval. It is found that following a sudden change in reactivity, the neutron level changes "promptly" to N, where[6]

$$\frac{N}{N_0} = \frac{1}{1 - \rho/\beta}.$$

The change in neutron level then approaches a stable exponential rise, $e^{t/T}$. The time required for the "prompt" jump to take place is $(l/\beta)/(1 - \rho/\beta)$ approximately. The change in N/N_0 for some time following a sudden change in reactivity for a reactor with uranium-235 fuel is shown in Fig. 5. The

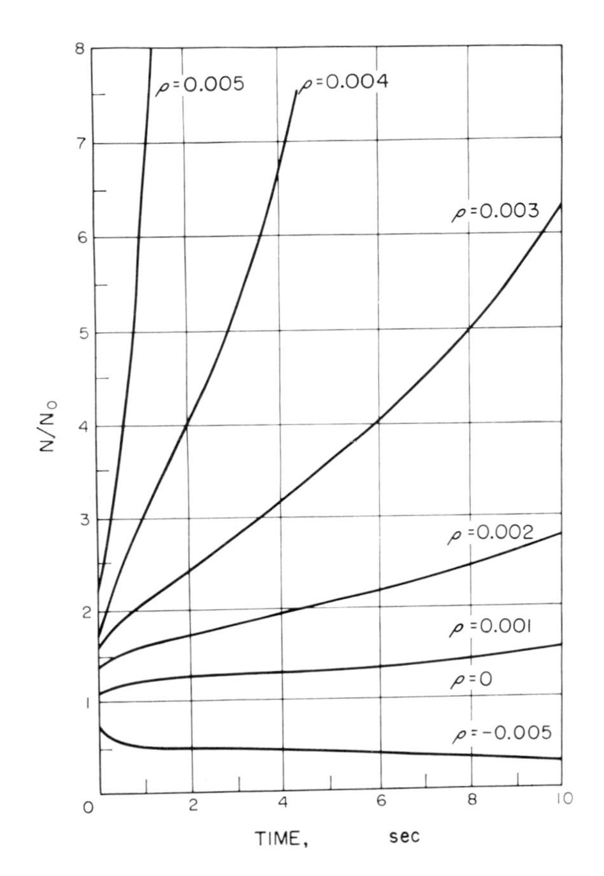

FIG. 5. (Reproduced from Chapter 3, *Reactor Dynamics*, Reactor Safety Monograph, by H. Soodak, Report NDA–2009–12.)

"prompt" jump in this figure appears as instantaneous. In Figs. 6–9 are shown the changes in the asymptotic reactor period T for changes in reactivity expressed as dollars for various neutron lifetimes for uranium-233, uranium-235 and plutonium-239. These curves are useful in many reactor hazards calculations even though the assumption of an instantaneous change in reactivity and a constant value for the reactivity following the change are limiting conditions for what might actually occur in a reactor where an appreciable time interval is required to change reactivity, and then the reactivity is a function of the reactor power.

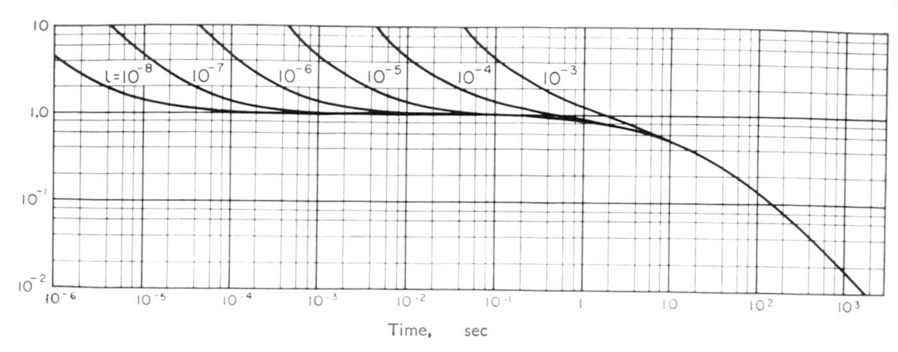

FIG. 6. Reactivity vs. asymptotic period (T) for various neutron lifetimes (l) for U^{233}. (Reproduced from Report ANL–5800 by permission of the United States Atomic Energy Commission.)

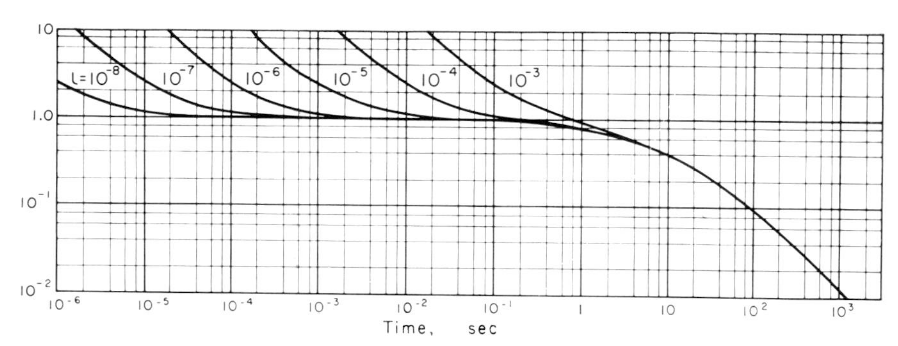

FIG. 7. Reactivity vs. asymptotic period (T) for various neutron lifetimes (l) for U^{235}. (Reproduced from Report ANL-5800 by permission of the United States Atomic Energy Commission.)

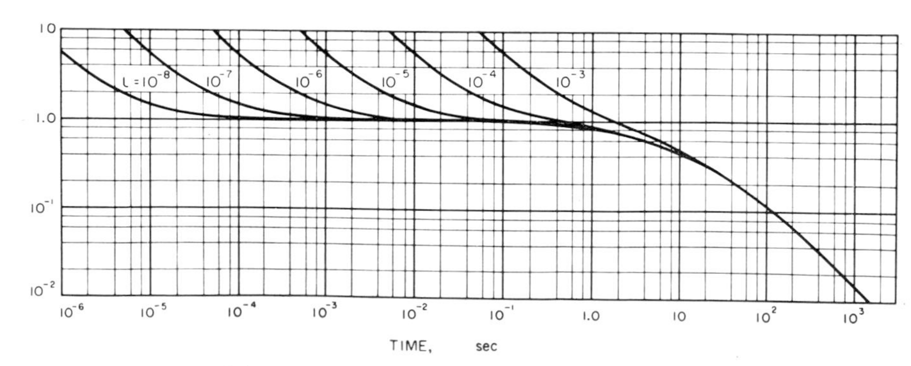

FIG. 8. Reactivity vs. asymptotic period (T) for various lifetimes (l) for Pu^{239}. (Reproduced from Report ANL–5800 by permission of the United States Atomic Energy Commission.)

It can be seen that when the reactivity exceeds appreciably the delayed neutron fraction β, its effect on the neutron reproduction time diminishes and for changes of reactivity of several dollars, the reactor period is determined by the neutron diffusion and slowing down times, which as has been noted above, may be very short with the result that the reactor period may become so small that the reactor cannot be controlled within human reaction times. Therefore,

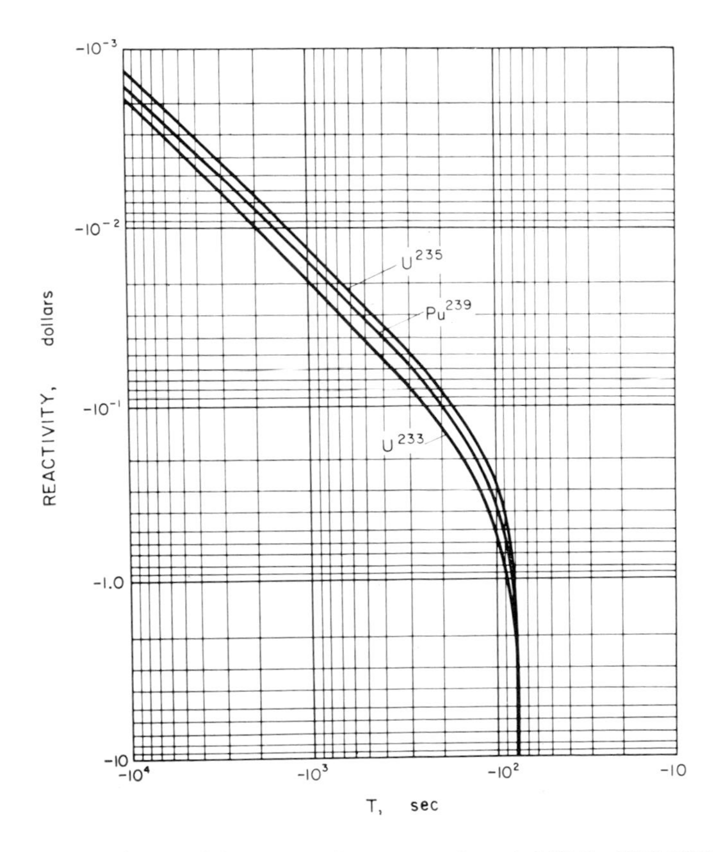

FIG. 9. Negative reactivity vs. negative asymptotic period (T) for U^{233}, U^{235} and Pu^{239}. (Reproduced from Report ANL–5800 by permission of the United States Atomic Energy Commission.)

the total amount of reactivity which can be inserted into a reactor under all possible conditions requires very careful study in hazards evaluation. It is also of interest to note that delayed neutrons are as effective in limiting the period of a fast reactor as for a thermal reactor for reactivity changes that are small as compared to β, the delayed neutron fraction. However, the short neutron lifetime in a fast reactor would make it very dangerous indeed under conditions of prompt criticality and this condition is, of course, the basis for a fission type nuclear bomb. Therefore, the total amount of reactivity in a fast reactor must be limited for safety and fortunately, this requirement does not seriously limit the operating time because the reactivity does not change rapidly with fuel burn-up as the result of the large fuel inventory required for a critical mass.

THE INHOUR

From the technology of the Hanford reactors has come a unit of reactivity expressed in inverse hours and called, therefore the inhour, *Ih*, and defined as that amount of reactivity which will make the stable reactor period equal to 3600 sec or 1 hr. The inhour is used where amounts of reactivity are involved that are much smaller than the fraction of delayed neutrons. For this limiting situation certain simplifications of the usual kinetic equations can be used. The usual equation relating reactor period T and reactivity is

$$\rho = \frac{l}{Tk} + \sum_{i=1}^{m} \frac{\beta_i}{1 + \lambda_1 T}$$

By making a number of simplifying assumption permitted by the assumed small value of reactivity it can be shown that

$$Ih \cong \frac{3600}{T}.$$

RAMP CHANGES IN REACTIVITY

A constant reactivity insertion rate is termed a ramp change and reactivity at any time t is given by the relation

$$\rho = \rho_0 + ct$$

where c is the fractional change in reactivity per sec $d\rho/dt$. If the control rods of a reactor are withdrawn at a constant rate an approximation to a ramp change results over those areas of control rod position where their effectiveness is a linear function of position. It can be seen, therefore, that the ramp functions describing reactor dynamics are a convenient mathematical model for describing approximately the condition of constant rate of control rod withdrawal.

Relations can be derived based on the one-group one-region model with further simplifying assumptions depending on the rate of reactivity change. For rates of reactivity insertion of a few dollars per sec and higher, the time scale of events is very rapid and as a consequence, the fractional change in delayed neutron emission is small and may be neglected. This is termed the "rapid rate approximation". For small rates of reactivity change of some cents per sec and less, the change in reactivity in the average neutron delay time is not large and as a consequence, the delayed neutron emission at any time is close to the value that it would have if the reactivity were held constant at its value at any time. Even with such simplifying assumptions for these special cases, the algebraic expressions become cumbersome and the use of an intermediate insertion rate greater than some cents per sec and less than some

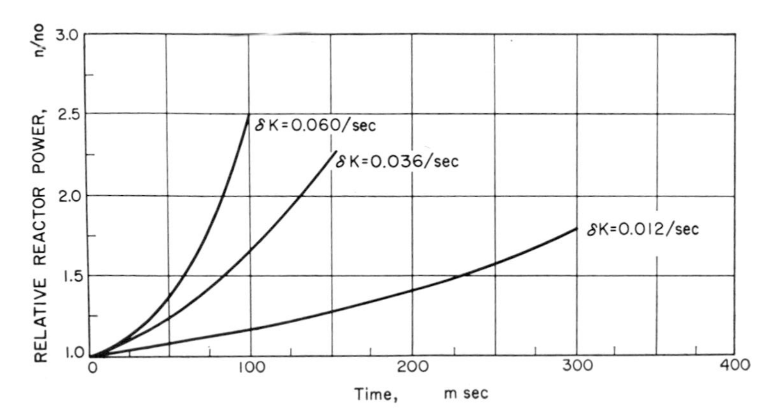

FIG. 10. Relative reactor power for k varying linearly with time. At time zero the reactor is critical. (Reproduced from Report WAPD–34, Vol. 1, by permission of the United States Atomic Energy Commission.)

dollars per sec requires direct numerical calculations. These problems have been solved for many cases with the aid of computers which make possible the inclusion of all the important delayed neutron groups and several groups of neutron energies.

VARIABLE REACTIVITY

It has been seen that a sudden increase in reactivity causes a "prompt" jump in the neutron density, N. Also, from kinetic equations it is found that a sudden change in $d\rho/dt$ similarly causes a prompt change in dN/dt. The actual inverse reactor period, α, is defined by the relation

$$\alpha = \frac{1}{N}\frac{dN}{dt}$$

and the change in α caused by the sudden introduction of an insertion rate, $d\rho/dt$, is

$$\Delta\alpha = \frac{d\rho/dt}{\beta - \rho}.$$

It can be seen that the reciprocal of $\Delta\alpha$ is the time required to reach prompt critical at the present rate of insertion of reactivity, and in an initially critical reactor, the insertion rate of b dollars per sec would cause the neutron level to begin rising with a period of $1/b$ sec. In an initially sub-critical reactor starting at x dollars below critical, the insertion rate of b dollars per sec would cause a period of $(1 \times x)/b$ sec. Assuming one group of delayed neutrons with an equivalent decay constant λ, it follows that

$$\alpha = p + \frac{d\rho/dt}{\beta - \rho}.$$

where there is no neutron source. Here p is the stable inverse period corresponding to the instantaneous value of ρ and is

$$p = \frac{\lambda\rho/\beta}{1 - \rho/\beta}.$$

As a result of the prompt jump in α caused by a reactivity insertion rate, instrumentation which measures the rate of change of neutron level or reactor power can provide for safety not only against a stable period that is too short for safety but, and perhaps of more importance, also protection is provided against a rate of rise of reactivity that would, if continued, bring the reactor up to prompt criticality and then result in periods that would be too short for safety.

THE XENON PROBLEM

In the startup of the first Hanford plutonium production reactor it was found that there was an unexpected decrease in reactivity associated with operations at high power levels. The study of this problem led to the discovery that xenon-135 formed as a fission product and from the decay of iodine-135 has a remarkably high thermal neutron absorption cross section of about 2×10^6 barns; and, therefore, this isotope can have a significant effect on reactors which operate at high neutron flux. Also the rate at which xenon-135 is destroyed by neutron absorption decreases with the neutron flux when a reactor is shut down while its production may continue at an appreciable rate by decay of the shorter half-lived iodine-135. The result is that the amount of xenon poisoning may increase significantly following reactor shut down, and may prevent the restarting of the reactor unless a large amount of excess reactivity is provided for over-riding the xenon poisoning during a period of up to a day following shut down and depending on the reactor neutron flux.

Xenon-135 is formed and removed by the following reactions

$$Te^{135} \xrightarrow{2m} I^{135} \xrightarrow{6.7h} Xe^{135} \xrightarrow{9.2h} Cs^{135} \xrightarrow{3 \times 10^6 y}$$

$$\rightarrow Ba^{135}(\text{stable})$$

$$Xe^{135} + n \rightarrow Xe^{136}(\text{stable})$$

Samarium-149 is another poison similar to xenon-135 in that they both may be removed by neutron capture at a rate equal to their production and, therefore, achieve an equilibrium concentration in a reactor operating at constant power. The following table gives values at equilibrium of the ratio of the poison cross section to the uranium absorption cross section for thermal neutrons.

Table IV[9]

Ratio of Poison Cross Section to Fuel Absorption Cross Section at Equilibrium for Thermal Neutrons

Thermal flux	Xenon	Samarium
n/cm² sec		
10^{12}	0.007	0.012
10^{13}	0.033	0.012
10^{14}	0.053	0.012
10^{15}	0.056	0.012

It is seen that the equilibrium xenon poisoning is not significant in very low flux reactions but becomes appreciable at a flux over 10^{12} and approaches a maximum value at a flux of 10^{15}. Since samarium-149 is stable, its equilibrium poison value does not increase with flux. In Figs. 11-13 are shown plots of the fraction of the xenon shut down value remaining at various times after shut down. At flux levels greater than 3×10^{13} it can be seen that the xenon concentration increases significantly and at a flux level of 10^{15} it would be difficult to operate a thermal reactor.

If sufficient excess reactivity is provided to over-ride the xenon poisoning and a reactor is started with a large excess of xenon poisoning over the equilibrium value, the neutrons will remove the excess xenon and a positive power coefficient of reactivity can result since an increase in power will cause the xenon in excess to be removed at a greater rate.

The neutron absorption cross section of xenon-135 decreases rapidly at neutron energies above 0.1 eV and this decrease can cause a positive component in the temperature coefficient in some cases. Although the temperature corresponding to this neutron energy is very high, as the temperature of a reactor is increased a larger fraction of the neutrons may have energies above this value. An intermediate reactor with the average energy of the neutrons causing fission in the range of 0.1 eV may be influenced by changes in xenon poisoning with neutron energy. A fast reactor operates with neutron energies above the level where the xenon absorption cross section is appreciable and, therefore, xenon poisoning is not significant. Calculations of the influences of neutron energies on xenon poisoning in thermal and epithermal reactors require the use of detailed multigroup procedures, since the nuclear properties are very sensitive to neutron energies.

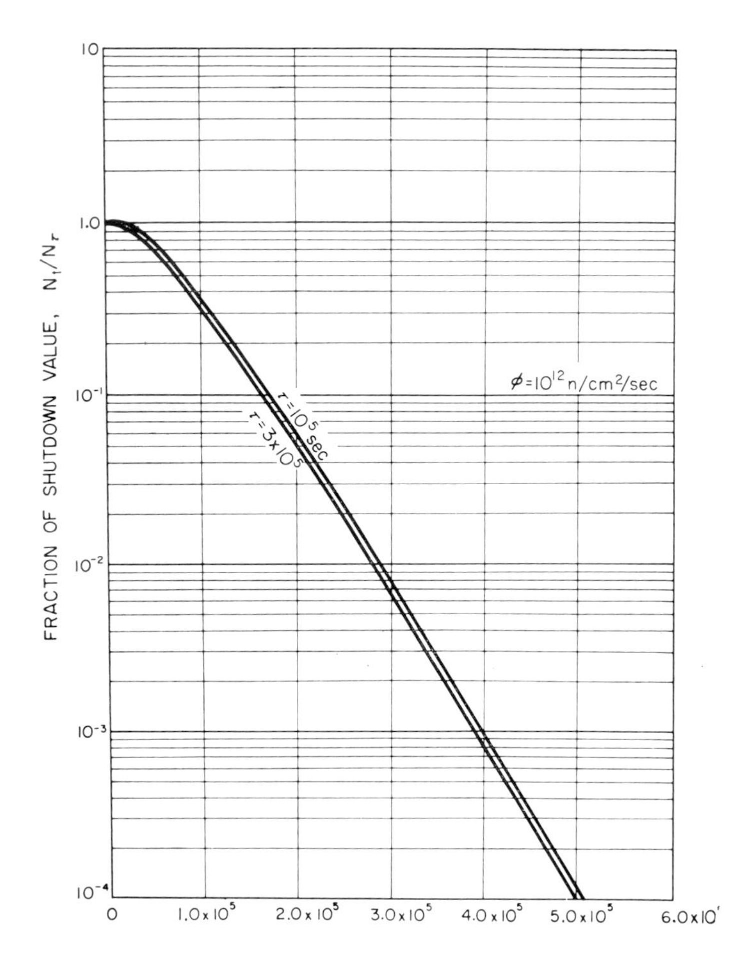

FIG. 11. Fraction of shutdown values of 9.2hr. Xe^{135} remaining at various times after shutdown. (Reproduced from "Uranium–235 Fission Product Production as a Function of Thermal Neutron Flux. Irradiation Time and Decay Time," by J. O. Blomeke, and Mary F. Todd, Report ORNL–2127.)

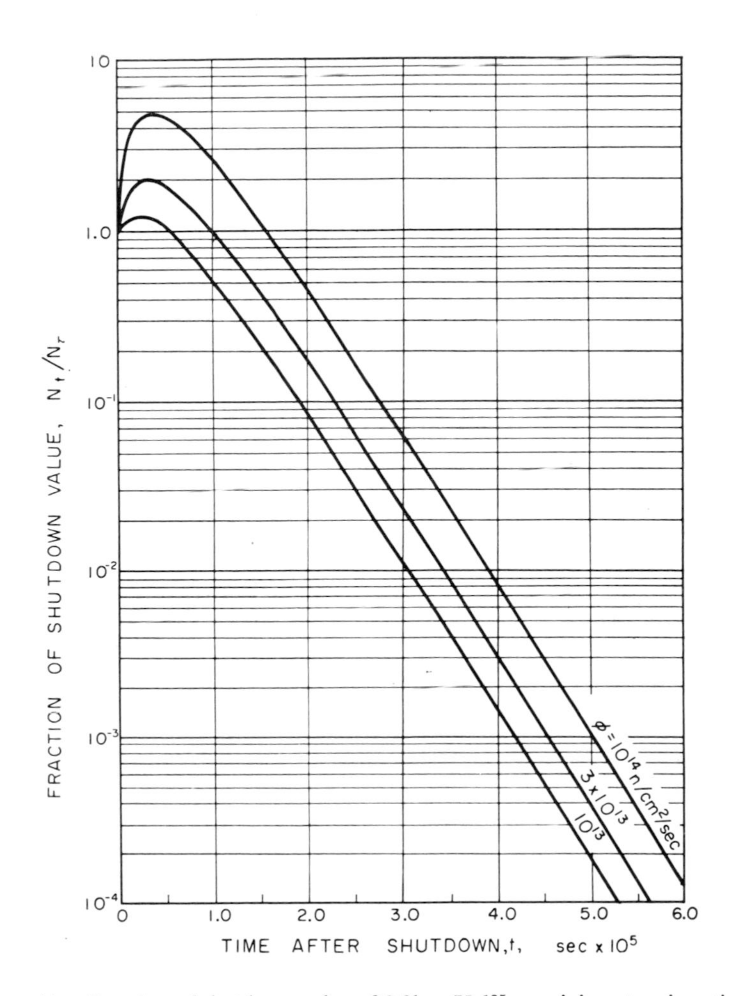

FIG. 12. Fraction of shutdown value of 9.2hr. Xe^{135} remaining at various times after shutdown. (Reproduced from "Uranium–235 Fission Product Production as a Function of Thermal Neutron Flux, Irradiation Time and Decay Time," J. O. Blomeke, and Mary F. Todd, Report ORNL–2127.)

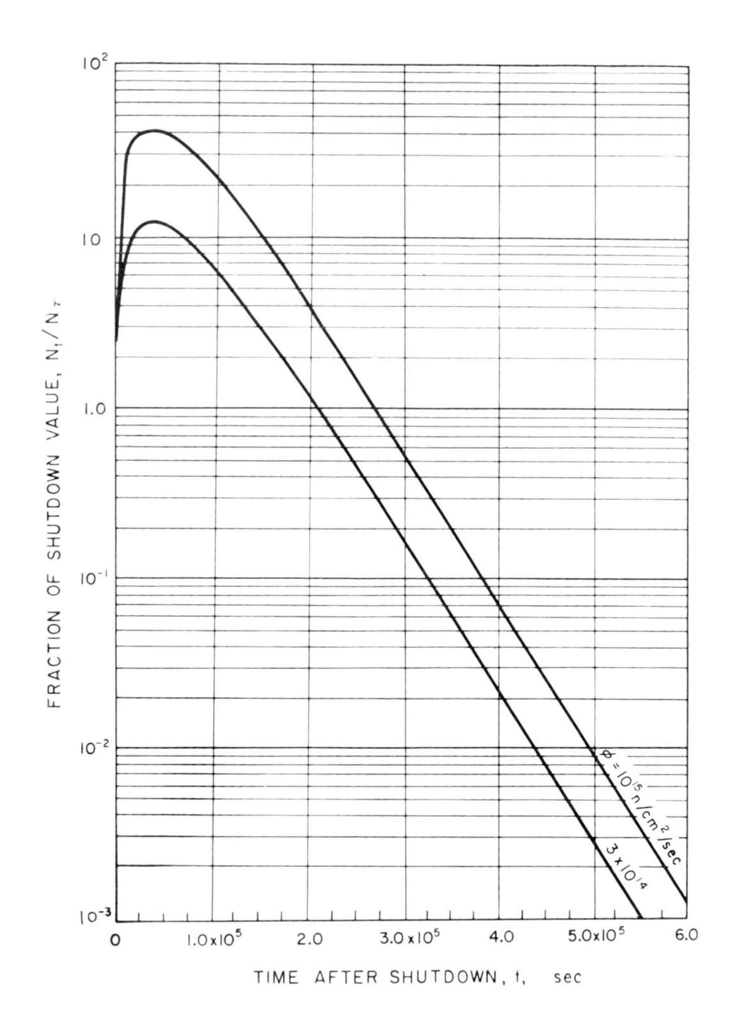

FIG. 13. Fraction of shutdown value of 9.2hr. Xe^{135} remaining at various times after shutdown. (Reproduced from "Uranium 235 Fission Product Production as a Function of Thermal Neutron Flux, Irradiation Time and Decay Time," J. O. Blomeke, and Mary F. Todd, Report ORNL–2127)

FEEDBACK AND STABILITY*

When a reactor is operating at a significant power level, the physical states of the components vary with the power level through changes in temperatures, pressures, densities and positions of elements. These changes due to power variations may cause reactivity changes or feedback. A power increase that results in a positive feedback or reactivity increase causes instability. Negative feedback is required for stability.

Feedback due to heating effects is absent in most criticality experiments when operated at a few watts (zero) power. However, there might be feedback in such a device due to instrumentation response, control rods, pump speeds, valve openings, etc. An important case of positive feedback occurs in high flux thermal reactors and is due to xenon-135 poisoning. An increase in neutron level causes a drop in the xenon-135 content due to burn-out by neutron absorption and a reactivity increase.

The reactivity at any time will be divided into components as follows:

$$f = \delta\rho_{ext}(t) + \delta\rho$$

$$\delta\rho = \sum_n \delta\rho_n$$

where $\delta\rho_{ext}(t)$ represents some external source of reactivity, and the $\delta\rho_n$ are the feedback reactivities due to changes in reactor components.

In order to describe adequately the feedback due to heating, it may be necessary to consider separately the temperatures of various parts of the reactor with reference to the effect of power on the temperatures and the effects of the temperatures on the reactivity.

As an example, consider the division of the reactor into only two parts, part one being all the fuel rods and part two being all the coolant in the core. Let T_1 and T_2 be the corresponding average temperatures of these parts. If the power of the reactor is proportional to the neutron density N, the effect of the power on the heating of fuel by fissions and the cooling of fuel by coolant can be described by an equation of the type:

$$\frac{dT_1}{dt} = -\frac{(T_1 - T_2)}{\tau_1} + b_1 N.$$

The net rate of heat removal from the coolant due to transport is proportional to $(T_{out} - T_{in})$ where T_{out} is the outlet coolant temperature and T_{in} is a fixed inlet temperature for the coolant. Assuming that $(T_{out} - T_{in})$ is proportional to $(T_2 - T_{in})$, a relation describing the heating of the coolant by fuel and the

* Material in this section is taken at length from *Reactor Dynamics*, by H. Soudak, Reference 5.

removal of heat by transport of coolant out of the reactor is:

$$\frac{dT_2}{dt} = \frac{(T_1 - T_2)}{\tau_{21}} - \frac{(T_2 - T_{in})}{\tau_{22}}.$$

The time constants, τ_1, τ_{21}, and τ_{22} depend on heat coefficients and flow rates. At equilibrium reactor level N_o the temperatures are:

$$T_{01} = T_{in} + \left(1 + \frac{\tau_{22}}{\tau_{21}}\right) b_1 \tau_1 N_0$$

$$T_{02} = T_{in} + \frac{\tau_{22}}{\tau_{21}} b_1 \tau_1 N_0 .$$

For small departures of the temperatures around their equilibrium values, the effects on reactivity may be assumed linear and:

$$\delta\rho_1 = a_1(T_1 - T_{01})$$

$$\delta\rho_2 = a_2(T_2 - T_{02}).$$

where a_1 and a_2 are the temperature coefficients of reactivity of fuel and coolant respectively.

For quasistatic (very slow) power changes, the temperatures at any time will be in equilibrium with the power level at that time. If it is assumed that the coolant parameters such as flow rate and inlet temperature are held constant so that τ_1, τ_{21} and τ_{22} remain constant, then quasistatic changes satisfy the relations:

$$\delta T_1 = \left(1 + \frac{\tau_{22}}{\tau_{21}}\right) b_1 \tau_1 \delta N$$

$$\delta T_2 = \frac{\tau_{22}}{\tau_{21}} b_1 \tau_{21} \delta N$$

where

$$\delta N = N - N_0$$

$$\delta T_1 = T_1 - T_{01}$$

$$\delta T_2 = T_2 - T_{22} .$$

As a consequence, a quasistatic change δN results in a reactivity change

$$\delta\rho_1 + \delta\rho_2 = a_1 \delta T_1 + a_2 \delta T_2 = \left[a_1 + \frac{\tau_{22}}{\tau_{21}}(a_1 + a_2)\right] b_1 \tau_1 \delta N.$$

Since by definition, the quasistatic power coefficient of reactivity is the reactivity change per unit fractional quasistatic change in N,

$$\left[\frac{\delta\rho}{\delta N/N_0}\right]_0 = \left[N_0 \frac{\delta\rho}{\delta N}\right]_0$$

where the subscript zero outside the brackets denotes that the changes $\delta\rho$ and δN occur quasistatically.

The net quasistatic power coefficient due to heating effects in the case being considered is then:

$$\left[\frac{\delta\rho_1 + \delta\rho_2}{\delta N/N_0}\right]_0 = \left[a_1 + \frac{\tau_{22}}{\tau_{21}}(a_1 + a_2)\right] b_1\tau_1 N_0.$$

The overall quasistatic power coefficient for the reactor is obtained by repeating the above steps for the other feedback mechanisms and then summing up over the separate coefficients. A positive power coefficient causes a monotonic building up of any small departure δN from the equilibrium level N_0, and thus in instability. A negative quasistatic power coefficient signifies that a small departure δN from the equilibrium level N_0 does not build up monotonically with time. It may be, however, that the departure oscillates and that the amplitude of the oscillation builds up with time. Thus, a negative quasistatic power coefficient does not insure stability.

The stability of the whole reactor system can be investigated by finding the natural periods of small vibrations of the reactor around the level N_0. The reactor is stable around N_0 if all the natural vibrations are damped. The natural periods are found by assuming that the departures of all quantities (δN, δC_1, δT_1, ρ etc.) from their values at level N_0 behave as e^{pt} where it is understood that p may be a complex number. For the example of feedback due to heating considered above

$$N = N_0 + \delta N$$

$$C_i = C_{i0} + \delta C_i$$

$$T_1 = T_{01} + \delta T_1$$

$$T_2 = T_{02} + \delta T_2$$

$$\rho = \delta\rho = \delta\rho_1 + \delta\rho_2.$$

The last equation makes use of the fact that the reactor is critical ($\rho_0 = 0$) when all quantities have the values corresponding to steady operation at level N_0. Assuming that all departures are differential quantities, and neglecting products of differentials, the above equations can be used to obtain "linearized" reactor equations. The departure quantities may be eliminated from the linearized equations by the use of the following relations to give an equation for p whose roots are the inverse natural periods:

$$\frac{d\delta N}{dt} = p\delta N$$

$$\frac{d\delta C_i}{dt} = p\delta C_i$$

$$\frac{d\delta T_1}{dt} = p\delta T_1$$

$$\frac{d\delta T_2}{dt} = p\delta T_2$$

The linearized basic neutron equations are:

$$\frac{d\delta N}{dt} = \frac{N_0}{\lambda}\,\delta\rho - \frac{\beta}{\lambda}\,\delta N + \sum_i \lambda_i \delta C_i$$

$$\frac{d\delta C_i}{dt} = -\lambda_i \delta C_i + \beta_i \frac{\delta N}{\lambda}$$

By use of equations for the departure quantities and the relation

$$\delta\rho = \left[\frac{\delta\rho}{\delta N/N_0}\right]_p \frac{\delta N}{N_0}$$

leads to the equation

$$Z(p) = 0$$

where

$$Z(p) = \left[\lambda p + \beta - \sum_i \frac{\lambda_i \beta_i}{\lambda_i + p}\right] - \left[\frac{\delta\rho}{\delta N/N_0}\right]_p.$$

The quantity

$$\left[\frac{\delta\rho}{\delta N/N_0}\right]_p = N_0\left[\frac{\delta\rho}{\delta N}\right]_p$$

is the power coefficient of reactivity, or the reactivity change per unit fractional change in N, for departures with time dependence e^{pt}. The quasistatic power coefficient

$$\left[N_0\frac{\delta\rho}{\delta N}\right]_0$$

is the value of the general power coefficient for p = 0. The general power coefficient

$$N_0\left[\frac{\delta\rho}{\delta N}\right]_p$$

is a function of p and is called the transfer function of the feedback mechanisms. The first bracketed quantity on the right side of the equation for $Z(p)$ may be called the neutron transfer function. The function $Z(p)$ is then the total transfer function. The roots of the equation

$$Z(p) = 0$$

are the inverse natural periods of the reactor for vibration around N_0, and if all the roots have negative real parts, then the reactor is stable for small vibrations around N_0. If any one root has a positive real part, the reactor is unstable. Real roots for this equation may be found by plotting each of the bracketed quantities of the equation for $Z(p)$ versus real p and noting the intersections. Plots versus real p of the values of

$$\sum_i \frac{\beta_i}{\beta} \frac{\lambda_i}{\lambda_i + p} - 1$$

are available in the literature. In cases where the feedback transfer function is a ratio of polynomials in p, the knowledge of the real roots may be used to reduce the degree of the equation. If the degree of the reduced equation is not too high the remaining (complex) roots may be found without undue difficulty. In cases where the feedback transfer function is transcendental or otherwise complicated, the existence or non-existence of roots with positive real parts may be determined without finding all the roots. A method for doing this is that of the Nyquist Stability Criterion.[10]

In the example considered above of feedback due to heating in a reactor, considering only the fuel rods and the coolant, the linearized temperature equations are

$$\frac{\mathrm{d}\delta T_1}{\mathrm{d}t} = -\frac{(\delta T_1 - \delta T_2)}{\tau_1} + b_1 \delta N$$

$$\frac{\mathrm{d}\delta T_2}{\mathrm{d}t} = \frac{(\delta T_1 - \delta T_2)}{\tau_{21}} - \frac{\delta T_2}{\delta T_1}$$

Here it is assumed that the inlet temperature T_{in} is constant and that the time constants τ_1, τ_2, and τ_3 are constant. If these are not constant, then their departures must be included. Thus, for example, the term

$$\frac{T_{01} - T_{02}}{\tau_{01}^2} \delta\tau_1$$

must be added to the right side of the equation for $\mathrm{d}\delta T_1/\mathrm{d}t$ in order to take into account the departure $\delta\tau_1$ of τ_1 from its value corresponding to operation at N_0. Also to be added is the linearized equation describing how τ_1 varies as N and temperatures vary. Using equations

$$\frac{\mathrm{d}\delta T_1}{\mathrm{d}t} = p\delta T_1$$

$$\frac{\mathrm{d}\delta T_2}{\mathrm{d}t} = p\delta T_2$$

in the linearized temperature equations leads to

$$\delta T_1 = \left(\frac{1}{A} + \frac{1}{AB}\right) b_1 \tau_1 \delta N$$

$$\delta T_2 = \frac{1}{B} b_1 \tau_1 \delta N$$

where

$$A = 1 + \tau_1 p$$

$$B = \frac{\tau_{21}}{\tau_{22}} + p\left(\tau_1 + \tau_{21} + \frac{\tau_1 \tau_{21}}{\tau_{22}}\right) + p^2(\tau_1 \tau_{21})$$

These equations together with the relation for the linear effects of temperature on reactivity

$$\delta\rho_1 = a_1(T_1 - T_{01})$$

$$\delta\rho_2 = a_2(T_2 - T_{02})$$

give for the feedback transfer function in this example

$$\left[\frac{\delta\rho}{\delta N/N_0}\right]_p = \left(\frac{a_1}{A} + \frac{a_2}{B} + \frac{a_1}{AB}\right) b_1 \tau_1 N_0$$

When a nuclear reactor is coupled to a power plant the analysis of the transient response and stability of the system must include the influences of the significant parts of the power cycle as well as the several reactor components. The resulting equations are usually too complex for convenient numerical solution by the above method, and, therefore, computing machines are used for studies of the kinetics and stability of complex systems. The analogue computer can be used for this purpose, and the assumption of linearized functions given above is a convenient simplification. The stability of the system and its natural frequencies can be investigated directly with an analogue computer by inserting various perturbations into the system. Special purpose analogue computers may be constructed for the solution of reactor problems and are termed reactor simulators.

CIRCULATING FUEL

If the reactor fuel is a liquid (solution, slurry, or fuzed salt) and is circulated through an external heat exchanger, some of the delayed neutrons will be emitted outside the core and therefore are lost. The fraction of delayed neutrons effective in reactor kinetics is therefore reduced in a circulating fuel reactor and the reactivity corresponding to prompt critical (one dollar) is reduced. The result is a reduction in reactor period for a given increase in reactivity in this type of reactor. Analysis of the equations describing the time behavior of a reactor with circulating fuel are available in the literature. In general the very strong negative temperature coefficient due to thermal expansion of the fuel out of the core makes homogeneous reactors very stable and the reduced effective fraction of delayed neutrons due to fuel circulation has not been found, in the reactors of this type which have been operated, significantly to affect safety.

REFERENCES

1. HENRY, H. F., Editor, Studies in nuclear safety, ORNL Report K–1380.
2. Reactor Physics Constants, ANL–5800.
3. HUGHES, D. J., & SCHWARTZ, R. B., Neutron cross sections, Report BNL 325, (Supplement Number 1).
4. GLASSTONE, SAMUEL, *Principles of Nuclear Reactor Engineering*, Van Nostrand, Princeton, New Jersey.
5. SOODAK, H., Reactor dynamics, Chapter 3, AEC Safety Monograph, Report NDA 2009–12.
6. BONILLA, C. F., *Nuclear Engineering*, McGraw–Hill, New York.
7. CONNOR, J. C., Pile kinetics for linear rate of delta *k* insertion—The worst accident solution, WAPD–RM–33, 1951.
8. BLOMEKE, J. O., and TODD, MARY F., Uranium-235 fission-product production as a function of thermal neutron flux, irradiation time and decay time, Report ORNL–2127, June 10, 1958.
9. SOODAK, H., Determination of reactor characteristics, Chapter 2, AEC Safety Monograph, Report NDA 2009–11.
10. Reactor kinetics, quarterly progress report, Oct.–Dec., 1956, Report NAA–SR–2134.

CHAPTER 4

CONTROL AND SAFETY SYSTEMS

INTRODUCTION

ADEQUATE means for safely increasing the neutron flux during start-up until the desired operating power is reached and of controlling the power are required for routine operations in addition to shut down mechanisms for normal and emergency conditions. The control and safety system must provide these functions and several others associated with the safety of the system, including automatic reduction of power or shut down under conditions of excess power, temperatures above established limits in the coolant or fuel elements, loss of power to instruments, low coolant flow, reactor period shorter than established limits, or for other reasons considered critical to the safe operation of the installation. Particular attention is, therefore, given to the method for controlling the reactor including the actuating mechanisms, the sensing devices for neutron flux and other variables indicating the conditions of operation, and the circuits which interconnect the elements of the control system with the interlocks and safety devices built into the control and safety system. The system may be quite simple for a low power research reactor of a type which has been proven to be inherently stable; however, for reactors designed for operation at high power levels, the control and safety systems incorporate multiple units in parallel with interlocks and many automatic safety provisions. Reactors designed for routine operation usually have automatic controls.

METHODS OF CONTROL

The factors which affect neutron multiplication can be used for reactor control as follows:

(1) Moving Fuel into or out of the Core

This method of control can be easily visualized for an aqueous homogeneous reactor where the concentration of fissionable material in solution can be increased until the desired power level is reached. This method of control would be convenient where the fuel solution is circulated through the reactor and an external loop. For small homogeneous research reactors which do not have a fuel circulating system, the amount of solution in the reactor core can be adjusted as a rough or shim control. In solid fuel reactors it may be inconvenient to provide adequate cooling for a movable fuel element in all positions of a movable element and this system of control is not usually used in power reactors.

(2) Moving a Neutron Absorbing Material into or out of the Core

A conventional control system for a thermal reactor would be an arrangement for moving rods containing boron or cadmium into and out of the active region of the core. Other materials with high neutron capture cross sections might be used such as gadolinium or hafnium.

(3) Changing the Amount of Moderator in the Core

A water moderated reactor such as the Canadian NRX natural-uranium heavy-water reactor, can be shut down by dumping the water from the core and shim control can be obtained by adjusting the level of water. The negative temperature coefficient of reactivity for reactors which have water or organic liquid moderators is obtained by having less than the optimum amount of moderator so that thermal expansion of the moderator and expulsion of liquid from the core causes a decrease in reactivity. A shut down mechanism is also provided by boiling of the liquid since the vapor displaces its volume of moderator.

(4) Movement of the Neutron Reflector

Removal of the neutron reflector will increase neutron leakage and decrease reactivity. This method of control has been used for a fast reactor since control by poison materials is not as effective as for thermal reactors at the high neutron energies in a fast reactor. In the Los Alamos Fast Plutonium Reactor (Clementine) a block of uranium metal could be dropped from the assembly thereby increasing neutron leakage and stopping the chain reaction. This device became inoperative due to distortion of components and it was considered safe to operate with this block of uranium held in position with jacks, since there was sufficient operating experience to give some assurance that the reactor could be safely operated without this additional safety feature.

(5) Increase in Core Surface

Since the neutron leakage is proportional to the core surface area an increase in surface area will increase the leakage and decrease reactivity. This increase in core dimensions will usually occur from thermal expansion and the resulting component of the overall temperature coefficient of reactivity associated with thermal expansion of the core structure is usually negative.

ROD SYSTEMS

The rate at which neutrons are absorbed by a rod of neutron absorbing material depends upon several factors in addition to the flux level in the vicinity of the rod. This rate of absorption is a function of the mass of absorbing material, its geometrical arrangement which is usually expressed as the surface area, the position of the rod in the core, the presence of the other control rods, and, of course, the neutron absorption cross section of the material averaged over the energy spectrum of the neutrons. These factors are related by complex functions and it is usually considered necessary for a new reactor design to determine experimentally the control rod effectiveness in a mock-up reactor simulating the nuclear characteristics of the reactor to be built. Also it usually is considered necessary to calibrate the control rod effectiveness by a series of experiments at the time of start up of a new reactor and after any significant modifications to the core. Cladding of the rods with a corrosion resistant material such as stainless steel, is provided to protect the rod and to prevent contamination of the system with particles of the neutron absorbing material. Although this system of control with poison rods is generally used, there are some disadvantages such as the loss of the absorbed neutrons which might have been used to produce a useful material, and a major change in the flux and power distribution within the active core due to the control rods. In the Materials Testing Reactor the movable fuel and poison methods are combined in the shim-safety rods which have cadmium in the upper section and fuel in the lower section. As the rods are lifted from the reactor the cadmium is removed and the fuel inserted.

The types of control rods for a typical reactor may be classified into groups as follows:

(1) *Shim rods* which have fairly large amounts of reactivity (several dollars) are used to shut down the reactor under controlled conditions and to maintain the reactor at near critical from hour to hour and day to day.
(2) *Safety rods* which are used to shut down the reactor in the case of emergency. These are sometimes the same as the shim rods. There should, however, be some safety rods that are always kept cocked except during shutdown. It may be argued that even during shutdown there should be some amount of reactivity held in reserve.
(3) *Control rods* which perform reactor control and regulation. There may be two of these—a coarse control rod typically containing a little over a dollar of reactivity, and a fine control rod, called the regulating rod, which contains less than a dollar of reactivity.

Reactors are kept at steady power operation by means of the regulating rod. This rod typically usually contains less than a dollar reactivity as noted above, and is capable of fairly rapid motion. Regulation may be provided either manually or automatically by means of a servo system. In either case it is convenient to have as the control signal the difference between the desired power and the actual power. The detector for the control device may be a compensated ion chamber or it is sometimes sufficient to control the reactor directly from the reactor power itself. In this case the usual detector is one of the power level indicators. In Table I are listed the rod systems used in some reactors. It is to be noted that the common practice is to use both safeties and shims. The usual technique is to have these rods held by magnetic clutches. Rods may be moved in or out of the reactors slowly by means of hydraulic or electric motors. In an emergency the magnetic clutches release the rods and they quickly go to their least reactive position under the action of gravity or some stored energy system such as a spring.

CONTROL DRIVE MECHANISMS

The control drive mechanisms are as varied in type and design as the reactors that have been built. The control rods have been moved manually in a few reactors; however, the control mechanisms for power reactors are highly developed devices which represent a major part of the cost of some

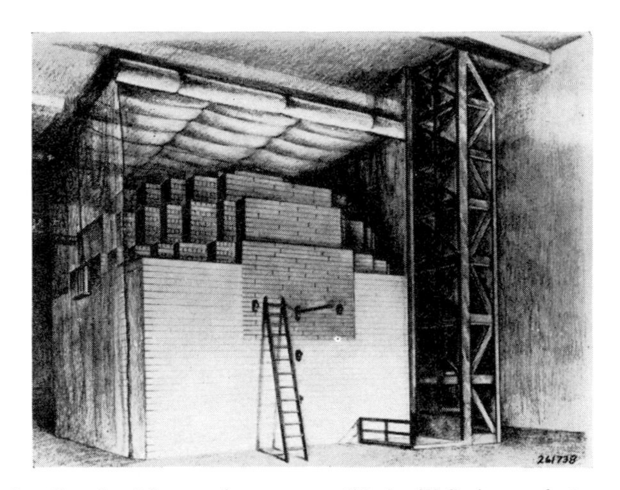

FIG. 1. Sketch of first nuclear reactor CP–1. (U.S. Army photograph.)

reactors. Certain types of rod drives have been standardized, particularly for the Naval Reactor Program, so that these complex mechanisms are produced with high reliability at minimum cost.

The control drive mechanisms must perform several functions in a reliable and reproducible manner including the following:

(1) Move the rod out of the reactor at a controlled rate which should be reversible at any time and with the motion of the rod limited at the extremes.
(2) Provide an indication of the position of the rod at any position with positive indications when the control element is fully inserted or fully withdrawn.

Table I

Reactor Rod Systems

Reactor	Shim Rods Type	Shim Rods Δk, %	Shim Rods Material	Safety Rods Type	Safety Rods Δk, %	Safety Rods Material	Coarse Control Rods Type	Coarse Control Rods Δk, %	Coarse Control Rods Material	Regulating Rods Type	Regulating Rods Δk, %	Regulating Rods Material
BNL	Poison	3.5	Boron	Same as shims						Poison	0.5	Boron
WBNS				Poison		Boral	Poison		Cadmium	Poison	0.1	Cadmium
Livermore L3				Poison		B_4C	Poison	1.4	B_4C	Poison	0.75	B_4C
BSR	Poison	5 to 7.6	PB–B_4C	Same as shims						Poison	0.8	Pb–B_4C
MTR	Poison-fuel	40	Cd–U	Same as shims						Poison	0.5	Cadmium
KAPL–NTR	Poison	6		Poison	1.5	Cadmium	Poison	1.5	Cadmium	Poison	0.1	Cadmium
CP–5	Poison	~40	Cd	Same as shims							0.4	

(Taken from "AEC Safety Monograph," by Brooks, W. L., and Soodak, NDA 2009–8.)

(3) Have stored energy (gravity, spring, inertia of a flywheel) to drive the rod rapidly into the reactor when other sources of power may be lost.
(4) Stop the rod at the end of high-speed insertion without damage (shock absorber, dash pot, snubber).
(5) Hold the rod from bouncing back out of the reactor following rapid insertion or prevent the rod from falling out of the reactor in the case of a mobile reactor which may be inverted.
(6) Prevent the rods from being blown out of the reactor by internal pressure upon loss of power or failure of a component.
(7) Provide a seal at the rod drive to prevent leakage for pressurized reactors.
(8) Have a release device (usually a magnetic clutch or a lifting magnet) which is fail safe so that loss of electrical power will cause rapid rod insertion.
(9) Provide for simultaneous movement of combinations or groups of rods.
(10) Prevent side motion or vibration of the rods under all conditions of coolant flow and rod position.
(11) Cool the rod including the part that is out of the core.
(12) In some cases where radioactive parts of the rod may be withdrawn through the shielding, special rod shielding may be required.

The specifications for the rod drive mechanisms will vary with the type of reactor but some general requirements which are common to several reactors have been cited. Safety rods are designed with an effectiveness large enough to keep the reactivity, k, less than unity by 0.1 to 0.2 under all operating conditions. About 5–10 min is usually allowed for safety rod withdrawal and the rate of withdrawal is uniform. For safety action in a "scram" a decrease in reactivity of 0.01 within 0.1 sec may be a design requirement. Release of the rod in a scram should occur with 0.05 sec after the release signal is received. The effective length of the shock absorber in stopping the rod should be limited to 5 per cent or less of the rod travel. The total effectiveness of the rods should be greater than the changes in reactivity due to temperature coefficients, fission product poisoning including the part of maximum xenon poisoning the system is to override, and the fuel depletion between loadings. The maximum rate of change of reactivity with the control rod drives should be about 0.01 per cent/sec although this value depends on the design of the system and the errors and time lags involved. The control rod motion should be reversible within 0.01 sec. For non-automatic regulating rod operation, the speed of motion is usually limited to a smaller value than may be permitted by an automatic regulating system depending upon the time response and the possible errors in the system.

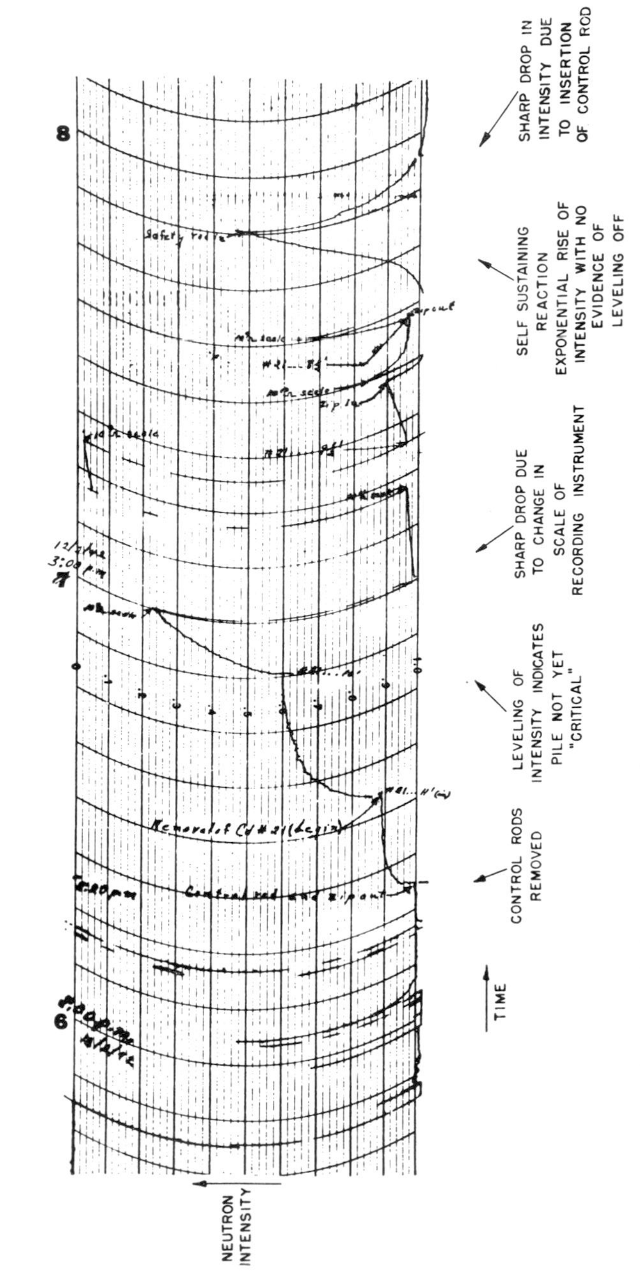

FIG. 2. Galvanometer recording of start up of CP–1. (Argonne National Laboratory.)

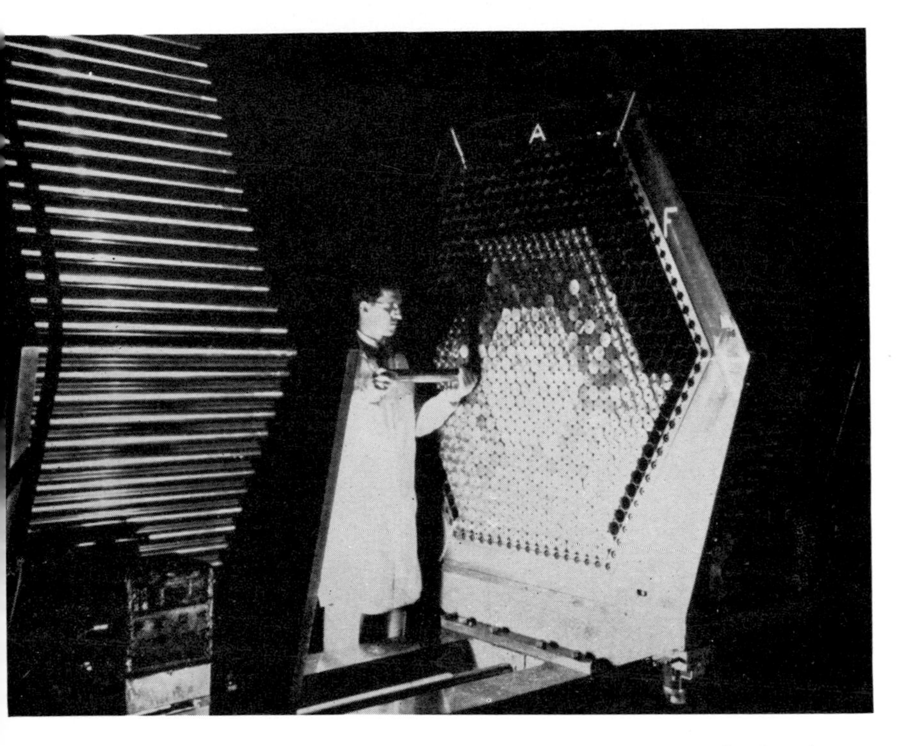

FIG. 3. Criticality experiment in preliminary pile assembly (Knolls Atomic Power Laboratory). (General Electric Power photograph.)

The control drive mechanism designed by the Argonne National Laboratory for their low power reactor (ALPR), as described in Report ANL-5744, is a typical design illustrating the arrangement of components. The mechanical drive incorporates the basic concept of transmitting rotary motion into linear motion by means of a rack pinion. A rotary shaft pressure seal is used. The pressure seal is of positive clearance, break-down seal ring type, which has controlled leakage and requires a minimum of maintenance. The leakage is bled to the pre-cooler and the condensate is returned to the tank. Approximately 0.1 gpm of water is bled continuously from each control drive housing or thimble. The system is driven by a reversible electrical motor. The external drive is positively engaged with the pinion shaft by means of a magnetic clutch. Failure of clutch current automatically inserts the rods rapidly into the core by gravity. Rapid insertion (scram) is accomplished within 3 sec over the full travel. The mechanism is so designed that a scram signal will not only release the magnetic clutch but the downward drive will further try to drive the rod down by positive action through a "free wheeling" clutch. In the event of a power failure, the control rod motor current is supplied by

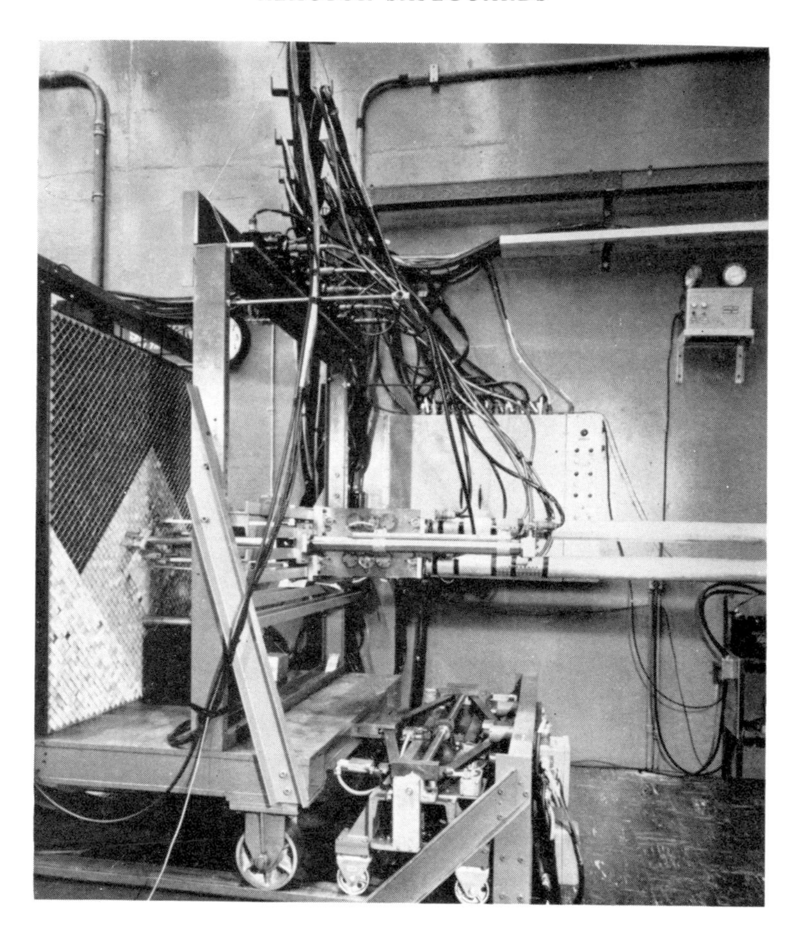

FIG. 4. Movable half of the FPR critical assembly (Knolls Atomic Power Laboratory). (AEC Schenectady Operations Office photograph.)

the emergency power system. The speed of normal rod travel is restricted by mechanical gearing. At the beginning of zero power tests, the speed was set at 3 in/min. Prior to power operation, the rate was adjusted by replacing gears so as to limit the reactivity insertion rate to approximately 0.01 per cent/sec, as determined from period measurements of differential rod worth. A torque-type shock absorber and stop arrangement, located within the drive mechanism, functions during rod drops. Position indicating devices are installed directly on the pinion drive to indicate position of the rods at all times within 0.05 in. over the full travel of 31 in. Since the drive is external, all motors and critical parts are readily accessible for maintenance or replacements.

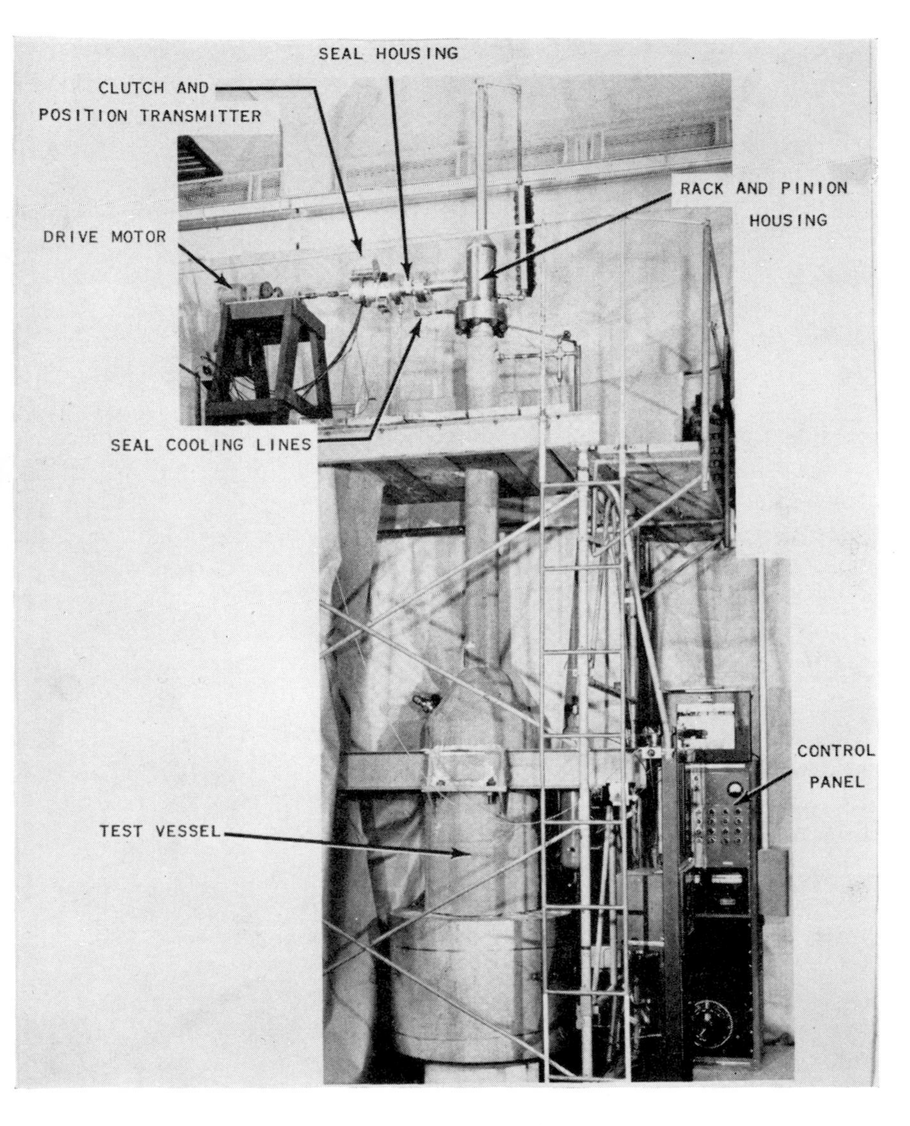

FIG. 5. Control rod drive test facility. (Reproduced from "Hazards Summary Report on the Argonne Low Power Reactor, ALPR," ANL–5744, Nov. 1958.)

In the development of this control drive mechanism, the test rig shown in Fig. 5 was used and an actual ALPR prototype mechanism was operated for 1550 hr at operating conditions with over 8000 cycles and 250 scrams being made after which visual inspection of the parts indicated satisfactory performance.

BACK-UP SHUTDOWN SYSTEM

In several reactors a completely independent system is provided in addition to the safety shim shutdown rods, to make the reactor subcritical. For reactors with a liquid moderator which is not essential for cooling the fuel elements, such as the NRX heavy-water-moderated-light-water-cooled reactor, the heavy water may be dumped into a storage tank thereby positively shutting down the machine. A back-up safety is provided for the Brookhaven Reactor in the form of a tank of chlorobenzene which can be forced into channels within the reactor thereby inserting considerable volume of neutron absorbing material. On some reactors steel balls containing boron may be fed from a hopper into channels leading to the interior of the reactor. For water cooled and moderated reactors a back-up shutdown system which provides for the addition of boric acid to the reactor water, has been included on some designs. In the case of ALPR a 120 gall storage tank is provided which contains boric acid in solution at a concentration of 100 g per gall of water. A manually operated pump is provided which has a capacity of 25 gall/hr when the reactor is operating at pressure, corresponding to a reactivity removal rate of at least $2\frac{1}{2}$ per cent/hr when the boric acid solution is pumped into the water in the reactor. When the reactor is at

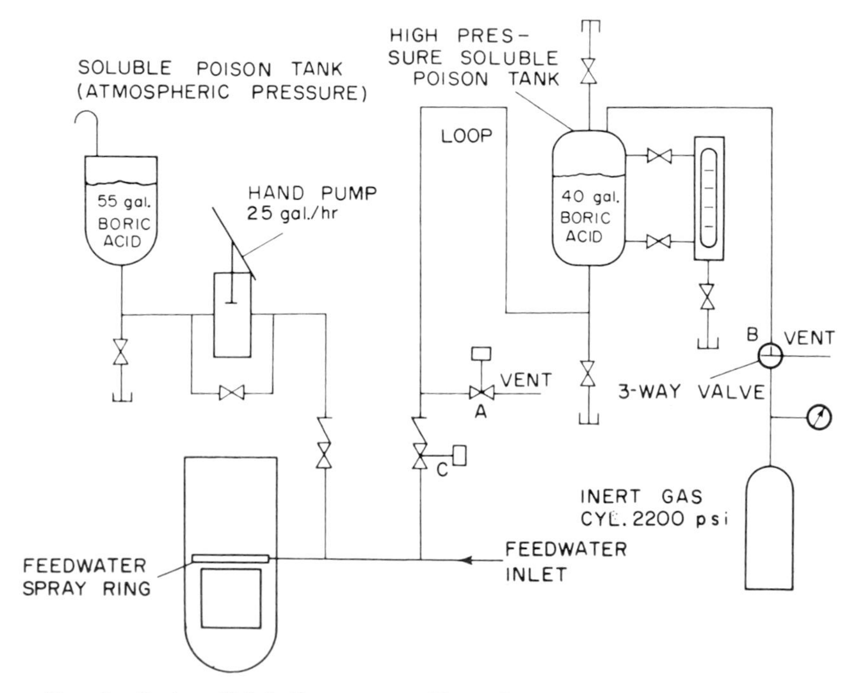

FIG. 6. Boric acid injection system. (Reproduced from "Hazards Summary Report on the Argonne Low Power Reactor ALPR," ANL–5744, Nov. 1958.)

atmospheric pressure, the solution may be introduced by gravity through a bypass hose into the reactor water. The boric acid can be removed from the reactor water by circulating it at room temperature through an auxiliary anion resin bed, Fig. 6.

REACTOR FUSES AND SAFETY DEVICES

The most difficult-to-guard-against possible sources of a serious reactor malfunction appear[3] to be failures of the control system and sabotage. All systems are subject to sabotage provided sufficient time is available for informed saboteurs to work. Therefore, the Reactor Safeguards Committee repeatedly recommended since 1950 that safety devices be developed which are self-actuating and self-contained without connections external to the reactor core and which would require the concerted efforts for some time period of more than one person to remove from the reactor core. The device should shut down the reactor on excessive power, temperature, or rate of change of power or temperature. The fuse should fail safe and have a short reaction time.

These requirements are difficult to fulfil and several possible systems have been considered as follows:

(1) An explosive charge containing a piece of fissionable material which overheats and ignites the charge thereby disassembling the reactor during a power surge. Such a device could have application to criticality experiments but use in a reactor is not attractive since the explosive would be decomposed by radiation and also it is undesirable to have an explosive material around nuclear facilities since the dispersal of radioactive material is to be avoided.

(2) Metal differential expansion elements with one of the metals perhaps containing fissionable material would detect changes in neutron flux. It appears that the response times of the designs that have been studied are too slow to be effective.

(3) Thermopiles using fission heat to actuate a safety system. Thermoelectric elements probably are too sensitive to radiation damage to be satisfactory; however, a thermionic device such as a cesium cell might be applicable.

(4) Radiation damage elements such as transistors might be used to actuate safety device; however, ionization sensitive devices seem to have inconsistent breakdown points.

(5) A material with a low Curie point might be used as the armature of a holding electromagnet so that on overheating, the magnetic properties would change and release the safety rod.

(6) A solder plug perhaps containing fissionable material that melts on overheating and releases a fluid under pressure.

Projects for the development of reactor fuses have been initiated principally at Atomics International Division of North American Aviation. This work under the direction of N. E. Huston is described in Volume II of the Proceedings of The Second United Nations International Conference on The Peaceful Uses of Atomic Energy, 1958[2]. Their safety devices ordinarily comprise the following components: a neutron sensor, a trip mechanism, a stored energy source and a stored poison. The sensor detects the danger situation and converts the measured quantity into a form of energy that will actuate the trip mechanism. The trip mechanism is a means of releasing the stored energy which propels the poison from its stored location to the location where it is more effective in absorbing neutrons. The solder-joint type of trip mechanism is considered feasible where the ambient neutron flux levels are greater than 5×10^{11}. Gas expansion is used for mechanisms with small energy requirements. A neutron absorbing material in gaseous or powder form is stored under pressure in a small "source" chamber which is separated from a larger "receiver" chamber located in the active region of the reactor core. The neutron absorber is either located outside the active core region or is in such a small chamber that the area presented for neutron absorption is small. In the event of an excursion the increased neutron flux triggers the solder-joint mechanism which releases the absorber into the receiver chamber thereby decreasing the reactivity of the reactor.

A safety device has been designed to be used in water cooled research reactors to limit the maximum operating power. The system shown in Fig. 7 was designed to fit into an unoccupied fuel element position, and is completely self-contained. This device contains between the storage and receiver chamber an intermediate chamber, the walls of which contain rupture diaphragms. A release tube leads from the intermediate chamber into the receiver and is sealed by a fission heater fuse which is soldered in place. In the "ready" state, air at atmospheric pressure is in the receiver, helium at 625 psi is in the intermediate chamber, and boron trifloride at 1250 psi is in the storage chamber. At the desired neutron flux level, the seal breaks releasing the helium pressure in the intermediate chamber, which produces an excessive pressure difference across the first diaphragm. The diaphragm bursts subsequently allowing the gaseous poison to expand into the receiver and produce the desired reduction in reactivity of the reactor.

Tests in the SPERT-I reactor were conducted to demonstrate the feasibility of this device. Five devices of this design were subjected to power transients which had periods in the range 110–10 msec. Rapid reactor shutdown was produced in each case with a large reduction in power and total energy release, as illustrated in Table II and Fig. 8. The total energy release was reduced in each case by a factor of 10 or more. Although these experiments have shown that research reactor safety devices are feasible, a number of tests must be conducted before their reliability has been fully demonstrated. The

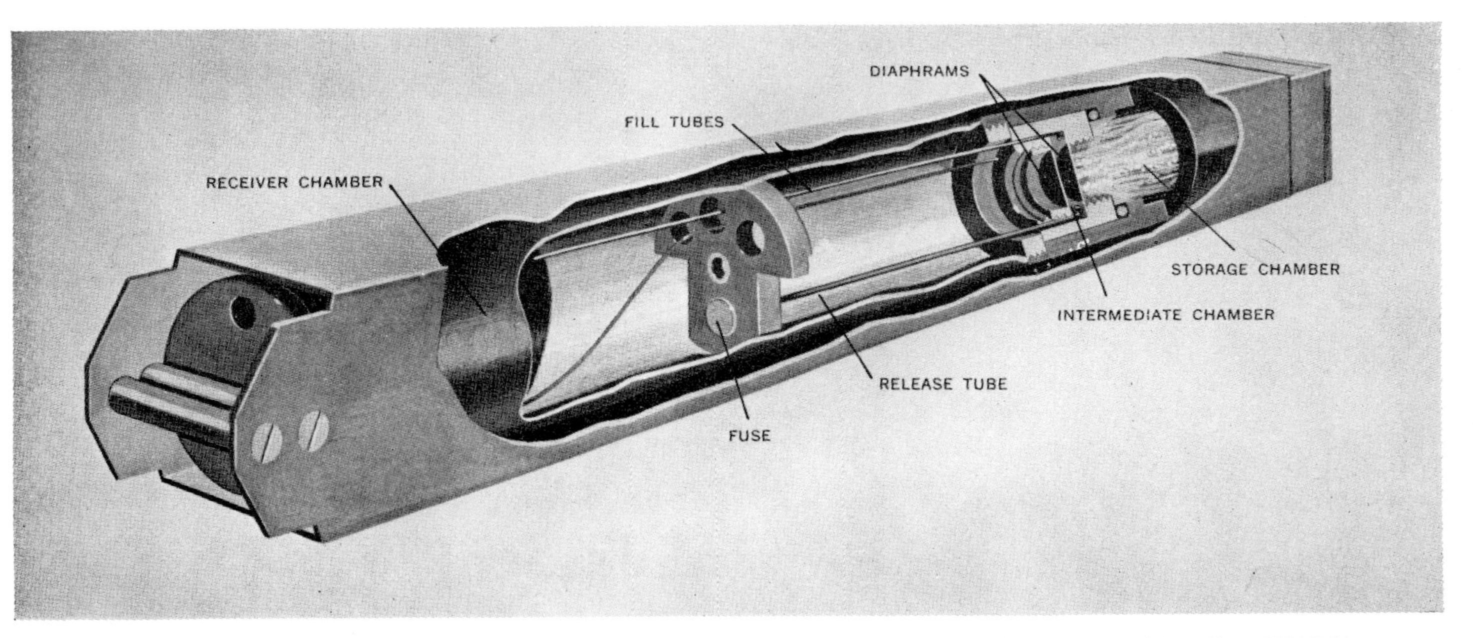

FIG. 7. Reactor safety device. (Reproduced from *Research and Development in Reactor Safety*, B. J. Garrick, Editor, USAEC.)

Table II
SPERT Test Data

Reactor period milliseconds	Reactor peak power (megawatts)	
	Without device	With device
107	16.0	8.15
53	68.2	13.8
28.6	244	22.3
12.3	1690	62.2
10.2	2210	86.5

(Taken from *Research and Development in Reactor Safety*, B. J. Garrick, Editor, USAEC.)

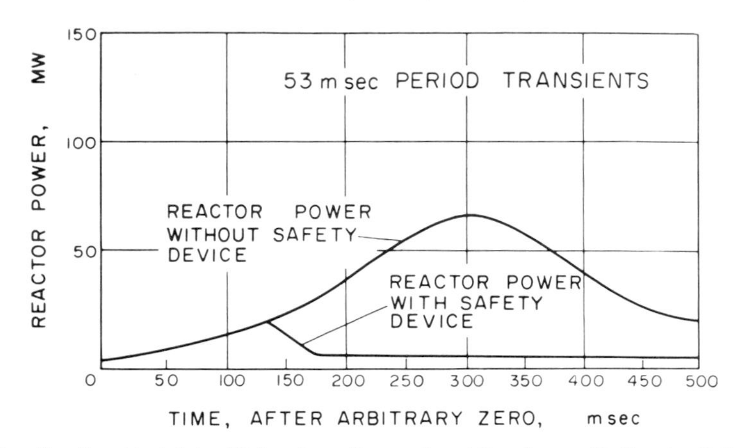

FIG. 8. Spert test data. (Taken from *Research and Development in Reactor Safety*, B. J. Garrick, Editor, USAEC.)

most promising device for power reactors appears to be a double gas chamber separated by a ruptureable diaphragm, with a fission heater in the source chamber. An increase neutron flux heats the gas in the source chamber rapidly producing a differential pressure which causes the diaphragm to rupture. Since a change in the ambient temperature increases the pressure in both gas chambers, such a fuse could be made relatively independent of ambient temperature.

REACTOR PROTECTION SYSTEMS

Nuclear instrumentation and associated automatic protective systems are provided to shut down the reactor in the event of excessive power or temperature or other conditions which endanger the reactor. This instrumentation may become so complex that the reactor cannot be operated for longer than short periods of time before its operation is interrupted by false signals or scrams. Also, in some cases, such as a submarine power plant, a sudden loss of power may create more danger than the condition in the reactor which is to be protected. Certainly a false shutdown is undesirable; however, failure of the safety system is dangerous. Therefore, combination relaying methods are used to improve the speed and reliability of the safety systems. For example, similar scram signals may be indicated by sensing elements connected directly to relays either electronic or magnetic or a combination of both. With at least three such assemblies provided, the relaying system may be designed so that, if two of the three measure a dangerous condition, a scram is initiated. Parallel systems are also used in which a fast-acting electronic relay is placed in parallel with the magnetic relay and, if failure of the electronic relay occurs, a magnetic relay will function. Such a system has the advantage of high-speed relay action in addition to the reliability of magnetic relays. To preclude the failure of one element or instrument in a measuring system, the readings of several elements may be converted to voltages and these voltages are compared by circuits called auctioning circuits, which are so arranged to send only the highest voltage to the shutdown sensing device. In addition, in some systems various degrees of safety action are taken according to the measurement of the condition which is abnormal. For example, in the reactor flux measuring system, if the indicated flux exceeds the set point by varying amounts the following actions will result:

20 per cent above set point:	Warning signal.
30 per cent above set point:	Shim-rod motion will be stopped and locked and no manual override for possible further out-motion will be possible until the causative condition is clear.
40 per cent above set point:	Slow-reversal will be initiated.
up to 50 per cent above set point:	Slow scram will be initiated.
over 50 per cent above set point:	Fast scram will be initiated.

In the design of a typical reactor for experimental power generation, the use of parallel circuits is indicated by the following:

A. Three channels with two channels indicating off-standard conditions will trip the safety circuits for abnormal pressure of the primary coolant, abnormal high-power level, liquid level in the steam generator, high pressure in the steam generator, low pressure of instrument air, and earthquakes of sufficient force.

B. Two channels with either channel indicating off-standard conditions will trip the safety circuits for abnormal flow of coolant in any process tube, electrical power failure, and short reactor period.

C. Two channels where both channels must indicate off-standard conditions to trip the safety circuit for high activity in the building ventilation air.

For the Argonne Low Power Reactor (ALPR), described in Report ANL–5744, abnormal system conditions which result in the release of the control rod drive clutchs and subsequent rapid insertion of all rods, together with abnormal conditions which result in enunciator operations, are listed in Table III. In some cases, (condenser pressure, reactor water level, reactor pressure) an abnormal condition first actuates the enunciator and, if allowed

Table III

Operating Conditions Resulting in Reactor Shutdown or Annunciation

Item	Operating conditions	Conditions causing annunciator operation	Conditions causing automatic reactor shutdown
1	Reactor neutron flux, high—Channel I	×	×
2	Reactor neutron flux, high—Channel II	×	×
3	Reactor neutron flux ion chamber high voltage supply, Channel I de-energized	×	×
4	Reactor neutron flux ion chamber high voltage supply, Channel II de-energized	×	×
5	Reactor water level, high or low	×	×
6	Reactor steam pressure, high	×	×
7	Main steam pressure relief valve, steam flow	×	×
8	Main steam safety valve, steam flow	×	×
9	Condenser pressure, high	×	×
10	Reactor period, short	×	×
11	Actuation of manual "Scram" pushbutton	—	×
12	Control power, de-energized	—	×
13	Rod drive clutch rectifier, failure	—	×
14	Bypass steam flow, high	×	—
15	Bypass valve discharge pressure, high	×	—
16	Condenser outlet air temperature, high	×	—
17	Condenser inlet air temperature, low	×	—
18	Feed pump discharge pressure, low	×	—
19	Purification water temperature, high	×	—
20	Air ejector after condenser temperature, high	×	—
21	Shield cooler outlet water temperature, high	×	—
22	Hotwell water level, high or low	×	—
23	Feed pump, automatic switchover	×	—
24	Reactor pressure, low	×	×
25	High pressure condensate tank water level, high	×	—
26	Low pressure condensate tank water level, high	×	—
27	Turbine oil inlet temperature, high	×	—
28	Thermal shield temperature, high	×	×
29	Shield cooler pump discharge pressure, low	×	—

× Yes — No

(Reproduced from Report ANL–5744, by permission of the United States Atomic Energy Commission.)

to progress further, causes reactor shutdown. Key-operated switches are provided for bypassing the reactor period and low pressure shutdown circuit.

In addition to the condition that all scram circuits be satisfied, interlocking circuits present initial withdrawal of control rods during reactor start-up unless (1) control power is switched and (2) all rods are completely inserted.

In this reactor the measurement of instantaneous reactor power level is accomplished through the use of fixed neutron flux detection channels as follows:

Channel I—An uncompensated, boron-lined ion chamber is used to drive a panel-type of microammeter and a sensitive moving coil relay in series. This channel furnishes rough reactor power information in the power range above the minimum of 1 per cent of full power and in addition, the relay can be adjusted to provide a high-flux shutdown at a pre-set power level.

Channel II—A second uncompensated, boron-lined ion chanber is connected to drive a multi-range amplifier-type microammeter and a sensitive moving coil relay in series. This channel will furnish reactor power information in the operating power range and below over a range of approximately five decades. In addition, a duplicate of the Channel I shutdown circuit actuation is provided.

Channel III—A gamma-compensated, boron-lined ion chamber is employed in conjunction with logarithmic period amplifiers to furnish log flux level and period information. The period circuit is so arranged to shut down the reactor in the event of a period shorter than a pre-set minimum (10 sec) when reactor power level is within 3 decades of full power.

Channel IV—A second gamma-compensated, boron-lined ion chamber is coupled to the power level indicator and recorder through a d.c. amplifier. This furnishes reactor power information over approximately 7 decades.

Channels V–VI—Two identical channels consisting of BF_3 proportional counters, pulse amplifiers and scaler-counters are provided. These channels are employed during core loading and initial reactor start-up for the detection of very low neutron levels.

In Fig. 9 are shown the effective ranges of the various channels of nuclear instrumentation. In Fig. 10 is shown the instrument block diagram for ALPR. In addition to the above channels, Channel VII, which is employed to monitor air ejector exhaust gamma-ray activity, is shown.

Depending upon the particular reactor, other quantities may be monitored and connected to trips. Examples of such quantities are:

(1) *Fuel element temperature.* The fuel element temperature in several places in the reactor may be monitored and the reactor shut down, if the temperature exceeds specified limits in any spot.

(2) *Rate of change of fuel element temperature.* The rate of change of the fuel element may be controlled to prevent damage to the elements due to excessive temperature. This limit on the rate of change of temperature may also apply during shutdown. For example, the Brookhaven Reactor is shut-down in such a way that the rate of change of temperature of the most rapidly changing fuel element is no greater than 2°C per min.

(3) *Coolant flow rate, coolant temperatures and pressure.* In addition to instruments which provide control and safety for the reactor it is necessary to have means to monitor the radiation level around the reactor. This will include instruments to count neutrons and gamma-beta rays. General survey instruments are required as well as hand and foot monitors for personal use.

A control system operating manual should be prepared including (1) a description of all control systems, (2) operating procedures to cover all normal and emergency circumstances, (3) a list of permitted modifications with circumstances which justify such modifications, and the appropriate operating procedures and (4) a list of absolutely prohibited modifications. Also, when the reactor is shut down, control systems should be locked to prevent accidental control rod withdrawal.

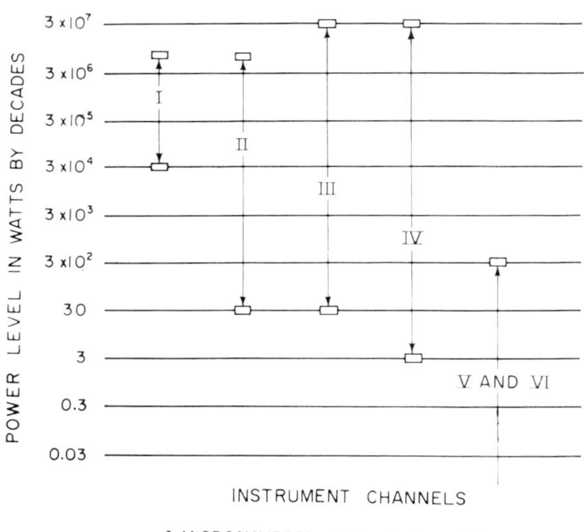

FIG. 9. Flux instrumentation. (Reproduced from Report ANL-5744 by permission of the United States Atomic Energy Commission.)

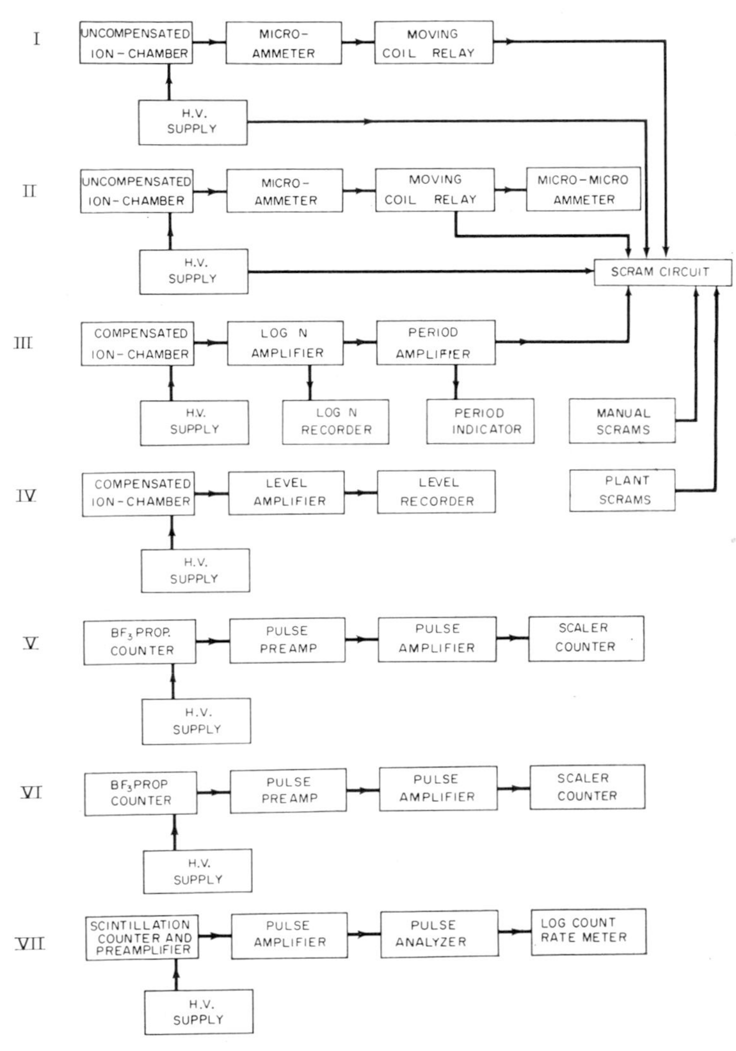

FIG. 10. Nuclear instrument block diagram. (Reproduced from Report ANL–5744 by permission of the United States Atomic Energy Commission.)

REACTOR START-UP PROCEDURES*

A source of neutrons adequate to produce a positive indication on at least one of the instruments of a reactor is required for control of the start-up procedure. In reactors that have been operated for some time, particularly reactors containing natural uranium fuel, a sufficient number of neutrons

* Taken largely from Brooks and Soodak, AEC Safety Monograph, Chapter 4, Reactor Control and Operation.

may be produced by spontaneous fission of uranium-238 or by radiation reactions with beryllium or other materials to produce an adequate neutron flux for start-up. However, in new cores and particularly reactors operating with enriched fuel it may be necessary to have a special neutron source such as a radium–beryllium or a polonium–beryllium source of the order of 10^6–10^7 neutrons per sec. This neutron flux must be "seen" by the neutron sensing instruments before the start-up procedures can be initiated.

The "source multiplication technique" is a procedure almost universally used in the initial approach to criticality with a new reactor or for criticality experiments. This technique is essentially the following:

1. The neutron source is placed in the unloaded assembly as near to the center as convenient.
2. A reading, such as a count rate reading, is taken on one or more of the sensing instruments and the inverse of this reading is computed.
3. The first fuel charge is placed in the assembly. The amount of fuel used is that which is definitely known not to be critical such as 10 per cent or less of the calculated critical mass.
4. Another reading of the instrument is taken and a curve is plotted of the inverse of the reading as a function of the amount of fuel added. Sometimes the curve is normalized so the first reading is unity.
5. A line is drawn through two points to intersect the abscissa. The point of intersection is the extrapolated critical mass. The next fuel addition is no more than half the difference between the loading and the extrapolated critical mass.
6. Steps 3, 4 and 5 are repeated and the curve is extended. When the multiplication becomes large enough for the effect of the control rods to be seen, points may be taken with and without the control and safety rods inserted. The difference in extrapolated critical mass in the two cases gives the control and safety rod effectiveness in terms of fuel mass. At this point fuel additions no greater than this amount are made. As the curve approaches zero, the safeties are left in the most reactive position.

START-UP AFTER CONTROLLED SHUTDOWN

The first step in the start-up of a reactor after a controlled shutdown is to test all controls and safety instrumentation, insofar as it is possible to do so while the reactor is shut down. At least one control instrument—usually a count rate meter working from an ion chamber—must give an indication with the source in place.

The reactor is brought to critical by whatever means are provided such as withdrawing shim rods, raising the water level, bringing the halves of a criticality assembly together, etc. This will normally be done in a safe and reversible manner with the source multiplication observed after each incremental increase in reactivity. The point at which the reactor passes critical may be observed in a number of ways:

1. The reading of a power level indicator will rise during a rod withdrawal but will be constant after the withdrawal if the reactor is below critical. Above critical the power level will continue to rise after the rod withdrawal has stopped.

2. A period meter will read a finite value during withdrawal of thc rod. The meter reading will return to infinity, however, when rod withdrawal ceases, if the reactor is subcritical but will remain on scale, if the reactor is above critical. The period read by the meter during a rod withdrawal is a direct measure of the rate of reactivity insertion caused by the rod withdrawal, and can thus be used to limit the rod's speed.

When the reactor has passed critical it is brought to full power at some stable positive period. It is customary to halt the power rise at some low power level such as 1 per cent of full power and to check the various instruments for operation.

The fact that reactivity is, of necessity, being inserted during start-up makes this time especially prone to accidents. For this reason elaborate precautions are taken during start-up. These are of two kinds—procedural and mechanical (interlocks).

Reactor operating teams usually consist of a crew chief or supervisor and one or more operators. Carefully planned procedures for start-up are worked out and followed. A start-up checklist is required to make sure that no point is overlooked.

In addition to these procedures, safety interlocks are provided in the control instrumentation wherever feasible, to prevent departure from the routine. The particular interlocks employed depend on the reactor but some examples of interlocks are:

1. Reactivity cannot be inserted unless at least one flux indicator gives a positive indication.

2. Reactivity cannot be inserted unless some safety rods are in their most reactive position and cocked.

3. The reactivity cannot be inserted unless coolant is flowing.

4. The reactor cannot be started unless the reactor room door is closed.

REFERENCES

1. Hazards summary report on the Argonne Low Power Reactor (ALPR), ANL–5744.
2. Fitch, S. H., Huston, N. E., Miller, N. C., and Saur, A. J., Integral safety devices for reactors, *Proc. 2nd United Nations Int. Conf. on the Peaceful Uses of Atomic Energy*, Vol. 18, pp. 186–192, United Nations, New York.
3. Garrick, B. J., Editor, *Research and Development in Reactor Safety*, USAEC.
4. Brooks, W. L., and Soodak, H., AEC Safety Monograph, Chapter 4, Reactor Control and Operation, NDA 2009–8.
5. Harrer, J. M., The Reactor Handbook, Vol. 2, Chapter 7.4, Nuclear reactor instrumentation and control, AECD–3646.
6. Schultz, M. A.: *Control of Nuclear Reactors and Power Plants*, McGraw-Hill, New York, 1955.

CHAPTER 5

CONTAINMENT

INTRODUCTION

NUCLEAR equipment which is considered to present undue hazards to populated areas or other installations is usually operated inside a structure providing somc protections against escape of radioactive material into the atmosphere. The usual concept of a containment structure is a steel sphere surrounding the reactor and designed to withstand the pressures which might be created in any plausible accident to the equipment. Protection against missiles formed by bursting pressure vessels or fittings may be considered necessary to prevent the puncturing of the containment vessel. There are other types of confinement including the filtration of the exit air from a building which is operated at slightly reduced pressure so that all leakage is inward. Sealed subterranean chambers might provide adequate containment. It can be seen that the design requirements for an appropriate containment means can be as varied as the types and applications of nuclear reactors and criticality experiments.

The need for containment results largely from the location of reactors near populated areas or adjacent to other important installations for reasons of economy in power distribution costs or for convenience in use as is often the case for research reactors. The major nuclear installations built by the Manhattan District of the Army Corps of Engineers were located so that containment was not considered necessary. The National Reactor Testing Station site was also selected on this basis. The need for better control of particulate material released from the Oak Ridge X-10 air cooled reactor when fuel elements failed, led to the development and installation of elaborate gas washing and filtration equipment. A potentially serious problem of contamination of the important research and development area was brought under control and the X-10 reactor might be considered to have some measure of confinement by filtration of the air from the reactor. The Brookhaven air cooled reactor was designed with an effective filtration system based on the Oak Ridge experience to permit safe operation on the site of the Brookhaven National Laboratory on Long Island, New York. The inlet air to the reactor is filtered through deep-bed pocket filters having an efficiency of over 85 per cent to remove dust which would become radioactive in passing through the reactor and add to the particulate material problem. The outlet filters for the hot air leaving the reactor are made of woven glass-fiber cloth guaranteed for service up to 500°F. The filters are designed to remove more than 95 per cent of all particles down to 5 microns and 25–30 per cent of smaller particles. When partially loaded, the filters will remove more than 95 per cent of particles down to 3–4 microns. These filters handle a total of 1,216,000 lb. of air per hr.[(1)] Since the air is drawn through the reactor by exhaust fans, any leakage will be into the cooling system. The exhaust air from the cooling system

FIG. 1. Submarine intermediate reactor facility, West Milton, New York. (General Electric Company photograph.)

is discharged up a stack 320 ft above ground level. The reactor with its control of leakage and filtration system might be considered to have a degree of confinement.

The steel containment pressure vessel for a sodium-cooled reactor built in the vicinity of West Milton, New York, established a precedence for the location of reactors near centers of population with reliance in part for safety on the containment structure to withstand any reasonably possible release of energy and to prevent the escape of radioactive material except at a rate sufficiently low so as not to constitute an off-site hazard. The site for this reactor installation was sufficiently large so that direct radiation from fission products released into the containment vessel was not considered a serious problem Since the presence of rock near the surface and the level of ground water would have made a deep excavation expensive, it was decided to build the steel sphere largely above ground. There were many problems of support, vibration, ventilation and allowances for rapid changes in atmospheric pressure and temperature that had to be carefully considered in the design of this structure. The possibility of a hydrogen–oxygen explosion in the structure was studied and such provisions as blast mats to prevent missiles

FIG. 2. Construction of submarine intermediate reactor facility, West Milton, New York. (General Electric Company photograph.)

from penetrating the containment vessel were included. Since this concept of containment was without precedence at that time such problems were studied exhaustively by the Knolls Atomic Power Laboratory, The Naval Reactors Branch and the Reactor Safeguards Committee.

The acceptance of containment instead of isolation with distance, has led to the construction of many reactors with this safety provision and therefore detailed studies of the design requirements for adequate containment structures have been in progress for several years. It is evident that the design of a containment structure must be based upon an analysis of the loads and forces it will have to withstand, the quantities of fission products that may be released from the reactor core and the maximum allowable rates of leakage and direct radiation from this radioactive material that can be permitted. These studies may include rates of release of energy in a reactor incident, chemical reactions between reactor materials, fire, shock-wave damage, missile penetration and even sabotage or war damages.

In considering the design of containment installations it is recognized that no arrangement gives absolute protection to the public. There is a general misconception that the large isolation areas at Hanford, Savannah River or the National Reactor Testing Station give absolute protection for off-site personnel. These sites were selected and planned on the basis of providing only a reasonable evacuation time for the off-site public in the event of a major release of radio-active materials. Similarly with containment vessels and other provisions such as confinement by filtration, only a degree of additional protection is provided, since there are always some ways, even though improbable, by which this containment can be circumvented. The most obvious and probable way in which the containment might be lost is through a door or port which has been left open. Also in time of war, bomb damage could destroy the containment and cause the release of radioactive material from a reactor. The nebulous problems of accident probability and damages are weighed against the cost of various degrees of containment, and as may be expected, many practical compromises are made.

MECHANISM OF RELEASE OF RADIOACTIVE MATERIAL

A reactor with clad solid fuel elements presents several barriers against the release of fission products. The accumulation of materials which are gaseous at the operating temperature will produce some internal pressure on the cladding depending on the types of fuel elements and the operating history. Also stresses will develop in the fuel element as a result of radiation damage and changes in physical properties. The fuel element may be attacked from the outside by corrosion and erosion. Failures resulting from the above causes are usually initially small (pin holes) and the rate of release of radioactive material slow and easily detected so that the faulty element can be replaced with only routine decontamination of the coolant required. Several hundred such failures have occurred with varying degrees of contamination which in a few cases have been expensive to clean up.

The type of fuel element failure usually considered in the design of containment systems is the overheating of the reactor core so that the fuel elements melt, vaporize or burn. Such overheating might result from a nuclear power excursion, loss of coolant while operating, or loss of coolant following shutdown with fuel elements heating from the radioactivity of the fission products. Once the assumption is made of excessive temperatures from any of the above causes, it is necessary to consider possible chemical reactions between reactor materials such as metal–water reactions or burning. Also excess temperatures may create pressures in the system and the rupturing of piping and vessels may result with the release of energy and production of missiles. Sudden releases of energy in the nuclear excursion, chemical reaction or rupture of equipment may produce shock waves which can damage structures or produce missiles by spallation. Various sequence of events that could lead to the release of radioactive material from a reactor core and containment structure are shown in Fig. 3, prepared by Brittan and Heap, Reference 2.

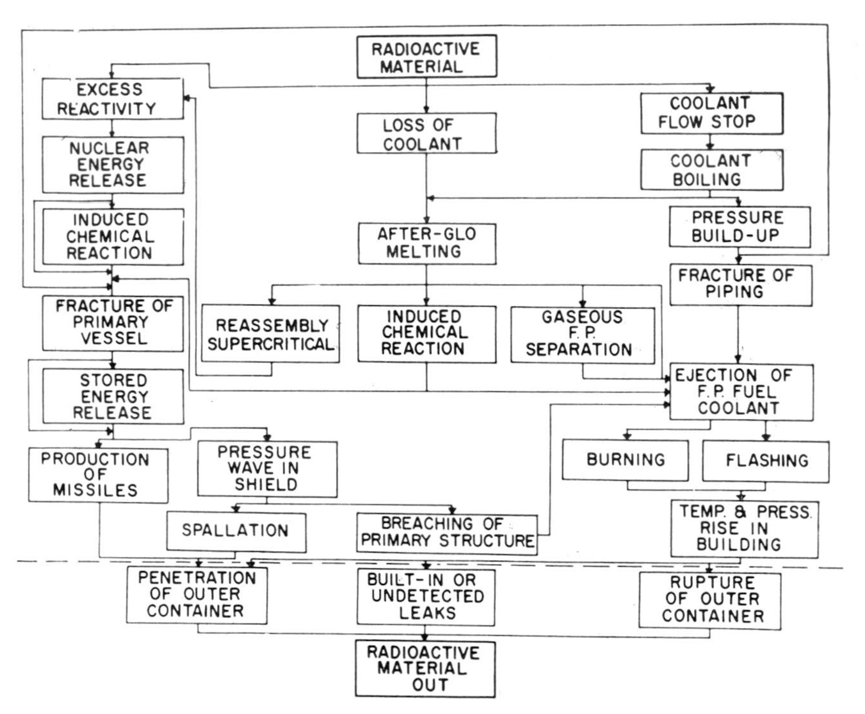

FIG. 3. Reactor containment problems. (Reproduced from "Reactor Containment" by R. O. Brittan and J. C. Heap, Vol. 11, *Proc. 2nd United Nations Int. Conf. on the Peaceful Uses of Atomic Energy*, United Nations New York.)

NUCLEAR EXCURSIONS

The power of a nuclear reactor will increase exponentially with time when the multiplication factor is greater than unity; and when the reactor is critical on prompt neutrons, very short periods of power increase of the order of a few thousands of a second may be obtained in thermal reactors. The sudden release of energy in a water moderated and cooled reactor has been found in the Borax tests,[3] discussed in more detail in Chapter 6, to result under some extreme conditions in the violent destruction of the assembly with ejection of fragments and components for a considerable distance. The disruption of the reactor was accomplished by a loud sound and a weak shock wave was felt at some distance. Although certainly not a nuclear letonation and perhaps best described as low order explosion equivalent perhaps to the explosion of only a few pounds of a high explosive, the event ertainly demonstrated that this nuclear reactor could release energy at a ufficient rate to result in destruction within a limited area around the reactor. The type of event is illustrated in Fig. 4.

FIG. 4. Final Borax 1 experiment at National Reactor Testing Station (Argonne National Laboratory.)

The necessary conditions for such a violent excursion are not well understood or easily predicted. Since most of the energy is released in a rapid nuclear power excursion during the last e-folding time or period, it can be expected that the period is a significant parameter. If the reactor period is shorter than the time required for the release of pressures generated in the core, it would be expected that augmented or shock pressures might be generated. A parameter for the time required for release of pressures is the velocity of sound and more specifically the time required for sound to travel across the core. The comment was made by the Reactor Safeguards Committee several years ago that unexpected and perhaps violent events may be expected when the reactor period becomes of the same order of magnitude as that required for sound to travel across the reactor core. See Appendix C.

There is some difficulty in predicting the velocity of sound in a liquid when vapor is present under boiling conditions, although theory would predict that

the presence of bubbles could decrease the velocity of sound in water by mor
than an order of magnitude. An experimental investigation of the phenomen
has been undertaken by Karplus at the Armour Research Foundation for th
Atomic Energy Commission[4]. Using air bubbles in water the data shown i
Fig. 5 were obtained. Although the results are not necessarily applicable t

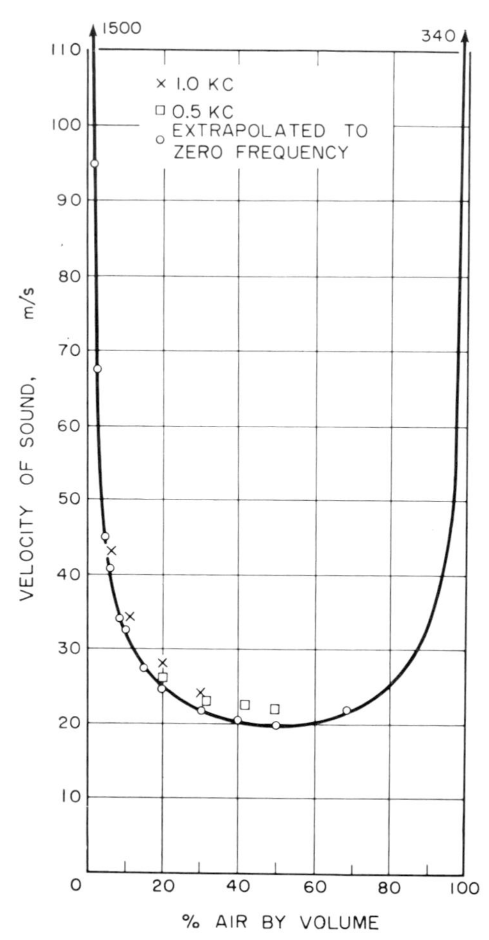

FIG. 5. Velocity of sound in water containing air bubbles. (Reproduced from ARF Report COO–248 by H. B. Karplus.)

steam–water mixtures, the magnitude of the effect is clearly indicated. Alsc
it has been verified that the velocity of sound is directly proportional to the
ambient pressure over an appreciable pressure range. As a consequence a
slow pressure rise in a mixture of gas bubbles and liquid tends to become a
shock wave rapidly. However, shock waves are expected to be severely
attenuated when passing through such a medium.

A Borax type reactor subjected to power excursions shorter in period than that which resulted in destruction of the core, might be expected to produce even higher peak local pressures and hence even more violent destruction. If it were possible to produce power excursions with periods very much shorter than a millisecond, at some sufficiently short period an appreciable fraction of the nuclear fuel would be fissioned before the core became disassembled by vaporization and expansion and it might be said that a nuclear explosion would occur. To produce such short periods requires that some material, either fuel or a nuclear poison, be moved an appreciable distance within a correspondingly short period of time. Fortunately, the forces required to move a component within such a short period of time are usually not available except through very unusual means. Therefore, the prediction can be made for almost every type of reactor that a nuclear detonation does not need to be considered. However it is important in reviewing the design of a reactor to consider means by which extremely heavy forces might occur which could affect the core. For example, forces could accumulate from the unreleased thermal expansion of core components. The structural failure of a core component subjected to such forces might permit the sudden movement of fuel elements in an undesirable manner.

Melt Down

A power excursion in a water cooled and moderated reactor with a period sufficiently long to permit transfer of most of the heat from the fuel elements to the water will usually be terminated without destruction by expansion of the water or its displacement by boiling. However, for a very short period most of the heat remains in the fuel element, thereby raising its temperature far above the coolant temperature and may lead eventually to melting of the fuel element as in the final Borax experiment. For a reactor whose coolant is not essential to the nuclear process such as a liquid metal or a gas cooled reactor, an inherent shutdown mechanism may not be present or sufficiently effective to prevent meltdown even in a slow excursion, if it is not detected and terminated by the control system.

When the rate of heat release, Q, in the fuel element is greater than the rate of heat transfer, U, from the element, the difference is the heat to be stored in the material of the fuel element by raising its temperature or in producing phase changes. If w is the weight of the fuel element, $\bar{C}_p$ is the average heat capacity including phase changes, and T_m is the melting point, the time t_m to reach the melting temperature from an initial temperature T_l is

$$t_m = \frac{w\bar{C}_p(T_m - T_l)}{Q - U}\,.$$

It can be seen that a reactor having a high rate of heat release and light weight fuel elements such as a power reactor with fully enriched fuel, will have a short time for emergency action before melting occurs, if the rate of hea release exceeds the rate of heat removal. Also a fuel element made of alum- inum will of course melt much more quickly than one made of a higl temperature material, other conditions being the same.

A reduction or loss of coolant flow from pump or blower failure, fault valve operation, or rupture of pipe lines are possible causes of meltdown anc must be carefully considered in reactor design and hazards evaluation Elaborate precautions are taken in the form of multiple pumps, dual cooling loops and emergency cooling systems to prevent the loss of cooling. In addition there are detection systems for reduction in coolant flow and in- creases in temperature.

The control system as described in Chapter 4 is designed to provide alarn signals and shut down the reactor in the event of a significant change in th heat transfer system.

Even after the nuclear chain reaction has been shut down there is appreci able release of heat by radioactive decay of the fission products and absorptio of the radiations released. In event of loss of cooling in a reactor, this deca heat may be significant and could lead to fuel element melting in some cases

The delayed heat problem is clearly a serious one. If some failure in th cooling system should occur, then even if the nuclear chain reaction is immed iately shut down by inserting control or safety rods, there will still be left substantial heat load which must somehow be disposed of. For example, i the fuel elements from the Materials Testing Reactor were suddenly remove from the reactor and left standing in the open air, they would melt down b themselves by delayed heat production.

DISSOLUTION OF FUEL ELEMENTS

Another mechanism by which fission products might be released from cla fuel elements is the dissolution of the fuel element by reactions with th coolant or some added chemicals. For example, if caustic were accidentl added in sufficient amount to the cooling water, aluminum fuel elements woul be dissolved rather quickly. There may be certain catalysts which could caus a rapid reaction between water and aluminum. However, the accidental addin of a chemical reactant in sufficient quantity to dissolve the fuel elements an release the radioactive material therein, is quite improbable and therefore i usually not considered in hazards evaluation.

METAL–WATER REACTIONS

The nuclear energy which might be released in a power excursion in mos reactor types is limited to less than 10^{-6} of the total energy from complet fissioning of all the fuel in the core. The nuclear reaction is usually terminate by expulsion of the moderator in water cooled reactors or by melting o vaporization of the core for extreme cases. In the final Borax 1 experimen the total energy liberated was only 135 MW-sec even though the peak reactor power was between 13×10^9 and 20×10^9 W.[6] For research reactors of the swimming pool type the estimated maximum credible nuclear power excursion is usually limited to some 1000 MW-sec. However, a complete chemical reaction between the aluminum in the core of such a reactor with water would produce 1680 MW-sec of energy and the combustion in air of the hydrogen thus formed would release about this same amount of additional energy.[2,7] It is seen therefore, that chemical reactions could yield about three times the energy released in the nuclear power excursion. Several studies have been made to determine the conditions required for chemical reactions between core components.

Table I

Heat of Reactions

Reaction	Heat of reaction cal/gram (of metal)
Nitroglycerin	1580
Black Powder	685
$Li + H_2O \rightarrow LiOH + \frac{1}{2}H_2$	6960
$Na + H_2O \rightarrow NaOH + \frac{1}{2}H_2$	1900
$2Al + 3H_2O \rightarrow Al_2O_3 + 3H_2$	3550
$Mg + H_2O \rightarrow MgO + H_2$	3190
$Zr + 2H_2O \rightarrow ZrO_2 + 2H_2$	1330
$3U + 8H_2O \rightarrow U_3O_8 + 8H_2$	437
Type 303 SS + $H_2O \rightarrow Fe_3O_4 + Cr_2O_3$	118
NaK(56%K — 44%Na) + $H_2O \rightarrow NaOH + KOH + H_2$	1473

(Taken from Report ANL–4503 by J. M. West and J. T. Weills.)

A reaction between a metal and water to produce hydrogen and a metal oxide may be exothermic depending on the metal and the temperature at which the reaction takes place. For the light metals the reaction is usually strongly exothermic, whereas for nickel and similar elements the heat of reaction is very small and the reaction may even be slightly endothermic. The heats of reaction are compared with the heat of explosion of nitroglycerin and black powder in Table I.[10]

In Figs. 6–8 are plotted the free energy of several metal–water reactions as a function of reaction temperatures. These figures are taken from Reference 12. It is seen that an increase in reaction temperatures changes the free energy of the reactions in different directions. The reactions of metals such as aluminum, magnesium and zirconium might be expected to become more violent with temperature; whereas for the alkali metals such as sodium the free energy change decreases and approaches zero.

Since aluminum is a common material of reactor construction and the reaction with water is strongly exothermic, the possibility of the aluminum–water reaction occurring in a reactor has been studied and debated. It is well known that aluminum is ordinarily not corroded appreciably by pure water below 350°F. A rapid, but not explosive chemical reaction, is inaugurated between saturated steam and common aluminum alloys. Molten aluminum in foundries is sometimes disposed of by pouring into water. However, there have been a few explosions when aluminum is poured into water and one explosion was particularly violent as evidenced by cracking a $\frac{3}{4}$ in. steel

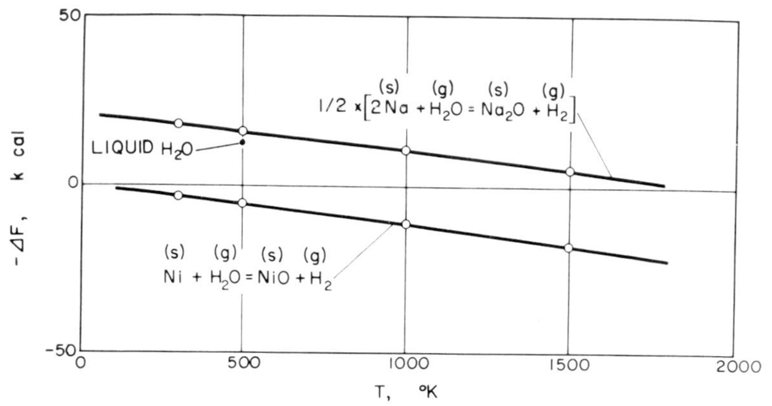

FIG. 6. Free energies of metal–water reactions. (Reproduced from Report NAA–SR–197 by W. C. Ruebamen, F. J. Shon and J. B. Chrisney.)

plate and hurling of a large piece of steel through 4 in. of timber.[8] This event led to an extensive series of tests, and the following facts were determined:

a. Explosions could not be obtained in water depths of less than 1 in. or of more than 30 in.
b. Aluminum poured from a 1 in. diameter opening will not explode, under given conditions, whereas aluminum poured through a 3 in. diameter opening will explode violently.
c. A grid which breaks up the falling stream is effective in preventing explosions.
d. The presence of soluble oil has prevented explosions whereas hydrated lime or rust greatly increased the tendency toward explosion.

Since the rate of a chemical reaction is proportional to the surface exposed, it seems necessary to have some means for dispersing the aluminum in the water to obtain sufficient surface area for an extremely rapid reaction. A hypothesis that has been made to explain the above experimental results, is that a pool of molten aluminum first collects on the bottom of the container and then the rapid generation of steam from water trapped under this molten pool as in lime or rust deposits, produces sufficient force to disperse the

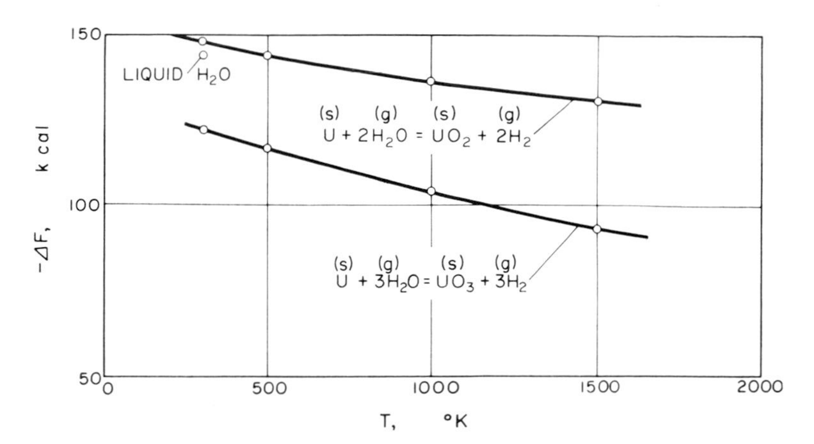

FIG. 7. Free energies of metal–water reactions. (Reproduced from Report NAA–SR–197 by W. C. Ruebamen, F. J. Shon and J. B. Chrisney.)

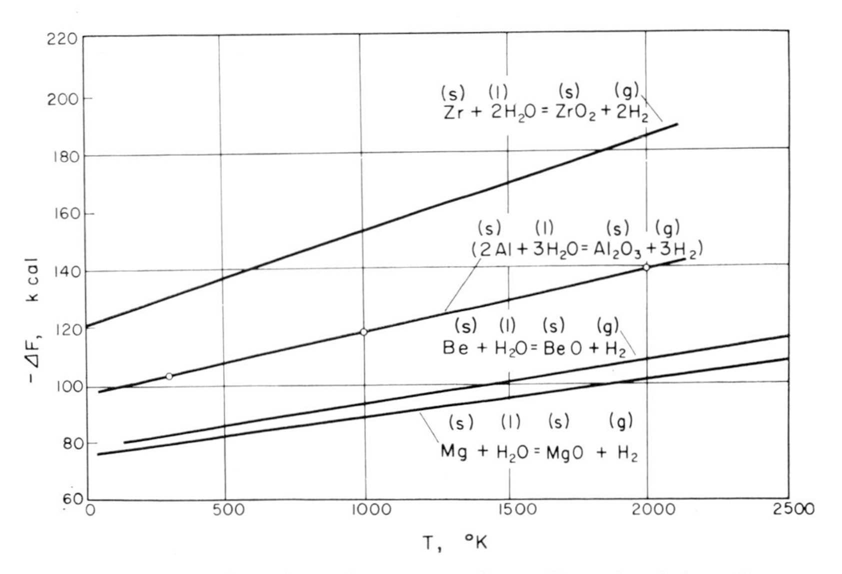

FIG. 8. Free energies of metal–water reactions. (Reproduced from Report NAA–SR–197 by W. C. Ruebamen, F. J. Shon and J. B. Chrisney.)

aluminum through the water. Steam explosions of this type have been observed with metals which do not react with water and such explosions can be of considerable violence. The dispersed aluminum may then react very rapidly with water but a means for initiating the reaction is required.

There have been several additional investigations of this reaction, and in the following paragraphs the results are reviewed for general information on this controversial problem. It will be noted that some of the findings seem contradictory which is not unusual for studies of such a complicated phenomenon.

When molten aluminum was forced by air pressure through a nozzle perforated by a large number of $\frac{1}{16}$ in. diameter holes, no detectable steam pressure or evidence of a chemical reaction was produced even though the aluminum was heated to 1000°C[10].

Shidlovski[11] reports the results of tests of mixtures of methyl alcohol or water with magnesium and aluminum. The aluminum or magnesium powder was placed in a lead beaker and moistened with the required quantity of water or methyl alcohol and the mixture was slightly compressed with a wooden pestle. In order to obtain homogeneous mixtures in the case of water–aluminum mixtures, 4 per cent of gelatine had to be introduced, while in the experiment with methyl alcohol the proportion had to be changed slightly by reducing the quantity of alcohol from that theoretically computed. For creating the initial impulse, a blasting cap was used and in many cases this was reinforced with a tetryl blasting charge. It was found that all mixtures were capable of explosive decomposition as evidenced by damage to the lead containers. The mixtures that proved to be most sensitive to the initial impulse were water and magnesium mixtures, which could be set off with the blasting cap alone. In the case of mixtures of water and aluminum an auxiliary blasting charge of tetryl was required. There was total destruction of the lead beaker and aluminum oxide powder and some metallic aluminum was found in the residue. It appears from these experiments and other data that reactions which produce more than 1000 cal/g of reactants and more than 500 l. of gaseous products per kilogram are suspect of possible explosive reactions.

Ruebsamen, Shon and Chrisney report in Reference 12 their experiments with electrically heated wires and foils of different metals immersed in water in a pressure vessel. Very rapid heating was accomplished by discharging a high voltage condenser through a sample and the extent of chemical reaction was calculated from measurements of the amount of hydrogen gas formed and the weight of metal lost. Reaction times were measured with an oscillograph. From these experiments it appears that there is a threshold amount of energy required to initiate the reaction which is possibly the amount of energy required to vaporize and disperse the metal in water. When sufficient initiating energy was supplied, substantial amounts of up to 100 per cent of the following metals and alloys were found to react; U, Zr, Al, Al–Li alloy and Al–U alloy; and under identical experimental conditions Ni did not react appreciably. Reaction times were of the order of 100 μsec.

In tests at duPont as summarized in Reference 9, crucibles containing molten metals were smashed under water with considerable force and confinement. Under these conditions, pure molten aluminum at 900°C did not react, but aluminum containing 1.0–7.4 per cent of lithium reacted slightly in some tests and violently in a few tests. Molten aluminum containing 5 per cent uranium at 900°C reacted violently in 3 out of a total of 20 tests. At higher temperatures, the activity of the metals did not seem to increase.

Higgins at Aerojet–General Corporation[9] conducted a series of tests with molten metals and water and in some tests a blasting cap was used to disperse the metal in the water. It was found that only with the blasting cap were violent reactions obtained with Zr, Zircaloy-2, 95 per cent Al–5 per cent Li alloy, and Mg. No appreciable reactions were observed with Ni, pure Al or type 321 Stainless Steel. Uranium and U–Mo alloy were tested without the blasting cap and no extensive reactions were observed. According to pressure–time measurements the reaction time for zirconium with water was approximately 1 msec and the expansion of the steam and hydrogen gas to atmospheric pressure took 4 msec. From observations of the damage done to the water container, it appeared that the reaction was similar to a low-order detonation.

Higgins and Schultz[13] report the extension of earlier studies at the Aerojet-General Corporation on the reactions of metals with water and oxidizing gases at high temperatures. The molten metals, zirconium, Zircaloy-2, aluminum, uranium, sodium–potassium alloy, and stainless steel were sprayed into water, and measurements were made to determine the explosive power, reaction times, reaction efficiency, and the effects of temperature and droplet size. It was shown that the percentage of reaction depends primarily on the particle size and on the temperature of the metal. The reactivity of molten aluminum is nil at temperatures up to 1170°C, but at temperatures above 1170°C, the reaction becomes increasingly more complete. At temperatures of 2200°C aluminum was found to be a more powerful reactant than zirconium. For both zirconium and aluminum there is a critical temperature above which a change in reaction rate occurs. An analysis was made of the final Borax-I experiment indicating that the fuel elements did not attain a temperature sufficiently high to react appreciable with the water present, but the safety factor was small. The reaction of stainless steel with water was surprisingly violent considering the small heat of reaction. From the reaction times measured it was concluded that these reactions could be described as deflagrations rather than detonations, but the metal–water reactions were slightly more brisant than black powder, as indicated by the rates of pressure rise.

In related experiments it was found that wires of zirconium did not burn in high pressure pure steam; however, when the steam contained as little as

0.5 per cent oxygen by volume, the burning of zirconium was self-sustaining.

Since the effects of radiation on metal–water reactions were uncertain and to determine the safety aspects of extending heat fluxes in the Boiling Heat Transfer Experiment to a point where plates would melt, the reactions of Al and Zr with water were investigated in the Materials Testing Reactor.[14] Test samples, containing sufficient U^{235} to cause melting were placed in autoclaves with water. These autoclaves were then irradiated from six to eighteen seconds in high flux reactor positions, and the pressure and sample temperatures were recorded. After irradiation the volumes of gas in the autoclaves were measured and the gas analyzed. A chemical reaction took place in all tests as indicated by the formation of hydrogen. No pressure transients were observed during the runs where the samples did not melt. Seven of ten submerged aluminum samples melted and transients up to 8000 psi and 23,000 psi were obtained during two of the runs where the sample temperature remained around 1400°F. No transients were observed during other aluminum experiments where the sample temperature went above 2200°F. Of the nine Nichrome V samples run, seven melted and transients were obtained during three of the runs with magnitudes up to 18,000 psi, 3000 psi, and 1000 psi, respectively. Six of seven zirconium samples melted with one sample producing a transient up to 10,000 psi. The violent reactions did not consume large portions of the samples or appear to have an effect on the per cent completion of the reaction although many sporadic pressure transients occurred. It was concluded from these tests, that all the fuel plates tested can be made to react violently with water although no enviromental conditions were found which would consistently result in a violent reaction.

Studies were made by Bostrom[33] to determine the rates of oxidation of Zircaloy 2 and Zircaloy B maintained at temperatures between 1300°C and 1860°C while submerged in water. Samples immersed in water were heated over a period of 5–10 sec to the desired temperature and maintained at this temperature. The times required to collect successive 250 ml volumes of hydrogen were then measured. The observed reaction rates are shown in Fig. 9 and 10 and are approximately those that would be expected from the extrapolation of existing data for zirconium in air. The reaction did not become violent or autocatalytic in nature even at temperatures above the melting point. This work was extended by Lustman[15] to determine the extent of reaction between zirconium-base alloys and water under conditions corresponding to a loss of coolant accident in the Pressurized Water Reactor. It was concluded from this analysis that less than 20 per cent reaction would be expected under conditions postulated for a loss of coolant accident. The amount of reaction within the core structure as estimated in this analysis was sensitive to even minor rates of cooling.

In the investigation reported in Reference 16, nuclear thermal transients were simulated by direct current electrical heating of metal specimens prepared

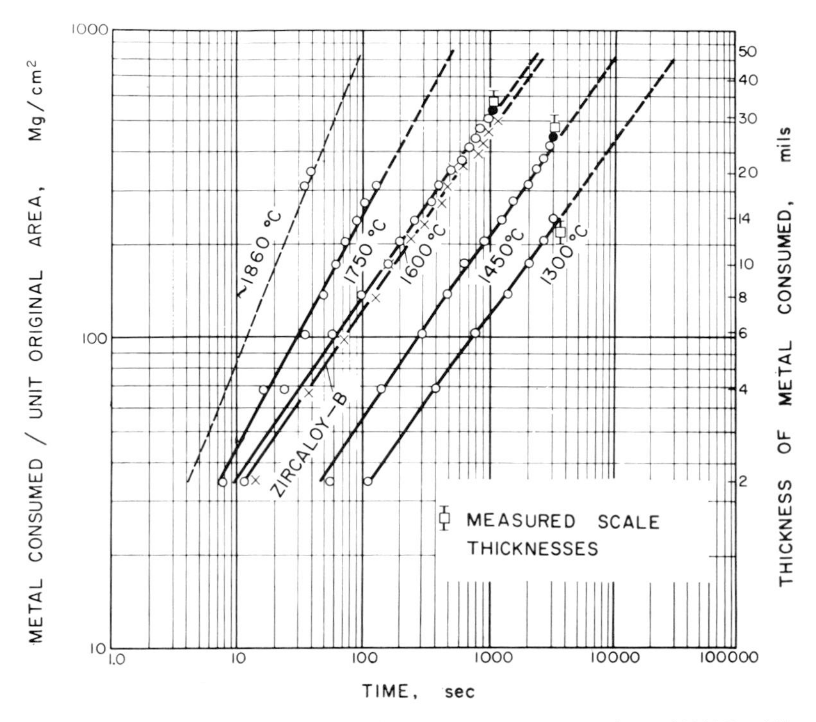

FIG. 9. Oxidation kinetics of Zircaloy–2 at temperatures above 1300°C. while submerged in water. (Reproduced from Report WAPD–104 by W. A. Bostrom.)

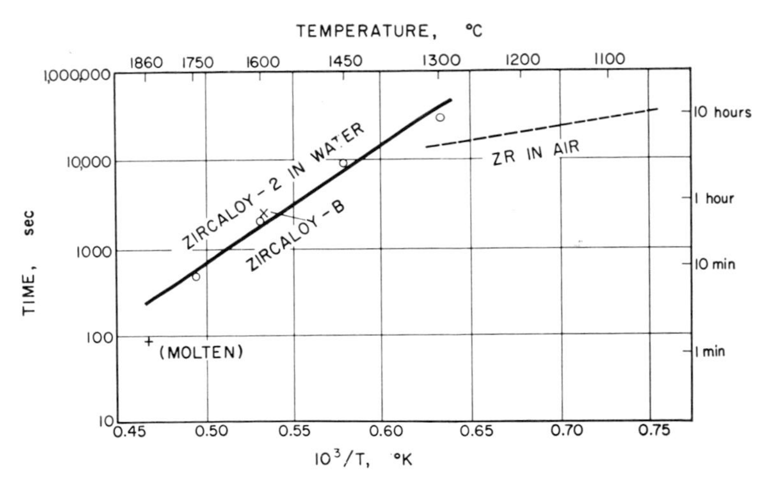

FIG. 10. Time to consume 0.087 in. thick plate as a function of temperature. (Reproduced from Report WAPD–104 by W. A. Bostrom.)

from Cu, Zn, Al, U and Zr. The pulse times were from 3 to 28 milli-seconds and the high current density vaporized part of the metal and pro-duced a phenomenon termed "electrical explosion". Peak over-pressures of about 16 psi were produced in the liquid as measured at a distance of $2\frac{1}{2}$ in from the specimens during the "electrical explosion". Evidence was not found of self-sustained metal–water reactions caused by fast transient heating of Ag, Al, Cu, Sn, U and Zn. Examination of unvaporized fragments of a heated Zr specimen, however, showed significant reaction with water and a penetration of 0.001 in. to 0.0015 in. of the metal took place in an estimated interval from 0.8 to 2 milliseconds when the glowing fragment was being quenched.

From the experiments reviewed above there is sufficient evidence to con-clude that aluminum, magnesium and zirconium and some of their alloys can react with water and release considerable energy in a manner which can sud-denly produce high pressures capable of damaging the installation; however, the conditions required for such a violent reaction are not well understood. It is noted that the containment for some reactors has been designed on the basis of withstanding the pressure generated by a chemical reaction of 50 per cent of the aluminum in the active core followed by combustion of the hydro-gen thus released. It was assumed that the heat released in these reactions turned water into steam which was injected into the building to increase the pressure, but without the shock forces associated with an explosion. A critical temperature of about 1200°C (2192°F) for an explosive aluminum–water reaction has been estimated. In the case of a pressurized water reactor with zirconium alloy fuel elements, a non-explosive reaction of 20 per cent of the zirconium with water has been used to evaluate the loss of coolant accidents.

A reaction between sodium and graphite is thermodynamically possible and it is known that sodium rapidly attacks amorphous carbon. However, when graphite is exposed to sodium the reaction proceeds only at elevated temperatures. For example Loftness[22] reports that graphite exposed to sodium vapor at 900°C showed a maximum weight loss of 60 per cent per week. However, it does appear that potassium, rubidium and cesium attack graphite and enter between the planes of the graphite lattice[23].

Sodium cooled reactors produce steam for power generation and are designed with an intermediate sodium or NaK loop so that the radioactive sodium from the reactor does not enter the steam boilers. In some designs there is an intermediate fluid such as mercury between the hot alkali metal and the water. However, the reaction of the hot alkali metal with water is possible in event of a major failure of equipment, and a tube failure permitting the water at high pressure to enter the sodium or NaK system has been postulated as a possible accident. Several studies of an applied nature have been made to simulate the tube failure in a heat exchanger and many tests on large scale equipment have been conducted at considerable expense. In some tests it was found that small leaks did not result in a violent reaction but tended to become sealed off by the formation of solid hydroxide. The failure of an entire tube in a heat exchanger has been simulated in tests. The injection of water at high pressures into the liquid metal resulted in a vigorous reaction and in a few tests heavy shock pressures were generated of sufficient magnitude to cause damage throughout the liquid metal system and within too short a time interval to permit the functioning of relief valves or vents. It has been found in some tests that the heavy shock pressures can be greatly reduced in magnitude or avoided by providing a free volume filled with inert gas in the liquid–metal loop. Relief valves are required to vent the hydrogen generated by the reaction and prevent excessive pressures in the liquid–metal system. The design requirements for adequate safety in a liquid–alkali metal steam generator require careful study and safety testing of equipment of new design should be considered.

COMBUSTION OF REACTOR MATERIALS

The burning of fuel elements in air is a possible mechanism for fission product release. This must be considered in many reactor designs where the fuel elements might attain high temperatures during a power burst or from fission product heating following shutdown, if adequate cooling is not provided. Ignition is said to occur when the temperature of a material rises spontaneously due to oxidation and a self-propagating oxidation reaction occurs at some elevated temperatures. Ignition will occur when the heating from an exothermal oxidation reaction exceeds the loss of heat from a point by conduction, convection and radiation cooling. By considering the mechanisms of oxidation and of heat loss, predictions of ignition temperatures have been made in Reference 28 which are in reasonable agreement with experimental values. The values of ignition temperatures of solid metals in Table II are taken from Reference 28.

Magnesium and its alloys are used as cladding in some reactors which are cooled with carbon dioxide. It is well known that magnesium will react with both carbon dioxide and with oxygen with the evolution of a large amount of heat. The oxide film which forms on magnesium protects the metal against rapid oxidation even up to temperatures of 450°C and ignition may require temperatures of 800°C[19]. It is noted that this ignition temperature is considerably above the value given in Table II thus indicating the variations in experimental values depending upon conditions.

For graphite in air it appears that the threshold for an appreciable oxidation rate of 1 per cent loss in weight in 24 hr is some 550°C. At 700°C nearly an hour is required for a 1 per cent weight loss. The threshold oxidation temperature in steam is about 700°C and in carbon dioxide is 900°C.[18] However once the metal becomes ignited, combustion of the magnesium and the hot uranium in the fuel element can be self-supporting.

Table II

Ignition Temperature of Solid Metals

Metal	Ignition temperature °F	Gas	Pressure atm.
Mild steel	2240 to 2330	air	1 to 7
W	2270 to 2350	air	1 to 7
Ti alloys			
RC–70	2880 to 2960	air, O_2	1 to 7
RS–70	2890 to 2940	air, O_2	1 to 7
RS–110–A	2860 to 2910	(a), O_2	1 to 7
RS–110–Bx	2850 to 2930	(a), O_2	1 to 7
Stainless steels			
430	2460 to 2490	(a), O_2	1 to 7
302	(b)	air, O_2	1 to 7
Cu	(b)	air, O_2	1 to 7
Ni	(b)	air, O_2	1 to 7
Ni alloys			
Inconel	(b)	air, O_2	1 to 7
Inconel X	(b)	air, O_2	1 to 7
Be alloys			
Berylco–10	1750 to 1760	air, O_2	1 to 7
Berylco–25	(b)	air, O_2	1 to 7
Mg	1171	O_2	1 to 10
Mg alloys			
20% Al	936	O_2	1
70% Zn	1004	O_2	1
25% Ni	934	O_2	1
20% Sb	1099	O_2	1
63% Al	862	O_2	1
Fe	1706	O_2	1
Sr	1328	O_2	1
Ca	1022	O_2	1
Th	932	O_2	1
Ba	347	O_2	1
Mo	1400	O_2	1
U	608	O_2	1
Ce	608	O_2	1
Al	(b)	O_2	1
Zn	(b)	O_2	1
Pb	(b)	O_2	1
Sn	(b)	O_2	1
Bi	(b)	O_2	1
Li	(b)	O_2	1
Cd	(b)	O_2	1
Na	(b)	O_2	1
K	(b)	O_2	1

(a) Does not ignite in air
(b) Melts before igniting

(Reproduced from Report NACA TN D–182 by W. C. Reynolds.)

The decomposition of water into hydrogen and oxygen in a reactor could lead to the accumulation of explosive mixtures particularly in aqueous homogeneous reactors. Water is decomposed by the absorption of radiation including neutrons and fission fragments. Therefore the rate of gas production in a homogeneous reactor requires careful consideration. A catalytic recombiner is usually provided on aqueous homogeneous reactors and the void spaces are well ventilated with a stream of inert gas. In heterogeneous reactors where fission fragments do not come in contact with the water the radiolysis is much less as it is produced only by gamma-rays and fast neutrons. Thermal neutrons do not decompose water except by a secondary process. The net rate of production of hydrogen and oxygen is the difference between the rate of formation and recombination. At a sufficiently high temperature and pressure the rate of the recombination can be sufficient to prevent an appreciable accumulation of the radiolytic gases.

Table III

Flash and Flame Temperatures of Polyphenyls

Compound	Melting point °F	Flash point °F	Flame point °F
Diphenyl	157	223	255
Ortho-terphenyl	131	340	390
Meta-terphenyl	194	405	445
Para-terphenyl	415	405	460
Santowax R	293	375	460

(Reproduced from Report NAA–SR–2323 by H. L. Sletten.)

Organic materials of interest as a reactor coolant or moderator are limited to the polyphenyls which could present some hazards particularly when hot. Flash and flame temperatures for some of the polyphenyls together with their melting points are given in Table III taken from Reference 20.

The solid polyphenyls are difficult to ignite by open flame, and when ignited the material burns with a sooty flame. The explosive properties of diphenyl dust have been studied and the minimum explosive concentration is 0.035 oz/ft^3. Ignition of the dust can be produced by open flame, by very weak electric sparks and by hot surfaces over 1200°F. The polyphenyls when heated to near the boiling point are easily ignited by flame. In the event of rupture of equipment containing hot polyphenyls under pressure, it is probable that the material will be sprayed into the atmosphere and the liquid may partially flash into vapor. The dispersed material will condense and solidify as a fine dust creating a potential explosion hazard, and precautions to prevent the ignition of the dust are required. All electrical installations in the area that might be exposed to this dust explosion hazard should be of the explosion proof type. Also there should be automatic warning systems for fire in the area.

The burning of sodium in air releases considerable energy which needs to be considered in designing containment for a sodium cooled reactor. Humphrey, as reported in Reference 21, has conducted experiments to determine the pressures which result from the sudden ejection of hot sodium into an enclosed space filled with air.

The dominant reaction that occurs when high-temperature molten sodium in a finely divided state is mixed with air is oxidation of the sodium to form sodium peroxide:

$$2Na + O_2 \rightarrow Na_2O_2$$

with a heat release of 124 kcal per mole of sodium peroxide formed. This reaction appears to proceed until all the oxygen is combined before additional sodium reduces the sodium peroxide to sodium monoxide:

$$Na + \tfrac{1}{2}Na_2O_2 \rightarrow Na_2O$$

with a heat release of 20 kcal per mole of sodium oxide. The presence of water vapor in the initial phase of the reaction results in the formation of sodium hydroxide:

$$Na + HOH + \tfrac{1}{2}Na_2O_2 \rightarrow 2NaOH$$

with a heat release of 85 kcal per mole.

Based on the energy derived from these reactions, the theoretically resulting atmospheric pressures and temperatures for contained sodium–air reactions have been calculated by Humphreys as shown in Table IV.

Table IV

Peak Pressures from Sodium Burning

Maximum Peak Pressures			
Na/O_2 Molar ratio	Oxygen consumed %	Theoretical peak pressure psig	Experimental peak pressure psig
2.54	87	119	78
2.54	85	119	75
3.05	99.9	127	85
3.05	99	127	79
4.07	97	138	75
4.07	80	138	74

(Taken from Study by J. R. Humphreys, Jr., Vol. II, Proceeding of the Second United Nations Conference on the Peaceful Uses of Atomic Energy, United Nations, New York.)

The experimental values cited above were obtained by the explosive ejection of a variable quantity of 400°C sodium into a sealed reaction vessel containing air. The sodium was ejected from an external reservoir into the reaction vessel by the detonation of a hydrogen–oxygen mixture. The volume of the reaction vessel was 2032 l. It will be noted that in all cases, the maximum pressures attained were substantially lower than those theoretically possible. This was attributed to incomplete sodium combustion and system heat losses. The reaction required 0.03 to 0.17 sec for completion, and during this interval, heat was lost to the reaction vessel surfaces before the maximum pressure could be developed.

The energy realized from the reaction of sodium with water vapor present at atmospheric humidity is less than 5 per cent of that potentially available from the sodium peroxide reaction; however, there is a fast, complete reaction of all water vapor in the system, and this process conceivably might influence the initiation and propagation of the dispersion process.

It should be noted that burning of the sodium from an exposed pool would be a much slower process than that studied experimentally. In such a slow process the absorption of heat into the container and equipment would reduce the peak pressure. A small leak in the system from which sodium is sprayed would also be a slow process compared with the explosive ejection of the molten metal studied. In a large containment building the surface to volume ratio is much smaller and therefore the fraction of heat lost to the container would be smaller and the peak pressure consequently slightly higher. However, it is doubted that this effect is of much significance.

Uranium Oxide Fuel

Fuel elements can be fabricated from uranium dioxide rather than uranium metal or its alloys and the oxide fuel elements seem to offer a number of advantages including minimum changes in properties and dimension during irradiation thereby permitting very high burn-up and also the chemical reactions possible between the metal and coolants including water or gases containing oxygen are avoided. The uranium oxide may be fabricated by cold pressing small pellets which are sintered at temperatures of about 1700°C. Pellets of densities above 95 per cent of theoretical can be made and the porosity is a minimum; however, even in this dense material the surface available for release of diffusing substances is appreciable.

Release of Fission Products

Lustman's studies are summarized in Reference 24 on the release of fission gases from uranium dioxide fuel in metal cladding which was not bonded to the pellets. In these experiments a number of clad specimens of various sintered densities were exposed in a reactor and the amount of the nobel fission gases, xenon-135 and krypton-85 were measured radio-chemically by puncturing the cladding and collecting the gases evolved. The data obtaine[d] was used to calculate the pressures in the fuel elements as a function o[f] temperature and burn-up. The following estimates of rate of diffusion wer[e] made as shown in Table V for pellets of 97.5 per cent of theoretical density:

Table V

Rate of Escape of Fission-Product Gases from Uranium Dioxide

Isotope	Temperature °C	Fraction escaping per second
Kr^{87}	350	0.275×10^{-4}
Kr^{87}	550	2.07×10^{-4}
Xe^{135}	550	5.61×10^{-5}

(Taken from "Release of Fission Product Gases from UO_2," by B. Lustman, WAPD–173.

The following total quantities of krypton-85 were released from the fue[l] during irradiation as shown in Table VI:

Table VI

Release of Krypton–85 from Uranium Dioxide

Burn-up MWD/T	Density % theo	Center temp.	Observed fractional release
2780	95	1815	23×10^{-2}
2060	95	1250	16×10^{-2}
1650	80	760	9.5×10^{-2}
350	95	925	5.8×10^{-3}
815	95	1600	1.6×10^{-2}
5870	95	925	1.1×10^{-3}

(Taken from "Release of Fission Product Gases from UO_2," by B. Lustman, WAPD–173.

From these measurements it was concluded that some free volume was required in the fuel elements to keep the internal pressure below 5000 psi.

MELTING OF REACTOR FUELS

Since little data has been available on the amount of radioactive material that might be released in a reactor incident, the assumption has sometimes been made of a 50 per cent release of all radioactive material into the environs. This value is probably very conservative, and therefore, studies were started to provide experimental information upon which to base better estimates. As part of this program, specimens of several types of fuel elements including stainless steel–uranium dioxide matrix clad with stainless steel (APPR type), zirconium–uranium alloy clad with zircaloy (STR type), and aluminum–uranium alloy clad with aluminum (MTR type) were heated above their melting points and the evolution of radioactive material measured. A summary of this work is presented by Parker and Creek in Reference 25 from which the following discussion has been taken.

The fission products of course include many elements among which are the rare gases, halogens and alkaline earth metals in addition to other classes of elements. It is necessary therefore to examine each class of elements and each element therein as a separate problem in evaluating the release of radioactive material. Fortunately much work has been reported on the properties of the fission products and from studies of pyrometallurgical processes for separating fission products from uranium, some predictions can be made of release upon melting. It can be expected that the rare gases, xenon and krypton, and the halides, iodine and bromine, will be released by vaporization. The alkali metals cesium and rubidium have high vapor pressures and would also be expected to evaporate. Molybdenum oxide has a high vapor pressure and this might separate as a slag and then disperse other radioactive materials

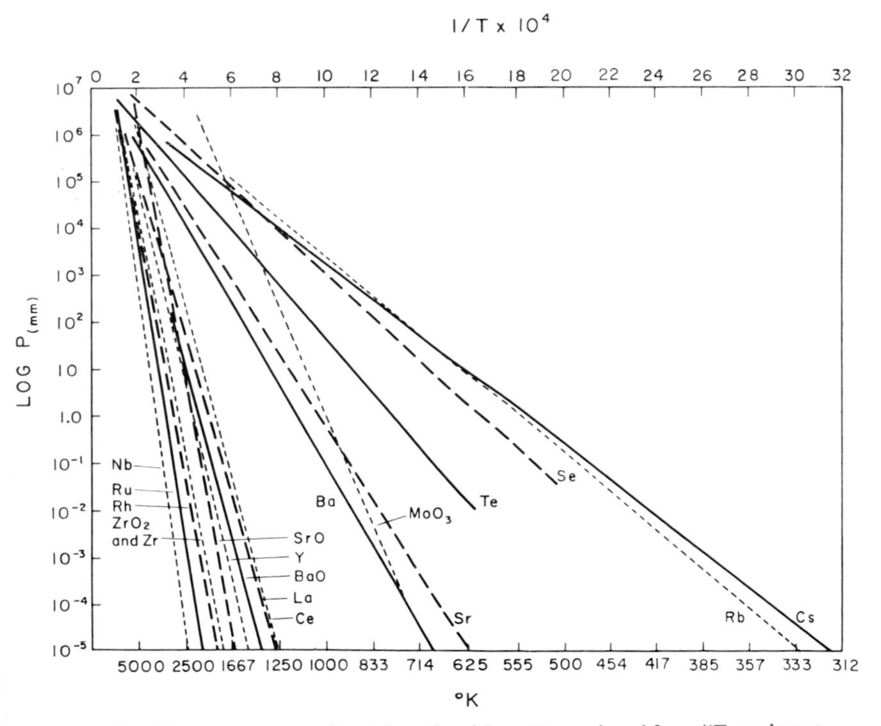

FIG. 11. Vapor pressures of metals and oxides. (Reproduced from "Experiments on the Release of Fission Products from Molten Fuels," by B. W. Parker and G. E. Creek, Report ORNL–CF–57–6–87.)

that diffused into this slag. It can be seen from the vapor pressure data in Fig. 11 that barium, strontium, selenium and tellurium will volatilize at sufficiently high temperatures in the absence of air which would form the less volatile oxides. Also some volatile compounds such as the halide salts may be formed between fission products.

There are also a number of processes which would tend to retain the fission products in the molten fuel. It seems that the rare gases are occluded in the fuel and are completely released only by purging of a metal. Some metals such as zirconium and niobium may form nonvolatile carbides. Fuel elements made with uranium dioxide will retain many fission products by occlusion in the oxide which has a very high melting point. Stable alloys may be formed by uranium fuel with such elements as tellurium and ruthenium. Also oxides of low volatility may be formed with barium and strontium particularly when heated in air.

In an experimental program, samples of the different types of fuel elements after irradiation were rapidly (10 sec or less) heated to above the melting point of the sample and the release of radioactive material identified. At trace concentrations of fission products, slow melting of the APPR plate at 1525°C in air or steam effected the release of 50 per cent of the rare gases, 33 per cent of the iodine, 9 per cent of the cesium, and traces of strontium. After 25 per cent burn-up, the cesium value increased to about 60 per cent. Aluminum alloy of the MTR type, also at trace concentration, upon melting at 700 °C released up to 2 per cent of the iodine, 10 per cent of the rare gases, and negligible portions of other fission products. Zirconium alloy of the STR type after 15 per cent burn-up, when melted at 1850°C, released up to 95 per cent of the rare gases, 90 per cent of the cesium, 60 per cent of the iodine, and only a trace of strontium. It was concluded that prolonged heating in air at temperatures in excess of the melting point results in the release of a large portion of the radioactivity. On the other hand, a moderate amount of heating in air or steam sufficient only to melt a specimen results mainly in the partial volatilization of the rare gases, the halogens, and the alkali metals. In the presence of air or water vapor, strontium and other fission products were not released. The percentage release of fission products into the atmosphere from molten reactor fuel is apparently proportional to the melting temperature and is affected mainly by the degree of oxidation, the concentration of fission products, and the type of fuel. The release of beta active fission products is summarized in Table VII taken from References 25 and 26.

In radiation exposure calculations for an assumed reactor incident involving melting of the core and some chemical reactions, the assumption of 50 per cent release of all fission products into the atmosphere is apparently unduly pessimistic. Based in part on evidence from tests on melting of fuel elements, much lower fission product release estimates were used in evaluating the hazards to the area surrounding the pressurized water reactor at the

Table VII

Calculated Beta Activity of the Fission Products

(Per Cent of Total by Element)

Irradiation time:		5 days				30 days				
Cooling time:		10 min	10 hr	10 days	100 days	10 min	10 hr	10 days	100 days	Melting behavior and % released
Group I-A	Xe	6.0	11.0	8.0	0.2	6.0	10.0	4.5	0.1	
	Kr–Rb	3.4	0.8	0.02	0.06	3.0	0.7	0.02	0.008	
	I–Br	9.0	15.0	8.5	0.1	10.0	13.0	5.7	0.04	High volatility
	Cs	3.5	0.001	0.03	0.09	3.0	0.001	0.04	0.24	10–100
Group I-B	Te–Se	7.0	4.0	4.8	0.05	6.8	4.0	2.2	0.1	Variable
	Mo	14.0	6.2	4.8	0.00	16.0	5.6	2.5	0.0	moderate volatility
	Tc	0.3	0.6	0.5	0.02	0.4	0.6	0.3	0.01	1–10
Group II	Ba	5.0	2.5	10.0	0.15	5.8	5.0	10.5	0.3	Volatile in vacuum
	Sr	6.0	4.5	3.3	11.0	5.8	4.2	4.5	8.5	nonvolatile in air
Group III-A	Y	11.0	10.0	3.3	14.0	10.0	8.0	5.5	14.0	Low volatility
	La	11.5	3.3	12.0	0.9	6.6	6.0	12.0	0.7	rare earths
	Ce	5.0	8.2	8.0	16.0	5.7	8.2	10.2	14.0	refractory
	Pr	3.6	3.8	11.0	8.0	4.5	6.0	12.0	10.0	< 0.1
	Nd	8.0	1.2	6.0	0.06	1.4	2.5	4.3	0.02	
	Pm	0.6	1.6	1.2	9.0	0.6	1.5	1.7	9.0	
Group III-B	Zr	5.8	13.0	7.0	15.0	5.5	10.0	5.7	14.0	Low volatility
	Nb	2.9	7.5	1.3	17.0	2.7	5.3	2.6	23.0	alloying and
	Ru	0.6	0.8	3.0	5.5	1.0	1.7	4.5	6.0	refractory
	Rh	0.6	2.2	5.5	6.5	1.3	3.3	5.5	6.0	< 0.1

(Reproduced from "Contributions of Chemical Elements to the Total Beta Activity of Fission Products," by F. J. Keneshea and A. M. Saul, Report NAA–SR–197.)

Shippingport Atomic Power Station as described in Reference 27. It was assumed that those volatile fission products significant in dosage calculation will escape completely from the portions of the seed core affected by melting or chemical reaction of zirconium with water. It was estimated that the time of release was short as compared with the time required for complete core meltdown. The elements which were included in this group were Br, I, Xe, Kr, Cs, and Rb. The non-volatile fission products such as the rare earths, Zr, Nb, Ru, Rh, and others of a similar nature were assumed to be not released due to occlusion in the zirconium oxide product of the zirconium–water reaction. It is expected that 90 per cent of the ZrO_2 reaction product would remain after the reaction as a coherent mass of oxide and that no more than 10 per cent of the reaction product would separate from this mass. These separable particles would not be extremely finely divided, so that entrainment would be difficult and fall-out of the particles, if entrained, would effectively prevent release of the non-volatile fission products associated with zirconium oxide. The alkaline earths, Ba and Sr, are particularly important in hazards evaluation and represent an intermediate case between the volatile and non-volatile fission products. An experiment on the release of fission products from a spent uranium fuel element, as a result of a slow uranium–water reaction, has shown that the release of barium and strontium is proportional to the amount of uranium reacted. Presuming that this experimental evidence applies to a zirconium–water reaction, a release of 5 per cent of the barium and strontium associated with the zirconium that reacts with water was used in the hazards calculations. Thus, with total core meltdown and a reaction of 20 per cent of the metal with water, the release of barium and strontium would be 1 per cent. In estimating the release of barium and strontium from the uranium dioxide blanket, the assumption that the release is proportional to the amount of zirconium reacting with water was made only as a matter of convenience in the calculations and there is little experimental evidence to support this assumption. The release of Mo and Te was neglected since their gamma emanations are of low energy and the combined effects of these elements on the radiation exposure is less than 10 per cent.

For the worst accident calculated in evaluating the PWR hazards, which would be the case of complete core meltdown with an open plant container, the isotopes of the elements which were assumed to be released constituted approximately 11 per cent of the total core fission products. Many of these isotopes would not contribute to the biological dosage because of short half-life and/or lack of significant radiations. Therefore, for this worst accident, only 2.77 per cent of the total core fission product activity in curies at shut-down is assumed to escape to the atmosphere and to contribute to the calculated dosage. Of this quantity, the isotopes of the volatile elements I, Xe, Kr, and Br constitute 2.4 per cent, Cs and Rb 0.34 per cent and the alkaline earths, Ba and Sr, 0.0092 per cent. Because of this selective release, the inhalation and external beta dosages assume a larger importance, compared to external gamma radiation, than would be the case, if the whole spectrum of fission products were assumed to escape.

SHOCK WAVES

The sudden release of energy in a nuclear excursion, rapid chemical reaction or release of internal energy of a compressed fluid by rupture of a vessel, could produce a pressure wave which travels out from the origin through solid and fluid media. This pressure wave could cause serious destruction of components near to the reactor including the containment structure and therefore, provisions may be required such as a blast shield to prevent breaking the container. Damage outside the immediate area of the reactor from the pressure wave would be negligible and therefore, need not be considered in evaluating public hazards except as related to the containment problem.

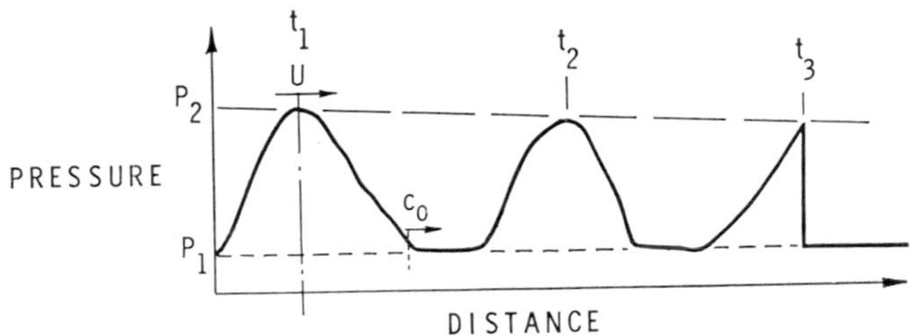

FIG. 12. The "shocking-up" process. (From "Containment of a Nuclear Power Excursion in the Air Force Nuclear Engineering Test Reactor" AFIT Report 58–12 by D. E. Asire.)

As a pressure wave or pulse moves away from the origin, the pressure profile for a very strong pressure wave goes through a rapid transformation to become a true shock wave provided the material has the characteristics which are common to most materials: (1) the volume decreases with increasing pressure and (2) the resistance to compression increases with increasing pressure. As a result of these properties the portions of the wave at high pressure have a higher local velocity of sound than at the wave front and therefore, the pressure peak overtakes the leading edge of the pulse to form the almost pure pressure discontinuity called a shock wave. This has been termed the "shocking up" process and is illustrated in Fig. 12.

The total increase in energy of a material as a shock wave passes through it consists in part of kinetic energy due to particle velocity and about an equal part for strong pressure waves of an increase in internal energy. The total increase in energy is the work done by compression at the peak pressure, P_2, from the initial volume, V_1, to the final volume, V_{2p}.

$$E_t = P_2(V_1 - V_2).$$

The kinetic energy part of this total energy increase, E_k, is

$$E_k = \tfrac{1}{2}U_2^2 = \tfrac{1}{2}(P_2 - P_1)(V_1 - V_2).$$

The internal energy gain per unit mass is the difference between

$$E_I = E_t - E_k = \tfrac{1}{2}(P_2 + P_1)(V_1 - V_2).$$

Now, if $P_2 \gg P_1$, the increase in internal energy is nearly equal to the increase in kinetic energy and the energy imparted by the shock wave is nearly equally divided

$$E_I = K_k\,.$$

It is seen from the above relations that the conditions behind the shock wave can be calculated, if the peak pressure or energy is known in addition to the relations between pressure and volume, and the equation of state for the material. In the case of a perfect gas it can be shown that air becomes much hotter behind a single shock wave than from an adiabatic compression to the same pressure. Since only the energy of an adiabatic compression can be recovered by subsequent expansion of the gas to the base pressure, a shock compression is a highly irreversible process. The part of the internal energy in excess of an adiabatic compression and which, therefore, cannot be recovered is termed "waste heat" and represents an important loss of energy from the shock wave. As it passes through several feet of water, a strong shock wave may thus lose 90 per cent of its energy. It is seen that the energy of a shock wave is divided about equally between internal energy and kinetic energy and that part of the internal energy is lost as "waste heat" in an irreversible compression.

Since the pressure behind a shock drops in passing from a solid into a gas by approximately the ratio of densities, an air gap would be expected to offer some protection from a strong shock wave. For a very thin air gap between solid material the drop in shock pressure in passing from the solid into the gas is only momentary and reflection of the pressure wave back and forth across a small gap soon builds the pressure almost to the level originally existing in the solid. The air gap must be large as compared to the positive phase length of the shock wave to provide a significant reduction in shock pressure. A small air gap can provide protection against direct impact of parts of a structure, since a shock wave imparts momentary velocity to a material. Isolation of the concrete structure and shielding of a reactor from the outer containment vessel by an air gap as small as 2 in. is believed to reduce possible damages to the outer containment vessel according to Reference 34.

A strong underwater shock wave in the vicinity of the center of explosion decreases in intensity with distance only as $1/R^{1.13}$. At greater distances in a gas or a liquid which is vaporized by the explosion, the energy decreases more rapidly and approximately as $1/R^3$. Air shock strengths can be estimated reasonably well by calculating the static pressure rise in the total volume of air behind the shock due to the explosion or excursion energy. Multiplication of this pressure by a factor of two will give the peak pressure at the shock front. This rule of thumb according to Reference 34 is applicable because of the rapidity with which a shock wave in air dissipates its energy.

In solids and liquids which are not vaporized and which have good absorption characteristics, the waste heat process will cause a decrease in energy or pressure as $1/R^7$. When the pressure has decreased below the dynamic crushing strength of the material, the decrease in pressure is much slower and about $1/R$. The importance of selecting suitable absorbing material for a blast shield can therefore be seen. Porous media such as low density concrete, wood, felted fibers, foamed ceramics, plastics, and elastomers are known to be valuable in the dissipation of energy and such materials have been used for this purpose in bombproof shelters, explosive magazines, chemical plants, and other installations where explosions are to be anticipated. Such materials also have been applied to the construction of blast shields for nuclear reactors to protect the outer container of the reactor against puncturing as the result of an explosion. The design of blast shields using such materials has been studied experimentally at Armour Research Foundation as reported by Napadensky and Stresau in Reference 31.

Some of the other properties of shock waves can be illustrated by considering the possible sequence of events following a sudden release of energy from a power excursion in the core of a reactor. A liquid-filled (water or sodium) reactor vessel with a gas-filled space between the top of the vessel and the liquid as in the EBWR or EBR 11 will be assumed. Following the release of energy a pressure wave will travel away from the core and become a shock wave after traveling a distance of less than the core dimensions in the case of a short duration power release. From experiments it appears that such a sharp, short duration shock is more destructive than a long duration shock of equal impulse value. In traveling through several feet of liquid to the reactor vessel about 90 per cent of the total explosion energy is dissipated as waste heat raising the temperature of the liquid only a few degrees. The shock wave travels through the liquid and upon meeting the wall of the pressure vessel is increased several fold by a reflection process. For an initial energy release equivalent to 300 lb of TNT, the pressure at the vessel walls may be several thousand atmospheres and far above the strength of the vessel. The vessel is ruptured, and the energy of the shock wave is released below and to the sides of the reactor vessel. A rarefaction wave travels back from the vessel walls providing eventual relief of pressure within the vessel and limiting the time duration of the pressure-loading on the cap structure. When the original shock wave traveling upwards reaches the liquid–gas interface above the core, an enormous drop in pressure occurs of the order of several hundred-fold.

This greatly reduced shock proceeds upward through the gas and is increased several fold in pressure by reflection from the top of the vessel. This reflected pressure, still greatly below the original pressure in the liquid, constitutes the pressure loading of the cap. This pressure is relieved by the arrival of the rarefaction wave from the bottom of the vessel and limits the time duration of the pressure-loading on the top structure. It should be noted that rupture of the top structure in this example is prevented by the gas space above the liquid. If the vessel were completely filled with liquid such an explosive release of energy would be expected to cause failure of the top of the vessel.

EBWR BLAST SHIELD

The design of the containment for the Experimental Boiling Water Reactor, EBWR, was based on a reaction of 25 per cent of the metal in the core with

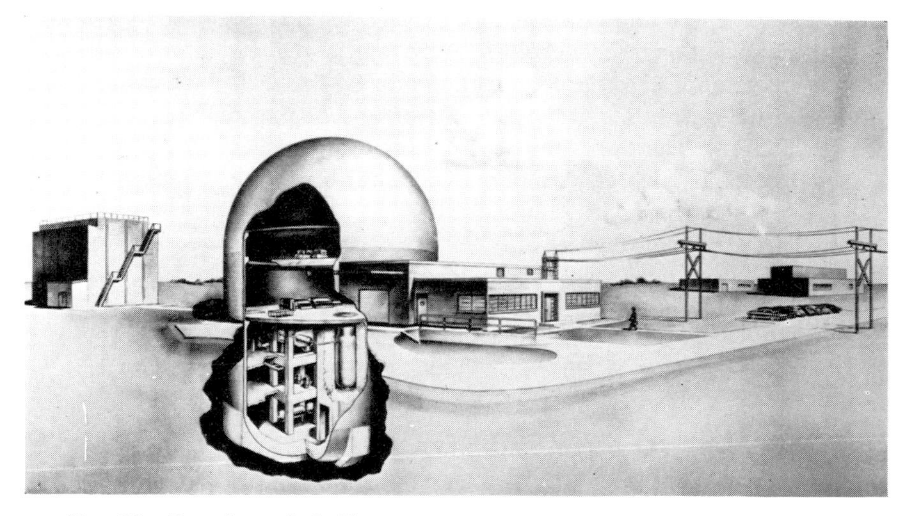

FIG. 13. Experimental boiling water reactor, Lemont, Illinois. (Argonne National Laboratory.)

water. The assumption was made that the release would be of the order of a TNT explosion with an energy release of 8 and 10^8 calories according to Reference 35. The following sequence of events was estimated by Porzel (Reference 34) to result:

(1) A shock wave pressure of 40,700 atm reaches the vessel walls below the water level in 1 msec with a duration of 2 msec and causes rupture of the vessel below the water level.

(2) The shock wave reaches the bottom head at about the same time and causes rupture from the peak pressure of 13,800 atm.

(3) The shock pressure is reduced by the steam space above the water level

to a value of 260 atm but has a duration of 6 msec. The parts of the vessel above water are not ruptured.

(4) The vessel cover which is also protected by the steam space also remains intact.

(5) The lower part of the biological shield is reduced to rubble.

(6) The bottom shield plug receives an impulse of 10 msec duration and is reduced to rubble.

FIG. 14. Steel containment building for EBWR Lemont, Illinois. (Argonne National Laboratory photograph.)

In the design of the Experimental Boiling Water Reactor it was found that the vertical columns for holding down the top structure and which are located within the concrete biological shield, would be subjected to lateral blast forces sufficient for destruction unless a specific blast shield were provided for protection.[30] A series of compression tests were conducted by Armour Research Foundation as described in Reference 30 to determine the optimum constituents of the blast shield. The final design consists of alternate layers of 3 in. of redwood and 1 in. of steel plate, repeated three times, and then alternate layers of 3 in. of Celotex and 1 in. of steel plate repeated twice. There was a reduction in pressure by approximately a factor of two through each layer of the softer material by conversion of a portion of the shock energy to heat energy. The residual material velocity is transformed into shock energy at each steel plate layer and the process is then repeated through the successive layers. By introducing five such steps, the shock wave pressure on the hold-down columns would be reduced to a value which the columns could sustain.

The Experimental Boiling Water Reactor, EBWR, was designed to hold down the remains of the reactor pressure vessel following rupture and blow-out of the bottom and sides below the water line. The upper portion of the large vessel is estimated from studies by the Armour Research Foundation in

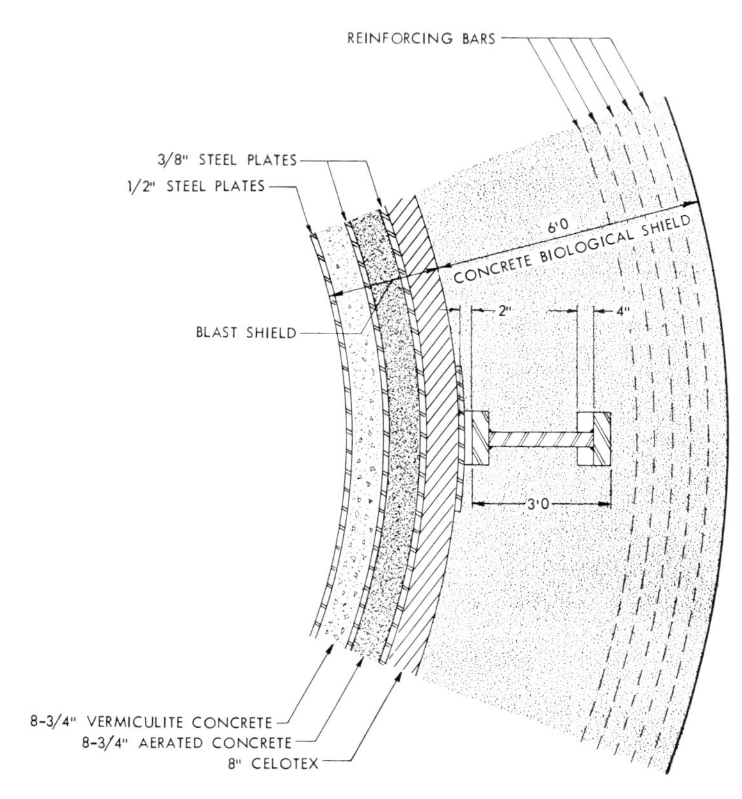

FIG. 15. Cross-section of blast shield.

Reference 34 to remain intact, and the upward force on this section induced by the shock wave was calculated at 21,000,000 lb. The force would be applied rapidly and for a very short interval, which implies high strain rate with accompanying low dynamic load factor. The system developed for restraining the upper portion of the reactor vessel has been termed the "hold-down" structure and is described in Reference 35. Three parallel hold-down beams span the top of the reactor vessel and are attached to two clusters of three hold-down columns on opposite sides of the reactor vessel. The columns, in turn, are anchored in the very heavy reinforced concrete basement floor. In order to transmit the upward force from the reactor vessel cover to the bottom flanges of the hold-down beams, a steel ring and a structural steel back-up plate serves as a biological and missile shield and distributes the upward force over a wide area on each of the three beams.

EXPERIMENTAL BREEDER REACTOR II BLAST SHIELD

From the design of the EBR II containment system the estimate was made that the maximum energy in a nuclear excursion would not be greater than that

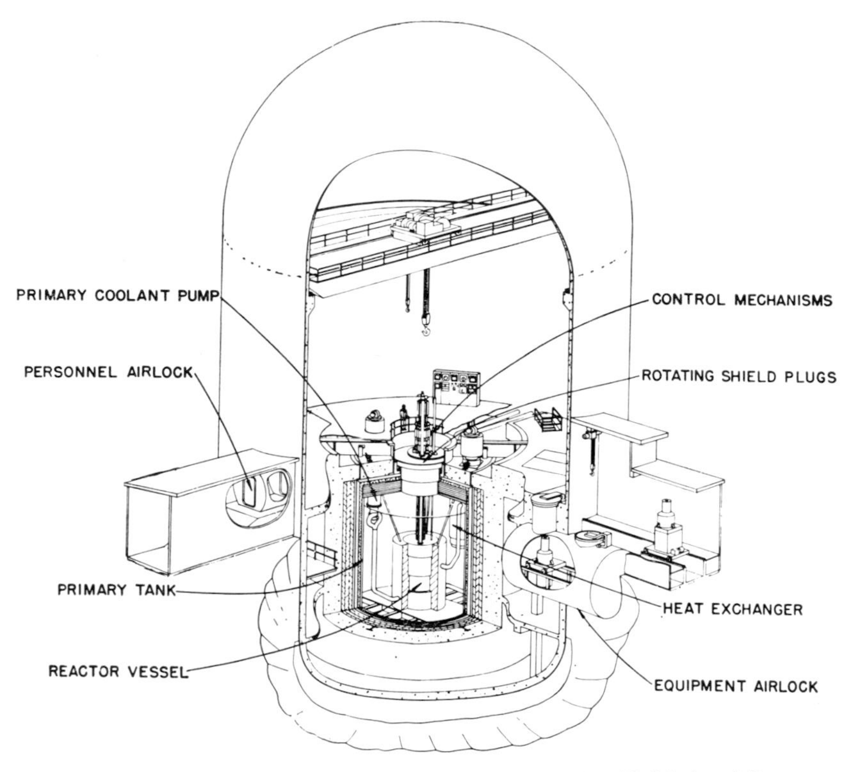

FIG. 16. Containment for Experimental Breeder Reactor II, National Reactor Testing Station. (Argonne National Laboratory.)

associated with a 300 lb TNT detonation. An additional maximum energy release equivalent to 10,000 lb of TNT in energy release from the sodium–air reaction was also estimated although this energy release would not affect the primary tank support structure and was therefore not considered in the design of the blast shield.

Some measurements were made for the design of the blast shield for the EBR II as described in Reference 31. Since space was available, the actual layers in the blast shield were made thicker than actually required. A change in absorbing material was made at each point where the transmitted pressure was estimated to fall below the material yield point and a laminated structure of three layers was selected. The first layer is vermiculite concrete (22 lb/ft^3) 8.75 in. thick which is estimated to be capable of reducing an incident shock pressure of 1530 psia down to 300 psia. The second layer of aerated concrete (16.5 lb/ft^3) is to reduce the 300 psia down to 75 psia across the thickness of

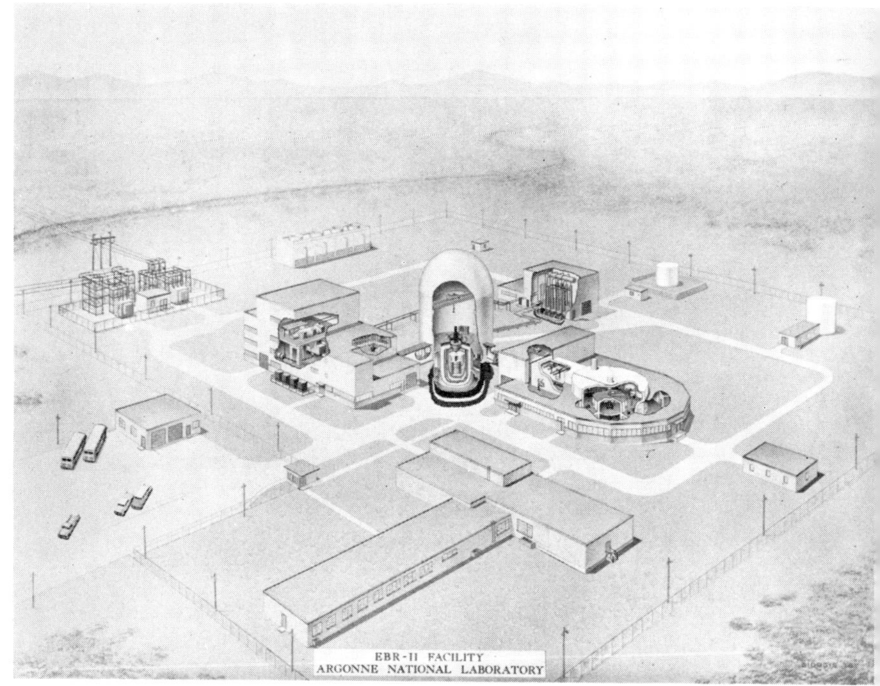

FIG. 17. Site plan for Experimental Breeder Reactor II at National Reactor Testing Station. (Argonne National Laboratory.)

8.75 in. The third layer is made of Celotex since a yield strength below 75 psia is required and in 8 in. of Celotex the pressure is to be reduced to 15 psia. The layers are separated by continuous cylinders of steel (3/8 in.) to prevent channeling of energy through the weaker or non-homogeneous regions of the concrete.

SPRAY

It has been seen that the material behind a shock wave is given motion and a velocity which depends on the material and the strength of the shock wave. At a liquid-gas interface, the velocity of the liquid may be sufficient to throw a spray up against the top container and exert a force on it. In the EBR II design, it is estimated in Reference 29 that the sodium velocity would be some 72 ft/sec and due to the reflection wave at the surface the particules would be thrown up at twice this velocity producing a dynamic pressure on the top structure of 3.53 atm.

In tests of a model of the Air Force Nuclear Engineering Test Reactor as described in Reference 32, a spray of water was observed to be thrown up and during the test simulating a 1000 MW-sec excursion, this spray dome was capable of throwing the model instrument bridge over the outer rim of the structure. From analysis it is estimated that water might be thrown up by the shock wave with an initial velocity of 30 ft/sec and reach a maximum

FIG. 18. Containment for Livermore Pool-Type Reactor. (University of California Radiation Laboratory photograph.)

height of 14 ft. The momentary pressure rise on any surface exposed to this spray of water near the surface would be 6 psi. If the liquid over the reactor core is very deep and confined by the walls to about the dimension of the bubble formed by the energy release, the expansion of the bubble may force a large slug of water upward with sufficient velocity to damage any overhead structure and the outer containment vessel that this large mass of water would strike. A low flat roof would be particularly vulnerable to such damage.

MISSILES

A shock wave traveling through a solid is reflected from a free surface as a tension wave. As described in Reference 5, as the leading edge of this tensile wave reflects into the material, it cancels the trailing edge of the on-coming compression wave. This tensile wave continues into the material until the magnitude of the original compression stress is approached. As this tensile stress builds up, it could become equal to the ultimate strength of the material and failure would occur. The result would be that a flat section or chip of the material would be thrown off at the free surface. The thickness of this fragment depends upon the profile of the original compression wave and the tensile strength of the material. The phenomenon is called spalling. The velocity with which the fragment is thrown out is twice the material velocity behind the incident wave. Multiple spalling from one strong, long-duration incident shock may occur since each spall presents a new free surface to what remains of the on-coming compression wave.

The maximum spall velocity from the Air Force Nuclear Engineering Test Reactor is estimated in Reference 7 to be some 15 ft/sec which is not a dangerous fragment for penetration of steel plate by relatively light pieces of concrete. A spall with this initial velocity would rise to a height of only 4 ft. The only heavy spalling that was observed from the model tests described in Reference 32 will be prevented in the actual reactor by capping the area with a steel plate.

The solid parts of a reactor immersed in liquid coolant are possible sources of missiles since a shock wave can accelerate material to a considerable velocity. The equation of motion of possible missiles being accelerated by shock wave would, according to Reference 36, include the following factors:

$$\frac{dU}{dt} = \frac{PA}{M} - g + \rho\frac{AU_s^2}{2M} - \frac{C_d AU^2}{2M}$$

where

U = velocity of missile
P = unbalanced shock pressure exerted on missile
A = effective area of missile
U_s = velocity of fluid immediately underneath the missile
M = mass of missile
ρ = density of fluid
C_d = drag coefficient, usually taken as unity.

In most cases the g and $C_d AU^2/2M$ are of negligible magnitude compared with the first term and may be ignored. The third term, $\rho AU_s^2/2M$, represents the force from the material velocity of the fluid on the underside of the missile. This term can usually be safely ignored also, because of its moderate magnitude and the extremely short time over which the force acts.

The unbalanced liquid pressure, P, acts for the time required for the shock wave to travel the length of the missile. This pressure decays exponentially with time while traveling the length of the missile, but for small missiles of dimension of a few inches, the pressure is assumed to be constant at the peak value. For an object of length of a few feet, an average pressure of half the nitial value was assumed in Reference 36. For a constant pressure, P, the equation of motion becomes

$$U = \frac{PAt}{M}$$

and from the value of pressure and time of application of the pressure in accelerating the missile which is the time for the shock wave to travel the length of the missile, the velocity can be calculated. After this initial acceleration the equation of motion of the missile is

$$\frac{d^2x}{dt^2} = -g - \frac{C_d A}{2M}\left(\frac{dx}{dt}\right)^2$$

where x is the distance traveled vertically.

Some estimates of missile velocities for the EBR II are given in Reference 29. This reactor is immersed in a large tank of sodium and there is $9\frac{1}{2}$ ft of sodium above the reactor cover. This cover is a cylinder 92.5 in. in diameter and 32 in. in height. The weight of the cover is 28,000 lb. It could be given a velocity of 35 ft/sec by an energy release equivalent to 300 lb of TNT in the reactor core. After traveling up through the $9\frac{1}{2}$ ft of sodium this velocity would be reduced to 11 ft/sec which would be further reduced to 6 ft/sec before striking the top structure which is designed to withstand this impact energy. Estimates were also made of other possible missiles and it is shown in Reference 29 that a solid steel sphere located 3 ft from the core would be given an initial velocity of 80 ft/sec which would be reduced to about 10 ft/sec after passing through the $9\frac{1}{2}$ ft of sodium. It is shown that other small spheres would achieve the same maximum velocity. In the EBR II design the control rod drive shafts which extend down through the tank to the reactor core exhibit the greatest potential in respect to missile generation. These shafts could be ejected with a velocity of 70 ft/sec. However, these rods are long and thin, and there would have to be nearly perfect alignment of the sodium shock force with the longitudinal axis of the shaft or else it would merely buckle or bind. Also, even though projected with a velocity of 70 ft/sec, this would not be sufficient for the rod to strike the top of the outer containment structure which is 96 ft high above the reactor.

Other studies of missiles from shock waves include the model testing of the Air Force Engineering Test Reactor. The missile of the greatest potential danger in the model was the core tank top plug, which was simulated by a simple bolted-down manhole cover. This almost broke loose in the last most energetic test sending fragments of bolts into the air with an initial velocity of 130 ft/sec in one case. The top plug in the actual reactor has been, therefore, designed not as a bolted flange, but to withstand impulse loading. This new top plug is a massive piece of concrete and steel, 2 ft thick and 8 ft in diameter. It is made up of a cylindrical layer of Barytes concrete sandwiched in between 3 in. thick steel plates and sealed with 1 in. steel plate. The bottom 3 in. plate also forms a segmented key that rotates into a horizontal keyway. The plug weighs about 25,000 lb and under the measured impulse of a 1000 MW-sec excursion would lift off its seat perhaps $\frac{1}{2}$ in., if in an unlocked position at the time. If the plug is locked, the 3 in. thick key will more than withstand any shear or bending stresses imposed by the impulse load. Other parts of the reactor installation were found from the model tests to constitute missile hazards. The thermal column roll door would be given a velocity corresponding to 18.5 ft/sec by impact from the graphite in the thermal column. Also the massive door at either end of each test cell would receive a direct air shock at least two times greater than the shock predicted in free air, because of the energy preference for channeling into the free air. The initial velocity of a 2800 lb model test cell door was estimated to have been 2.6 ft/sec.[7]

A valve or thermometer well might break loose and be propelled by the jet of escaping fluid. If the object remains in the jet of fluid a considerable velocity can be attained before striking the wall of the vapor container. The jet, which has mass and velocity, imparts impulse to the missile from which the increase in velocity of the missile from an initial state of rest can be determined. Pressurized water, initially at 1200 psi and 450°F, would drop very rapidly to saturation pressure of 423°F and the saturated fluid expanding into the vapor container would reach a peak velocity within a few feet from the orifice of rupture. The angle at which the jet spreads at right angle to the direction of flow is apparently between 12° and 60° and calculations were made with both values, to determine the limits of this uncertainty. The rate of release of fluid in the jet depends on the area of the rupture and the velocity of the issuing fluid. It was assumed tearing out a 2 in. valve would be associated with an opening approximately 3 in. in diameter. It was estimated that some 629 lb/sec of fluid would flow through this opening. A 2 in. valve designed for 1500 psi service would weigh some 50 lb and have a cross section area of 30 in^2. This missile might be propelled on traveling 40 ft in the jet to a velocity between 700 and 500 ft/sec depending on the angle of spread of the jet assumed. It was concluded that the velocity of other missiles accelerated in this way would not exceed these values.

Missiles can be produced from reactor systems operating at high internal fluid pressures as the result of several types of failures. If the head is released from a pressure vessel, the head could become a missile propelled by the jet of gas or vapor released from the vessel. Also in some cases the vessel itself could, if free, attain considerable velocity by discharge of the fluid at high pressure through the open end like a rocket. Other possible sources of missiles include valves, thermometer wells and pipes which might be released by failure of welds. This problem of missiles from a pressurized water system was examined carefully in the design of the Army Packaged Power Reactor and much of the following is taken from the Hazards Summary Report for that reactor.[36]

The maximum possible velocity for self propelled missiles was examined by considering a 2 in. schedule 160 pipe initially filled with water at 1200 psi and 450°F. The available energy in the fluid when expanded to atmospheric pressure would be 38 BThU/lb of fluid. If the pipe is accelerated through the

FIG. 19. Aerial view of Army Packaged Power Reactor, Fort Belvoir, Virginia. (Corps of Engineers, Dept. of the Army photograph.)

entire travel which is limited to approximately 40 ft within the vapor container, a velocity of 435 ft/sec could be attained. Similarly, if such a pipe were filled with 425 psi saturated steam at the instant of detachment, the available heat energy for expansion to atmospheric pressure is 255 BThU/lb, and a velocity of 510 ft/sec could be attained within the containment vessel. From these conservative estimates, a maximum velocity of 500 ft/sec was taken for self-propelled missiles.

MISSILE PENETRATION

The penetration of concrete and steel by missiles has been correlated on a semi-empirical basis by the military establishments and several formulas are available. The following were used in Reference 36 in evaluation of the possibility of missile penetration of the container for the Army Packaged Power Reactor. The Ballistic Research Laboratory formula for concrete is:

$$P = \frac{6Wd^{1/5}}{d^2} \times \left(\frac{V}{1000}\right)^{4/3}$$

where

P = penetration of missile, in.
W = weight of missile, lb
d = diameter, in.
V = velocity, ft/sec.

The Ballistic Research Laboratory recommends that the calculated penetration in concrete be multiplied by a factor of 1.3 to insure no complete penetration due to the cracking of concrete ahead of the point where the missile stops. In the case of the container for the Army Packaged Power Reactor, since the container has 2 ft of concrete inside the $\frac{3}{4}$ in. steel container wall, the use of the 1.3 factor was not considered essential.[36] The calculated penetrations of several typical missiles are shown in Table VIII.

Table VIII

Calculated Concrete Penetration

(*Taken from Reference 36*)

Missile	2 in. steel valve	Thermometer well	2 in. steel bar	2 in. steel pipe
Weight, lb	50	5	21	30
Area, in^2.	18	1.3	3	4.4
Diameter (equiv.)	4.8	1.3	2	2.4
If velocity = 500 ft/sec penetration, in.	7	7.5	14.5	15
If velocity = 700 ft/sec penetration, in.	10	11	21	22

Since it is considered that 1 in. of steel is equivalent to 12 in. of concrete in a compound wall, the wall of the vapor container would be equivalent to 33 in. of concrete and no penetration is indicated even when the above values are multiplied by the factor 1.3.

The Ballistics Research Laboratory formula for penetration of steel as given in Reference 36 is:

$$t^{3/2} = \frac{\frac{1}{2}mV^2}{K^2 \times 17{,}400 \times d^{3/2}}$$

where t = wall thickness, in.
m = mass of missile, slugs
V = velocity of missile, ft/sec
K = a constant depending on the grade of steel and $\cong 1$
d = the diameter of the missile, in.

The missile penetration through steel was calculated from this formula as follows:

Table IX

Calculated Steel Penetration

(*Taken from Reference 36*)

Missile	2 in. steel valve	Thermometer well	2 in. steel bar	2 in. steel pipe
If velocity = 500 ft/sec penetration, in.	1	.8	1.5	1.5
If velocity = 700 ft/sec penetration, in.	1.5	1.3	2.3	2.3

Another formula for the penetration of concrete is used by the Navy and as given in Reference 36 is:

$$D = KA_pV'$$

where

D = depth of penetration, ft

K = a coefficient depending on the nature of the concrete and is 0.00799 for massive concrete, 0.00426 for normal reinforced concrete, such as would be used in building construction, and is 0.00284 for specially reinforced concrete.

$$V' = \log_{10}\left(1 + \frac{V^2}{215{,}000}\right)$$

V = velocity of the missile, ft/sec

A_p = weight of missile per ft^2 of projected area, lb/ft.

The following missile penetrations were calculated using this formula:

Table X

Calculated Concrete Penetration

(*Taken from Reference 36*)

Missile	2 in. steel valve	Thermometer well	2 in. steel bar	2 in. steel pipe
Weight, lb	50	5	21	30
Area, in.^{2}	18	1.3	3	4.4
A_p, lb/ft^2	400	550	1000	1000
If velocity = 500 ft/sec penetration, in.	8	10	19	19

The values thus calculated correspond closely to those from the Ballistic Research Laboratory formula multiplied by the factor 1.3.

In the "PWR Hazards Summary Report"(43) the possibility of damage to the container by a flying fragment or "missile" resulting from failure of a component, pipe or vessel in the primary coolant system is considered since the simultaneous loss of integrity of the primary coolant system and the plant container might permit release of radioactive material to the outside atmosphere. It was concluded that brittle failure of a part of the primary coolant system must occur in order to release a high velocity fragment or component as a missile, since a ductile failure would probably take the form of a tear or split in the pipe or component wall, or a crack at a joint without shearing off the pipe or component. Careful attention has been given the design, fabrication, and testing of all parts of the primary coolant system, so as to exclude the possibility of brittle fracture. Therefore, special design precautions were not considered necessary for protection against internal missiles. Since the container is surrounded by concrete and is mainly underground, protection against external missiles is provided.

The brittle fracture referred to above is a cleavage type of fracture that occurs without appreciable plastic deformation in ferritic type steels which normally exhibit excellent ductility in a conventional tensile test. Investigations of failures by brittle fractures have established that three conditions are necessary and sufficient as follows:

(1) A stress raiser of some type such as cracks or sharp notches, that result from design or fabrication.

(2) A stress sufficiently large to cause localized yielding in the vicinity of the stress raiser.

(3) A service temperature sufficiently low so that very little plastic deformation occurs before fracture.

The operating procedures for the PWR plant specify that the reactor vessel, the pressurizer and the boiler, which are made of ferritic steel, will never be pressurized at temperatures so low as to be in the brittle range (i.e. never below 120°F), and also the reactor will not be critical below these temperatures.

Excessive exposure to high neutron flux can cause a reduction of ductility and an increase in the nil ductility temperature of carbon or molybdenum steels. Experiments were performed at fast flux levels orders of magnitude greater than experienced by the PWR pressure vessel and at various temperature levels and it was concluded that the SA-302 steel in this vessel will not suffer significant impairment of ductility through irradiation damage. The PWR piping in the primary system is Type 304 stainless steel.

In further support of the view that brittle failure producing missiles can be excluded, the results of a survey of accident experience with vessels has been in Reference 43. The pertinent results of the survey are:

(1) No record was found of a sudden major rupture or brittle failure in large vessels operating within design limitations at PWR operating temperatures.
(2) Cracks have occurred in large, heavy wall stainless steel pipes but, so far none have resulted in opening of a major rupture.
(3) Record was found of two cases of rupture in large diameter (6–12 in.) pipes but, these were caused by severe graphitization in carbon–molybdenum steel pipe. These cases are not applicable to the PWR plant which uses stainless steel in the primary coolant piping.
(4) Small (1–3 in.) stainless steel lines have ruptured, but these ruptures have usually occurred as axial splits.

MISSILES FROM ROTATING MACHINERY

There is some experience with missiles produced by failure of high speed rotating machinery such as the turbine rotor disintegration which occurred at Ridgeland Avenue Power Station near Chicago. From the results of the analysis presented in Reference 35, it appears that the most dangerous fragment which might emerge from the steam turbine casing in the Experimental Boiling Water Reactor installation would be one of the heaviest wheels traveling at a velocity of 212 ft/sec. This analysis indicates that a 12 in. thickness of concrete or $1\frac{1}{4}$ in. of mild steel plate would contain this fragment. The containment shell of the EBWR has a 12 in. thick reinforced concrete liner missile shield above the main floor. The ceiling slab at this height completes the concrete envelope. This slab has a $\frac{3}{8}$ in. thick steel plate on the underside and is 12 in. thick. There is no concrete liner for the shell above the ceiling slab, but wherever openings occur in the concrete envelope, other provisions were made for missile shielding. A 12 in. thick concrete wall protects the main air-lock and emergency air-lock, and $1\frac{1}{4}$ in. thick steel plates protect the bolted freight door and the large welded temporary opening, which was left accessible for future major equipment changes. A doorway through the concrete wall in front of the main air-lock is closed by means of a $1\frac{1}{4}$ in. thick plate. These plates are either bolted in place or hinged depending on their size and frequency of use.

SIZE OF BREAK OR LEAKS IN COOLING SYSTEMS

In some hazards summary reports the possible breaks or leaks in the primary cooling system have been divided into three catagories according to size, for convenience in considering different types of accidents. Although the following sizes of leaks apply to pressurized water reactor systems, similar situations may be found with other types of power reactors.

A small break or a crack is defined for a pressurized water reactor as one with a flow area equivalent to a hole of approximately $\frac{1}{4}$ in. diameter or smaller. Cracks have occurred in large piping and shown up as steam leaks and similar small breaks and other leaks can be expected to occur more frequently than large breaks. The normal charging system can usually make up the loss from a small break or crack.

A medium size break is defined as one with a flow area ranging from that of a hole $\frac{1}{4}$ in. in diameter to a hole as large as 6 in. in diameter. Surveys of industrial and shipboard experience of piping failures indicate that the probability of pipe failure of a type which will open a hole in the pipe equivalent to the pipe internal diameter, rapidly diminishes as the pipe size increases above 2 or 3 in. in diameter, however, this experience indicates that failure of 2 and 3 in. pipes should be considered as possible. In pressurized water systems, the emergency water injection system is usually designed to keep the core from melting in the event of a medium size break.

A large break is defined as one with a flow area larger than 6 in. in diameter. Usually the case of circumferential fracture of the largest pipe in the primary cooling system with flow out both ends of the broken pipe is considered in designing the containment system. The occurrence of a fracture of such a large pipe that is carefully designed and fabricated of ductile material is

FIG. 20. Containment for General Electric Test Reactor at Pleasanton, California. (General Electric Company photograph.)

improbably as indicated by studies of industrial pipe failure experience, and containment design based upon such an assumed accident is probably very conservative.

DESIGN OF CONTAINMENT STRUCTURES

A reactor which might otherwise present an undue hazard to the public can be enclosed within a container which is designed to be sufficiently gas tight under the conditions of the maximum credible accident to confine the radioactive material and limit the hazard. In many installations the containment structure also serves to house the reactor installation. In some cases the gamma radiation from radioactive material released inside this container could present some hazard to the local area and therefore the container may be lined with concrete or other means for radiation shielding may be provided.

FIG. 21. Containment for Industrial Research Laboratories Reactor, Princeton, New Jersey. (Industrial Research Laboratories photograph.)

A containment structure may therefore be designed to serve as a pressure vessel, laboratory or power plant building, and emergency radiation shield. Since considered an important public safeguard, the design, construction, testing and maintenance of containment structures require careful attention and conformity with all applicable codes and regulations. Although the design of the containment for each installation must be adapted to the reactor type as well as the plant location, the standards being prepared under the Nuclear Standards Board of the American Standards Association will represent a major contribution to the design of reliable containment which is not unnecessarily costly.

Steel Containment Vessels

Most reactor containers in the United States have been of steel construction particularly where appreciable pressures might be involved and low leakage rates required. Design and construction of large steel pressure vessels is well understood and a containment vessel can be constructed with assurance that specifications can be fulfilled and the required protection provided.

Fig. 22. University of Munich Research Reactor Gerehing, Germany. (AMF Atomic Division of American Machine and Foundry Company photograph.)

Loads that might be applied to a container require careful consideration. The maximum internal pressure resulting from the release of heat and vapor or other gases are the basis for the design. The weight and velocity of missiles, if any, define the requirements for missile shielding. In some cases protection against pressure waves may be required. The design must also provide for all internal and external loads including wind forces, snow load, changes in atmospheric pressure, weight of cranes, etc.

Although a containment vessel may prevent the release of radioactive material to the atmosphere, a steel containment vessel would not appreciably reduce the gamma radiation from radioactive material released from the reactor core into the structure. Local shielding would not reduce the hazard since the radioactive material might be distributed throughout the volume of a large containment vessel. The only methods of reducing this direct radiation hazard are distance and shielding materials such as earth, water, or concrete. For research reactors which operate at moderate power, a suitable shield may consist of an uncapped cylinder of concrete. For reactors operating at high power in populated areas, a shielded roof is usually necessary. In Reference 40 the external hazard to be expected from gamma radiation through the walls of the containment structure with various types of shielding is considered in some detail.

The design and construction of steel containment vessels for pressures greater than 15 psig are based on the ASME Unfired Pressure Vessel Code, Section VIII. Several Code cases regarding nuclear reactor containment vessels are applicable including Nuclear Case Interpretations No. 1224, 1226, 1228, 1235, 1238. The design and construction of containment vessels for internal pressures less than 15 psig may follow the American Petroleum Institute Standard 620 for large, welded, low-pressure storage tanks. Other applicable codes include the ASA Standard A58.1–1955 "Building Code Requirements for Minimum Design Loads in Building and Other Structures", and National Fire Codes of the National Fire Protection Association 1958. The Pacific Coast Building Officials' Conference—Uniform Building Code provides design requirements for areas where earthquakes must be considered.

Concrete Containment Structures

A windowless concrete structure was recommended in Reference 39 for the University of Michigan Research Reactor which was designed to operate intermittently at power levels up to 1 MW. The building described in this report has concrete walls with a minimum thickness of 12 in. reinforced on both faces to control shrinkage. The specifications provide for proper density of concrete and careful control of placing and curing. The roof is of poured concrete, and special attention was given to constructing the joints and they were held to a minimum in number. The inside surface of the building was treated with a thin film of Gunite and finished with 3 coats of aluminum paint. All exterior openings were held to a minimum and are of a type which would permit quick and positive closing and sealing.

A concrete containment structure provides protection to the CP-5 research reactor at the Argonne National Laboratory. The walls of the structure were constructed by continuous pouring to minimize the cracks and a separate expansion chamber was provided to take care of pressure changes. However,

FIG. 23. Containment for Argonne Research Reactor CP–5, Lemont, Illinois. (Argonne National Laboratory photograph.)

this type of construction has not been used for power reactors in the United States since concrete structures are difficult to make gas tight under high internal pressures. Also the design of large concrete pressure vessels is not well established. For these reasons, steel pressure vessels have been preferred; however, concrete containment structures have a capacity to support cranes and any overhead water tanks. Concrete constructions should conform to the "Building Code Requirements for Reinforced Concrete, AC1–318–56" and "Report on Tentative Recommendations for Prestressed Concrete of AC1–ASCO Joint Committee 323". Other applicable codes include the Pacific Coast Building Officials' Conference—Uniform Building Code, which contains requirements for withstanding earthquakes.

UNDERGROUND CONTAINMENT

The underground location of a reactor within a rock formation would be nother alternative to the steel containment vessel, and the underground construction of nuclear power reactors is considered in some detail in Reference 37. It appears that effective containment in case of an accident can be obtained in an underground rock excavation, particularly if provided with leakproof lining. The underground building is not subject to any extreme

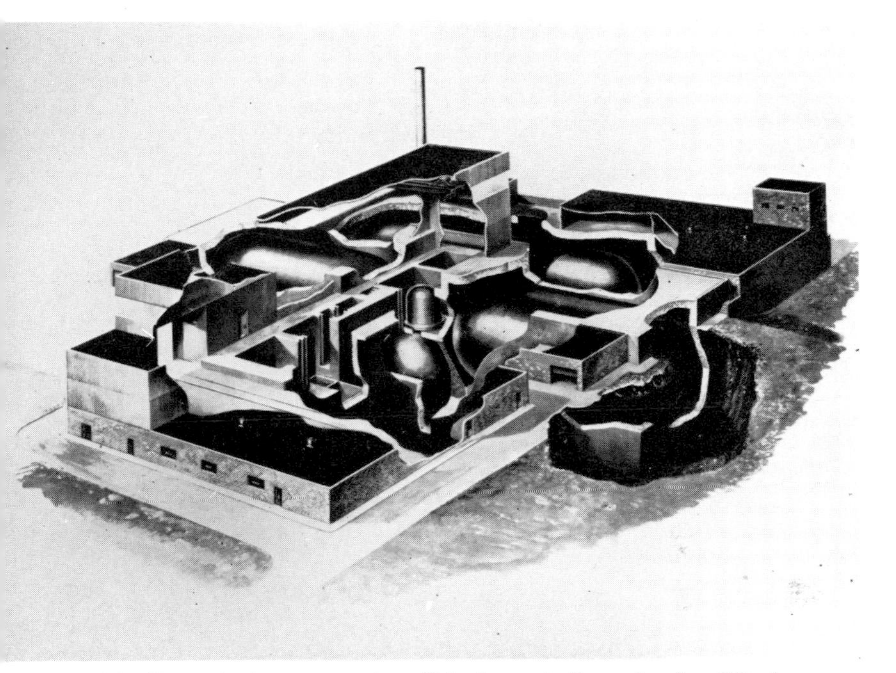

FIG. 24. Pressurized water reactor, Shippingport, Pennsylvania. (Westinghouse Atomic Power Div. photograph.)

changes in temperature, or disturbing surface phenomena such as hurricanes, blizzards, etc., and is relatively invulnerable to earthquake shock. Superior protection against the effects of enemy attack (such as blast and radiation) is possible in underground construction.

Granite, limestone, sandstone, and salt formations usually permit wide spans, high ceilings, and regular pillar arrangements, if needed. However, where cost is not an important factor, industrial plants can be installed successfully in virtually all types of geological formations. However, in general, there should be at least 50 ft of sound rock over the excavation, if much pressure must be contained. Too great a depth may cause high stresses around the opening due to weight of the overlying rock.

Some of the disadvantages of underground construction include th greater plant cost which for a typical EBWR installation is estimated i Reference 37 to be 3–7 per cent of the total cost of the plant. Also it may b difficult to determine in advance the exact condition of the rock structure such as heavy flow of water in the rock strata.

Underground reactor construction has been favored in some countrie where the methods of underground excavation have been highly developed

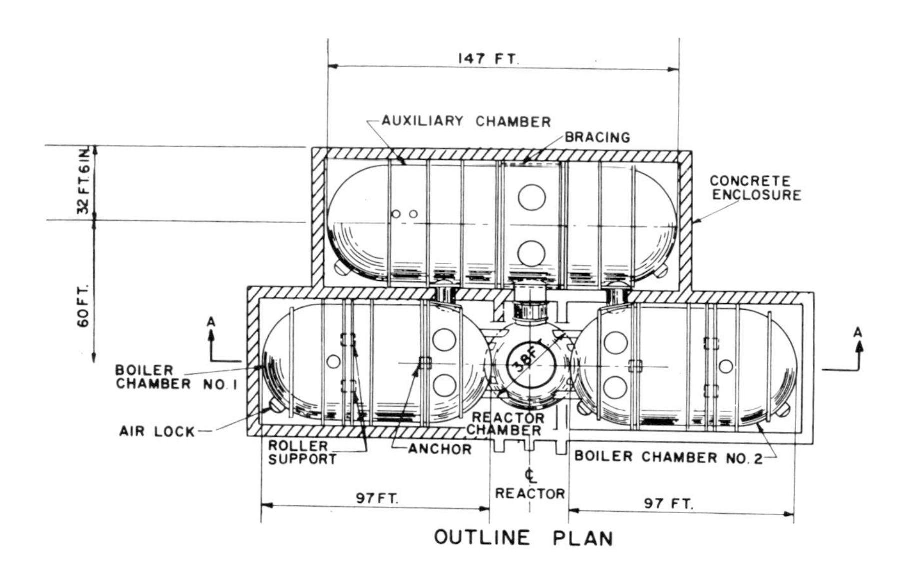

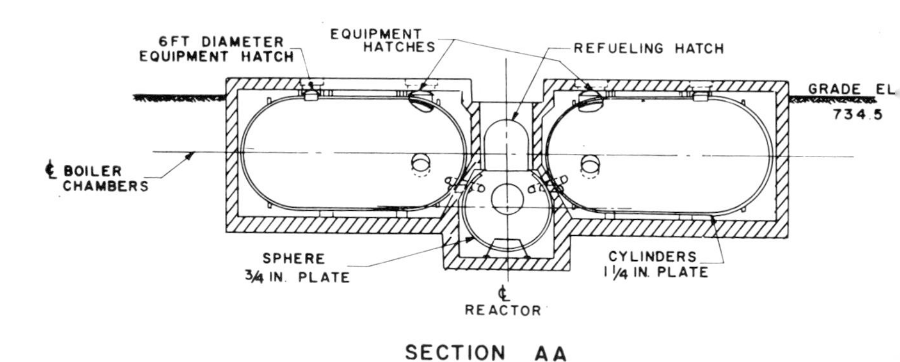

FIG. 25. Outline of Shippingport Containment Chambers. (Reproduced from "Vapor Containers for Nuclear Power Plant," by C. T. Chave and O. P. Balestracci, Vol. 11, *Proc. 2nd United Nations Int. Conf. on the Peaceful Uses of Atomic Energy*, United Nations, New York.)

The Swedish research reactor in Stockholm is located 85 ft underground in solid granite. Also the Halden Reactor in Norway is housed in a rock excavation. A tunnel of approximately 165 ft will lead into a large room of 100 ft length, 32 ft width and approximately 40 ft height. The reactor and some of the associated equipment will be mounted in a pit below the floor level of this room. The pressure locks are designed for about 40 psi over-pressure.

The Shippingport Container

The design of the vapor container installed for the Shippingport plant is reviewed in Reference 42 from which the following description has been taken. The system consists of interconnected horizontal cylinders and a sphere. This design resulted from certain specific criteria which are not expected to apply to future projects. In order to provide flexibility in core designs, the plant was designed so that an entire core could be removed from the reactor and handled in a fuel canal filled with water. Handling of individual fuel elements rather than an entire core, as has been specified for some other plants, would have made the fuel handling system a secondary consideration in the design of the vapor containers.

Another requirement was that the reactor be housed in an underground substructure and completely shielded by earth or concrete to protect personnel from radioactivity released by a nuclear incident. It was also specified that the plant be designed for normal operation with no personnel inside the container, but limited access during operation was required to allow adjustment of valves and instruments in certain areas containing auxiliary equipment. The requirement of underground installation led to the use of horizontal cylinders to obtain the required volume without excessive depth of excavation.

The ASME Code for Unfired Pressure Vessels, under which the containers were built, specifies a maximum plate thickness of $1\frac{1}{4}$ in. for nonstress-relieved vessels. The vessels were limited to a maximum diameter of 50 ft in order to keep the plate thickness within specified limits for vessels of 50 psig design. An auxiliary chamber of large volume was installed, in addition to the container holding the major equipment, to provide the total volume needed for pressure limitation in the event of an accident.

The use of multiple containers resulted in a more compact arrangement than would be possible with a single container of 50 ft diameter. A single container would also have incorporated a central depression to accommodate the reactor at a level below the steam generator and to provide space above the reactor for the fuel canal. It was found that excessive stresses would result at the points of juncture of the depressed reactor and the cylinder.

In the final design, shielding is provided by free-standing concrete enclosures. Concrete lining of the container with earth backfill against the shell was considered, but proved impracticable. This design would have required a network of tunnels for access of piping and conduit and would have made impossible the periodical shell inspection specified in the Pennsylvania Pressure Vessel Code. In addition, construction time would have been increased because the lining could be placed only after the vessel was completely fabricated and tested, and installation of equipment would have been deferred until the lining was completed.

The final design consists of four interconnected vessels. The steam generators and demineralizers are housed in two horizontal cylinders 50 ft in diameter and 97 ft long, with hemispherical ends. Concrete shielding is provided inside these containers to allow access to the demineralizers or to either of the two steam generators when one is shut down and the other operating. The reactor is enclosed in a spherical chamber 38 ft in diameter, located between the steam generator containers. An 18 ft diameter cylindrical refueling hatch is located at the top of the reactor container. The pressurizer and other auxiliary systems are installed in the auxiliary chamber, which is a 50 ft diameter by 147 ft horizontal cylinder with hemispherical heads. This auxiliary chamber is located beside the steam generator containers. It is connected to the reactor container by a 12 ft diameter cylindrical tube and to each of the steam generator containers by 8 ft diameter tubes. This arrangement permits rapid equalization of pressure between the containers if fluid should be released from the system. The total effective container volume of 470,000 ft^3 is adequate to limit the pressure rise to 52.8 psig upon release of all primary water in the system.

The individual containers are located in four adjacent concrete shielding enclosures with walls of 5–7 ft thickness, which provide the necessary shielding of the primary plant components. These enclosures also serve to provide weather protection for the container steel, access to the container and piping and structural support for the earth shielding. The vapor containers and concrete enclosures are located below grade with only the top portion of the shielding enclosure projecting above grade.

As noted above, all equipment and piping handling high pressure primary coolants were designed for conservative stresses in order to avoid a requirement for extensive installation of concrete structures for missile protection.

Each of the steam generator containers has two 10 ft diameter and one 6 ft diameter bolted hatches for access of equipment, and two air locks for access of personnel without violation of the integrity of the containers. The auxiliary chamber has two 10 ft diameter equipment hatches and two air locks. Penetrations of the container shells for piping and electrical services are designed to be vapor tight under all operating conditions. A heating and ventilating system maintains the container atmosphere at a temperature between 50°F and 122°F under all operating conditions.

The Army Package Power Container†

Several container shapes were considered, including vertical and horizontal ylinders and spheres for the APPR-1 container. Provisions for removal of a hole core and limited access during operation were not required in this esign. An upright cylinder with hemispherical ends proved best suited for nclosing the primary equipment with economy of space. This design also nade it possible to install secondary systems and auxiliary facilities at a ninimum distance from equipment within the container, so that lengths of iping, conduit and cables could be kept short.

Stability of support for the upright cylinder was achieved by locating the ower head containing the reactor and heavy concrete foundations below rade. This arrangement also provides additional shielding for the reactor hile permitting construction of the major portion of the container in the pen. A spherical container would have required more space and would have nade necessary location of the spent fuel pit at a greater distance from the ontainer and at a greater depth below grade. The cylindrical shape also implified the problem of supporting the weight of the inner concrete lining pecified.

The location of the reactor in an important military reservation near the ational capitol resulted in a particularly conservative design. It was conidered necessary to allow approach to the container from any direction vithout exposure of personnel to excessive radiation. To meet this requirenent, the primary shield is supplemented by a vapor container lining and by n external biological shield. The lining is of reinforced concrete 2 ft in hickness and covers all free interior surfaces of the shell, except at access penings and points of penetration of pipe and conduit. The external biologcal shield consists of a 3 ft layer of concrete placed against the container shell nd extending to the upper bend line at a height of about 28 ft above ground levation.

A further function of the concrete lining is to provide missile protection or the container shell. The three principal vessels within the container, the eactor, steam generator and pressurizer, are prevented from becoming nissiles by anchoring them to the concrete floor in the base of the container vith restraints capable of withstanding the high thrusts resulting from upture of a lower head. Small metal objects cannot acquire sufficient velocity o penetrate the concrete lining and container as noted above. The $2\frac{1}{2}$ in. hick steel covers for access openings also are proof against penetration by uch missiles. The covers are anchored to the concrete lining for added mpact resistance. The concrete lining also would serve to limit the shell

† Taken from Vapor containers for nuclear power plants, by C. T. Chaye and O. P. Balestracci. Vol. II *Proc. 2nd United Nations Int. Conf. on the Peaceful Uses of Atomic Energy*, United Nations, New York.

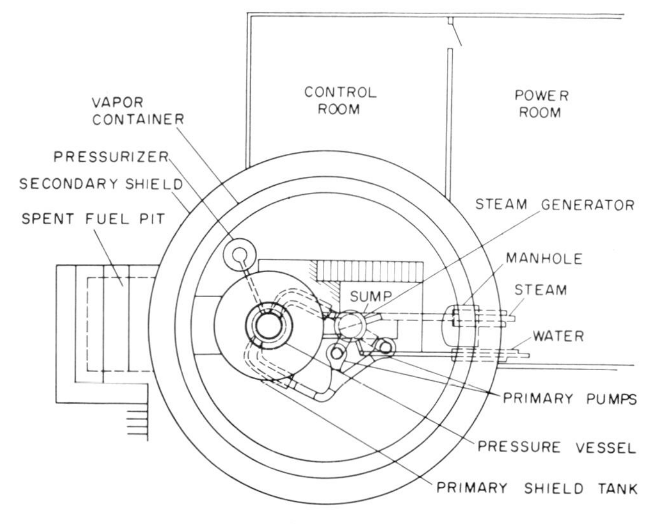

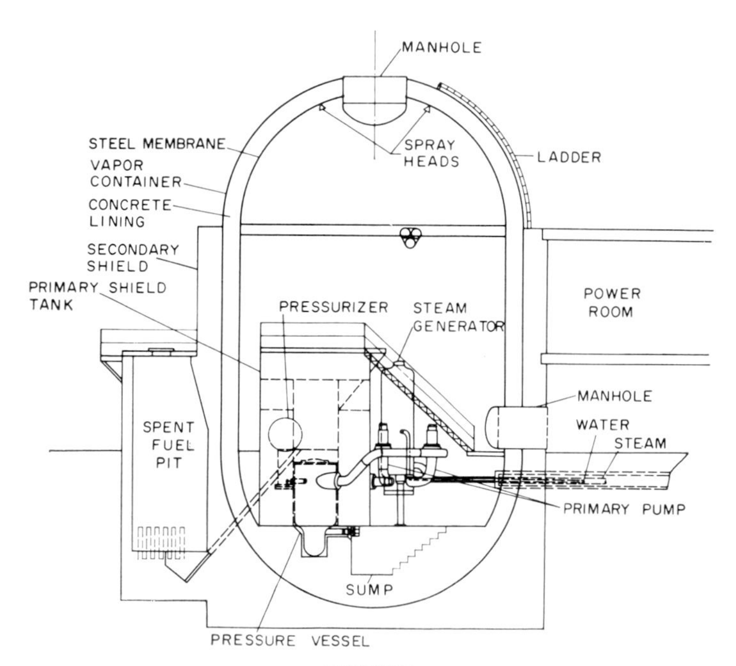

FIG. 26. Army Packaged Power Reactor (APPR–1) vapor container (Reproduced from "Vapor Container for Nuclear Power Plant," by C. T. Chave and O. P. Balestracci, Vol. 11, *Proc. United Nations Int. Conf. on the Peaceful Uses of Atomic Energy*, United Nations, New York.)

temperature rise following an accident, so that secondary shell stresses due to temperature effects would be negligible.

The concrete lining is covered by a 10 gauge steel membrane to protect the porous concrete and to facilitate decontamination. A steel lining seemed advantageous at first, but the erection problems experienced were so formidable that this design will probably not be repeated in future plants. Nevertheless, the steel liner did serve its purpose when a safety valve leak during operation required decontamination of the container.

The maximum credible accident postulated for the plant included simultaneous release of all pressurized fluids from the primary and secondary systems. This would create a maximum pressure of 66.3 psig within the container, which has a net volume of 37,000 ft^3. Container shell openings, reinforcements and other construction details were designed in accordance with ASME Pressure Vessel Code. The vessel does not conform to the code in some respects. For example the requirements of the code for a pressure relief device could not be met without violating the purpose of the container. Also the maximum stresses in the shell that would result from the "maximum credible accident" could not be held by any practical design to the limits specified by the code. For the accident postulated, the maximum stress imposed on the container shell was calculated to be 16,400 psi. Steel plates of the specification used has a minimum yield strength of 32,000 psi, which is almost twice the calculated stress; however, the code specifies a maximum working stress of 13,500 psi for the vessel design conditions.

The limit specified for container leakage was 0.075 ft^3/hr at an internal pressure of 66 psig. A leakage test was conducted along all welds in the container shell using a modified mass spectrometer with helium tracer gas. No measurable leaks were detected and this was accepted as evidence that the leakage was below that specified.

The Yankee Plant Container

The following description of the vapor container designed for the Yankee Atomic Electric Company Plant at Rowe, Massachusetts is also taken from Reference 42. This design represents an evolution of ideas from previous installations. Here a spherical container was selected, since it is more economical than an upright cylinder for a large plant. Elevation of the container above grade made it possible to avoid excavation costs and expensive construction methods. A hydraulic lift is used to handle fuel between the elevated container and a fuel pit having a minimum submergence below grade. Shielding and missile protection is provided by a detached concrete structure rather than by a concrete lining, which would have been expensive to install in a spherical container. Because of the remoteness of the Yankee plant location, it was considered unnecessary to provide an external shadow shield for protection of the public in the event of a nuclear incident. Minimizing of external shielding requirements contributes further to the practicality of a spherical vapor container.

The final design of the Yankee plant container evolved through several stages. The first design considered was a single horizontal buried steel cylinder with hemispherical ends, lined with concrete 2 ft thick. The reactor and shield tank were to be located at the center of the container. Two horizontal steam generators were to be placed at either side of the reactor and separated from it by concrete partitions. Design of lining and supports for the container would have been difficult. Placement of equipment in the container would have involved a formidable craneway structure extending over the top of the shell so that apparatus could be lowered through a manhole. These difficulties and the excessive estimated cost of the design concept led to its abandonment.

The second concept was a group of three above-grade vertical cylindrical shells with hemispherical heads. The reactor was to be located in the central container and each of the other containers was to house two steam generators separated by concrete walls. While this arrangement was considered less expensive than a single buried cylinder, it did not eliminate certain disadvantages. Handling equipment within the containers would have required a craneway above the top heads, or other expensive means. Either concrete linings or detached internal concrete structures with a shadow shield completely around the three containers would still have been necessary. An additional problem was the maintenance of rigid support while allowing lateral expansion of three vessels having common openings.

Advantages found for a single spherical container included lower fabrication costs, easier analysis of stresses at discontinuities and penetrations, and the possibility of using a crane to handle all large equipment within the container. Moreover, space was available for installation of individually shielded compartments for four loops arranged radially. The individual compartments would limit the extent of an accident and would allow maintenance to be carried out on a shutdown loop while others were operating.

In its first design, the sphere was supported by a short conical skirt, and equipment was handled through a top opening by means of an expensive crane arrangement. The design was greatly improved by elevating the entire container and placing the equipment access hatch on the underside of the container so that equipment could be handled by an internal polar crane from a railroad track below the sphere. The conical support for the container shell would have given superior stress distribution, but this design was abandoned in favor of columnar supports spaced around the equator or the sphere when it was found that no fabricator could handle economically the course of beveled thick plate required at the junction of the skirt and shell.

Another variation studied was a single 115 ft diameter horizontal aboveground cylinder with hemispherical ends. Equipment would have been

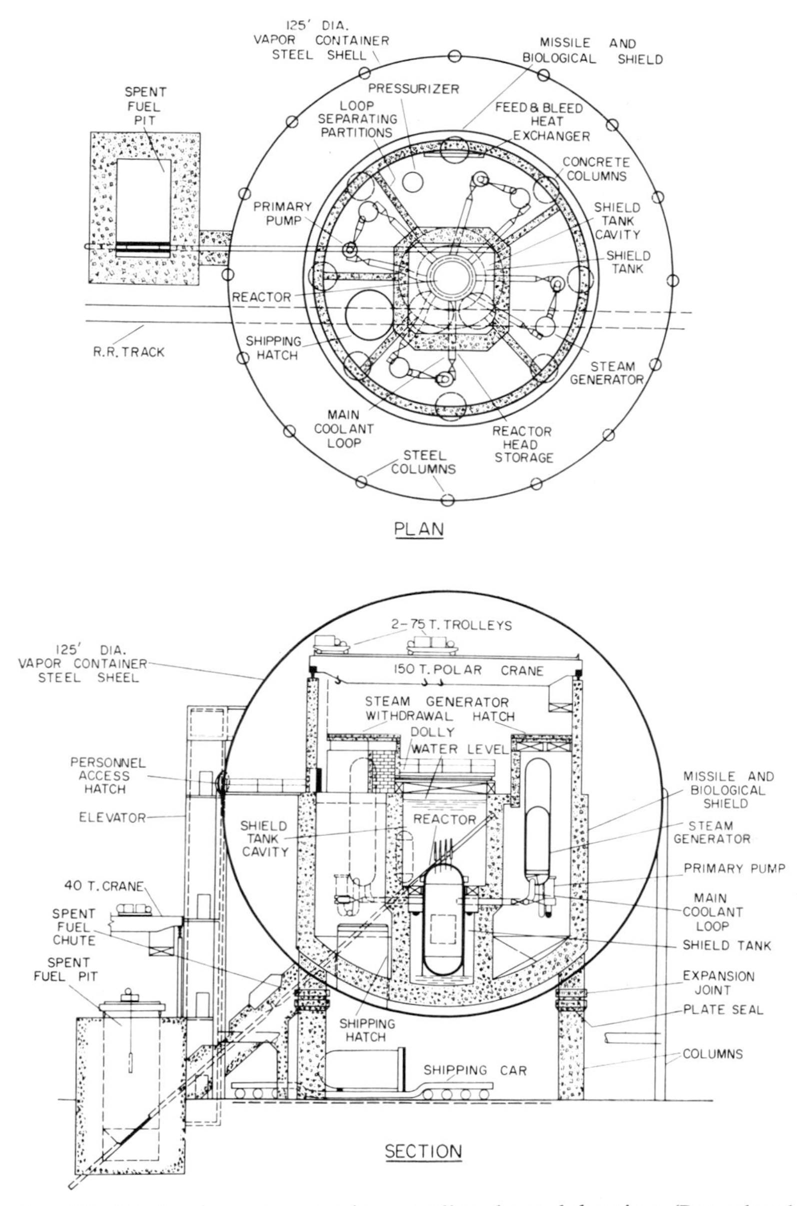

FIG. 27. The Yankee vapor container—outline plan and elevation. (Reproduced from "Vapor Containers for Nuclear Power Plant," by C. T. Chave and O. P. Balestracci, Vol. 11, *Proc. 2nd United Nations Int. Conf. on the Peaceful Uses of Atomic Energy*, United Nations, New York.)

Table XI

U.S. Reactor Containment Data

Symbol	Name	Location	Nuclear Thermal Power kw	Type
EBWR	Experimental boiling water reactor	Lemont, Illinois	20,000	Boiling water—Power
VBWR	Vallecitos boiling water reactor	Pleasanton, California	30,000	Boiling water—Power
Dresden	Dresden nuclear power station	Morris, Illinois	626,000	Boiling water—Power
APPR–1	Army package power reactor No. 1	Fort Belvoir, Virginia	10,000	Pressurized water—Power
PWR	Pressurized water reactor (Shippingport atomic power station)	Shippingport, Pennsylvania	231,000	Pressurized water—Power
Yankee	Yankee Atomic Electric Co. Reactor	Rowe, Massachusetts	392,000	Pressurized water—Power
Indian Point	Consolidated Edison Thorium Reactor	Indian Point, New York	500,000	Pressurized water—Power
GETR	General Electric Test Reactor	Pleasanton, California	33,000	Water moderated—Testing
Enrico Fermi	Enrico Fermi Atomic Power Plant	Lagoona Beach, Michigan	300,000	Fast breeder—Power
EBR–II	Experimental Breeder Reactor No. 2	NRTS, Idaho	62,500	Fast Breeder—Power
HRT	Homogeneous Reactor Test	Oak Ridge, Tennessee	5,220	Aqueous homogeneous—Power
West Milton	General Electric Co. KAPL Reactor Test-facility	West Milton, New York		Sodium cooled intermediate—Power
MITR	Massachusetts Institute of Technology Reactor	Cambridge, Massachusetts	1,000	Heavy water—Research
PRTR	Plutonium Recycle Test Reactor	Hanford, Washington	70,000	Heavy water—Testing
CP–5	Argonne Research Reactor	Lemont, Illinois	2,000	Heavy water—Research
LPTR	Livermore Pool Type Reactor	Livermore, California	1,000	Tank—Research
NASA	National Aeronautics Space Administration Reactor	Sandusky, Ohio	60,000	Tank—Testing
AFNETR	Air Force Nuclear Engineering Test Reactor	Dayton, Ohio	10,000	Tank—Testing

U.S. Reactor Containment Data

	Diameter (ft)	Length or height (ft)	Distance grade to bottom (ft)	Plate thickness (in.) Main	Top	Bottom	Material	ASME code stamp	Gross volume (ft³)	Free vol (ft³)
EBWR	80	119	−56	0.625[a]	0.375[c]	0.625[d]	A201–A300	No	497,500	400,000
VBWR	48	110	−35	0.875[a]	0.436[c]	0.436[c]	SA212		160,00	
Dresden	190		−39	1.25[b] upper			A201–A300	Yes	3,600,000	2,880,000
				1.40[b] lower						
APPR–1	32 (ID)	52	−10	0.125[a]	0.375[c]	0.375[c]	SA212		52,700	37,000
	36 (OD)	64	−20	0.875[a]	0.500[c]	0.500[c]				
PWR	2–50[e]	97		1.25[a]			A201–A300	Yes	635,000	473,000
	1–50[e]	147	−52	1.25[a]						
	38[b]			0.75 & 1.00[b]						
Yankee	125		+24	0.875[b] upper			A201–A300	Yes	1,020,000	840,000
				1.250[b] lower						
Indian Point	160		−75.5	0.89[b] upper			A201–A300	Yes	2,140,000	
				1.03[b] lower						
GETR	66	105						No	322,000	230,000
Enrico Fermi	72	120	−45	1.03[a]	0.52[c]	1.03[d]	A201–A300	Yes	415,800	
						1.25 knuckle line				
EBR–II	80	139	−47.5	1.00[a]	0.50[c]	1.00[d]	A201	No	595,000	450,000
HRT	54 × 30.5 × 19 parallelepiped						A285 (partly clad)	No	28,100	25,000
West Milton	225		−38	0.9[b] upper			A201–A300	No	6,000,000	
				1.1[b] lower						
MITR	74	50	−20				A283	No	258,700	
PRTR	80	121.5	−46.5	0.375[a]	0.625[c]	[d]	A201–A300		504,500	375,000
CP–5	70	48	0				Concrete	No		
LPTR	80	52	0	0.250[a]	0.3125[c]			No	257,500	
NASA	100	109	−56	0.750[a]	0.750[d]	0.750 irregular	A285–54T	No	520,500	
AFNETR	80	160	−48	0.563[a]	[c]	[d]	A201–A300	No		

(a) Cylindrical
(b) Spherical
(c) Hemispherical
(d) Hemiellipsoidal
(e) Axis of cylinder is horizontal
(f) Hydrostatic

(Reproduced from Report ANL–5948 by permission of the U.S. Atomic Energy Commission.)

Table XIII

U.S. Reactor Containment Data

	Coolant properties assumed for calculation of stored energy			Energy released in accident yielding design pressure on container, 10^6 BTU			Positive design pressure psig	Negative design pressure psig	Strength test pressure (pneumatic) psig
	Pounds and type	Initial pressure psia	Initial temperature °F	Stored	Chemical	Nuclear and other added net during accident			
EBWR	16,800 H_2O	600	488	8	3.2	None	15	0.5	18.75
VBWR	35,000 H_2O	1000	545	20	None	None	45		56.25
Dresden	376,000 H_2O	1000	545	200	None	None	29.5	1	37
APPR–1	12,000 H_2O	1500	600	5	None	2.4	63.3		30
PWR	145,000 H_2O	2000	523	78	None	4.0	52.8		70
Yankee	186,000 H_2O	2150	518	94	None	None	34.5		40
Indian Point	163,000 H_2O	1500	500	80	None	None	27.5	1	34.4
GETR							5	0.2	6.35
Enrico Fermi	Sodium—non pressurized			Designed on sodium–air reaction plus shutdown nuclear heating			32	2	40
EBR–II	Sodium—non pressurized			None	18	None	24	1	30
HRT	3,910 D_2O	2000	572	0.35	None	None	30		32[f]
West Milton	Sodium—non pressurized						19.75	0.8	25
MITR	Non pressurized D_2O				2.2	0.13	2	0.1	
PRTR	18,100 H_2O or D_2O	1050	504	{ — / 9	7	3.5 } Two bases	15	0.58	18.75
CP–5	Non pressurized D_2O			Designed on barometric pressure differences			2″ Hg		
LPTR	Non pressurized H_2O			Designed on barometric pressure differences			2	2″ H_2O	
NASA	Non pressurized H_2O				1.4	0.13	5		
AFNETR	Non pressurized H_2O						13		16

[f] Hydrostatic test

(Reproduced from Report ANL–5948 by permission of the U.S. Atomic Energy Commission.)

Table XIV

U.S. Reactor Containment Data

	Operation assumed before incident	% release of fission products to container	Gamma source strength history (assumed) for contained fission products, curies (t in seconds)	Specified[g] leak rate	Observed leak rate	Leak tests used
EBWR			$1.74\times10^{8}t^{-0.21}$	0.25	0.05	h, i
VBWR				1.00	<0.02	j
Dresden				0.5	0.016	h, j
APPR–1	9,120 MWD					k (for welds)
PWR	33,750 MWD	10%		0.1	0.25	h, j, k
Yankee	392 MW∞	100%	$3.26\times10^{9}t^{-0.21}$	0.2		h, k, i
Indian Point				1.0		
GETR				2.0	<2.0	
Enrico Fermi	300 MW∞	100%	$3\times10^{9}t^{-0.21}$	0.06		h, j
EBR–II	8,100 MWD	50% of f.p. 50% of Pu^{239}	$5.4\times10^{8}t^{-0.21}$	0.125		h, i
HRT	5 MW∞	100%	$6.6\times10^{6}t^{-0.21}$	0.2	<0.2	k
West Milton				0.5	<0.1	h, i
MITR		20% gaseous of which 35% unstable	$0.5\times10^{6}\Gamma(t)$	2.0	1.25	
PRTR	14,000 MWD	25%	$0.17\times10^{8}t^{-0.22}$(<$10^{4}$ min)	0.27		
CP–5	18 MWD∞	10% Halogens 100% Noble gases	$5.5\times10^{2}t^{-0.21}$			
LPTR		100%		3.0		
NASA	60 MW∞	100%	$5.46\times10^{8}t^{-0.21}$	0.3		
AFNETR				0.1	<0.1	

(g) % contained volume per 24 hours at design pressure
(h) Soap solution
(i) Direct pressure temperature measurement
(j) Reference system
(k) Freon or helium injection

MWD = Megawatt days
MW∞ = Megawatt–infinite operation

(Reproduced from Report ANL–5948 by permission of the U.S. Atomic Energy Commission.)

needed to allow handling of machinery within the container by a traveling crane. Although this concept was estimated to cost little more than a spherical container, it was considered less practicable.

The resulting design of the vapor container is an elevated sphere 125 ft in diameter. The minimum thickness of the shell plate is $\frac{7}{8}$ in. An equipment access hatch is located near the bottom of the container shell, and a spur track allows positioning of heavy equipment under the hatch so that the equipment can be handled by a polar crane within the container. Refueling operations are carried out by using an inclined hydraulic chute connecting a partly submerged spent fuel pit with the flooded space above the reactor core.

The vapor container encloses all pressurized parts of the main coolant system. The maximum accident postulated for this plant was the loss of coolant from the open ends of a severed steam main within the container. An accident of this nature would produce a maximum pressure of 34.5 psig in the container which has a net volume of 840,000 ft^3. The ductile austenitic stainless steel used for the main coolant system and the stainless-clad carbon steel reactor vessel are not considered feasable sources of missiles. Nevertheless, missile protection is afforded by the concrete structures surrounding the reactor and main coolant system. The bolting down of components to prevent them from becoming missiles, as was done for APPR-1, is not practicable for the large reactor and steam generators used for the Yankee plant.

The testing of the container includes a pressure test at $1\frac{1}{4}$ times the maximum working pressure and bubble tests at the test pressure, a halogen gas tracer test at all welded seams, and a pressure reduction test.

REFERENCES

1. *Research Reactors*, Government Printing Office, 1955, p. 409.
2. Brittan, R. O. and Heap, J. C., Reactor containment, *Proc. 2nd United Nations Int. Conf. on the Peaceful Uses of Atomic Energy*, Vol. 11, United Nations, New York.
3. Dietrich, J. R., Experimental determination of the self regulation and safety of operating water-moderated reactors, *Proc. 1st United Nations Int. Conf. on the Peaceful Uses of Atomic Energy*, Vol. 13, United Nations, New York.
4. Karplus, H. B., The velocity of sound in a liquid containing gas bubbles, ARF Report, COO-248.
5. McCullough, C. R., Mills, M. M. and Teller, E., The safety of nuclear reactors, *Proc. 1st United Nations Int. Conf. on the Peaceful Uses of Atomic Energy*, Vol. 13, United Nations, New York.
6. Dietrich, J. R., Experimental investigation of the self-limitation of power during reactivity transients in a sub-cooled, water-moderated reactor, AECD-3668.
7. Asire, D. H., Containment of a nuclear power excursion in the Air Force Nuclear Engineering Test Reactor, AFIT Report 58-12.
8. Russell, A. S., Aluminum Company of America, private communication.
9. Higgins, H. M., A study of the reaction of metals and water, AECD-3664.
10. West, J. M. and Weills, J. T., ANL-4503, October 1950.

1. SHIDLOVSKI, H. M., Explosive methyl alcohol–water mixtures with magnesium and aluminum, *Zhur. Prik. Him.* **19**, 371–77 (1946).
2. RUEBSAMEN, W. C., SHON, F. J., and CHRISNEY, J. B., Chemical reaction between water and rapidly heated metals, NAA-SR-197, 27 October 1952.
3. HIGGINS, H. M., and SCHULTZ, R. D., The reaction of metals with water and oxidizing gases at high temperatures, IDO-28000.
4. ELGERT, O. J., and BROWN, A. W., In-pile molten metal–water experiments, (IDO-16257), Abstracted in TID 3073.
5. LUSTMAN, B., Zirconium–water reactions, WAPD-137.
6. BENDLER, A. J., ROROS, J. K., and WAGNER, N. H., Fast transient heating and explosion of metals under stagnant liquids, AECU-3623.
7. KINDSVATER, H. M., Investigation of rate of reaction between water-reactive metals and water, Aerojet Research Laboratories, Note No. 50, 1946.
8. CURRIE, L. M., HAMISTER, V. C., and MACPHERSON, H. G., The production and properties of graphite for reactors, National Carbon Company.
9. SALESSE, M., Safety of magnesium canning for CO_2-cooled reactors, *Nucleonics*, **16**, No. 2123-4.
20. SLETTEN, H. L., Organic moderated reactor experiment safeguards summary, NAA-SR-2323.
21. HUMPHREYS, J. R., Jr., Sodium–Air reactions as they pertain to reactor safety and containment, *Proc. 2nd United Nations Int. Conf. on the Peaceful Uses of Atomic Energy*, Vol. 11, United Nations, New York.
22. LOFTNESS, *et al.*, NAA-SR-126.
23. MONTET, G. L., Reaction of graphite with sodium, AECU 3142.
24. LUSTMAN, B., Release of fission gases, from UO_2, WAPD-173.
25. PARKER, G. W., and CREEK, G. E., Experiments on the release of fission products from molten reactor fuels, ORNL-CF 57-6-87.
26. KENESHEA, F. J., and SAUL, A. M., Contributions by chemical elements to the total beta activity of fission products, NAA-SR-187, June 1952.
27. VALENTINE, R. F., Hazards to the area surrounding PWR due to atmospheric diffusion of radioactivity, WAPD-SC-548.
28. REYNOLDS, W. C., Investigation of ignition temperatures of metals, NACA TN D-182.
29. MONSON, H. O., and SLUYTER, M. M., Containment of EBR-11, *Proc. 2nd United Nations Int. Conf. on the Peaceful Uses of Atomic Energy*, Vol. 11, United Nations, New York.
30. PORZEL, F. B., Some hydrodynamic problems in reactor containment, *Proc. 2nd United Nations Int. Conf. on the Peaceful Uses of Atomic Energy*, Vol. 11, United Nations, New York.
31. NAPADENSKY, H., and STRESAU, R., Explosives test evaluation of blast shields for nuclear reactors, ARF D 132D11-1.
32. BAKER, W. E., and PATTERSON, J. D., Blast effects tests of a one-quarter scale model of the AFNETR, BRL Report No. 1011, Aberdeen Proving Ground, Maryland, Ballistic Research Laboratories, March 1957.
33. BOSTROM, W. A., The high temperature oxidation of zircaloy in water, WAPD-104, 19 March 1954.
34. PORZEL, F. B., Design evaluation of BER (Boiling Experimental Reactor) in regard to internal explosion, Report ANL-5651, January 1957.
35. HEINEMEN, A. H., and FROMM, L. W., Containment for the EBWR, *Proc. 2nd United Nations Int. Conf. on the Peaceful Uses of Atomic Energy*, Vol. 11, United Nations, New York.
36. Alco Products, Inc., Hazards summary report for the Army Packaged Power Reactor, APAE-2, July 1955.
37. BECK, C., Engineering study on underground construction of nuclear power reactors, AECU-3779, April 1958.
38. BRITTAN, R. O., Reactor containment including a technical progress review, ANL-5948.
39. LUCKOW, W. K., MESLER, R. B., and WIDDOES, L. D., The nuclear research reactor at the University of Michigan, AECD-3669.

40. GELLER, L., and EPSTEIN, R., A general method for evaluating containment shieldin under normal and emergency conditions, *Proc. 2nd United Nations Int. Conf. on th Peaceful Uses of Atomic Energy*, Vol. 11, United Nations, New York.
41. THOMPSON, T. J., BENEDICT, M., CANTWELL, T., and AXFORD, R. A., Final hazard summary report to the Advisory Committee on Reactor Safeguards on a researc reactor for the Massachusetts Institute of Technology, MIT-5007, 1956.
42. CHAVE, C. T., and BALESTRACCI, O. P., Vapor containers for nuclear power plants *Proc. 2nd United Nations Int. Conf. on the Peaceful Uses of Atomic Energy*, Vol. 11 United Nations, New York.
43. PWR hazards summary report, WAPD-SC-541, September 1957.

CHAPTER 6

SAFETY FEATURES OF WATER REACTORS

INTRODUCTION

WATER has good nuclear characteristics as a neutron moderator and also excellent heat transfer and thermodynamic properties; and, therefore, many of the interesting types of reactors for research, production and power generation are water cooled or water cooled and moderated. The technology of water reactors has developed rapidly and several water cooled and moderated power reactors have been built. It was found that a water-moderated reactor with solid fuel elements can be designed to have a strong-negative-temperature-coefficient of reactivity provided that less than the optimum ratio of hydrogen atoms to fuel atoms is present in the core since heating and expansion of the water reduces the value of this ratio. The remarkable effectiveness of water expulsion from the reactor core in limiting a nuclear power excursion was inadvertently demonstrated in the ZPR-I criticality incident described later. This experience contributed to the development of the concept of the boiling water reactor where the volume of steam voids in the core controls the reactivity. It has been found that steam formation occurs in a sufficiently short time to control substantial sudden increases in reactivity although at very short reactor periods, excessive temperatures would be expected because of the limitation on heat-transfer rates between the surface of the fuel element cladding and water particularly under the conditions of steam blanketing.

It has also been found that a homogeneous solution of a uranium salt in water as a fuel makes possible the design of a reactor with a very strong negative-temperature coefficient of reactivity. The decrease in density of the moderator and fuel increases neutron leakage as the solution volume increases but also arrangements can be made for solution to be expelled from the active core upon expansion thereby removing fuel from the core. Bubble formation from radiolysis of the solution and boiling further increases these effects. Also since the fuel and moderator occur as a solution there is no delay in transfer of heat as in a solid-fuel reactor.

The safety features of the water-moderated reactor may not be present in reactor designs using water only as the coolant. For example, in a water-cooled, graphite-moderated reactor with natural uranium fuel, the water may be a significant neutron absorber so that removal of the cooling water from the core substantially increases reactivity. Loss of cooling water by failure of a water pipe or pump failure could therefore lead to a serious accident, if the control system should not function. Also local boiling in a process tube may expel water and increase power locally so as to cause the condition to spread through the core volume. It was for this reason that the water-cooled, graphite-type of reactor was not selected when the decisions were made on Windscale production reactors. The gas-cooled graphite reactors which were selected and built do not have this inherent instability.

REACTOR EXPERIMENTS

The testing of water reactors to determine safety characteristics was limited to experiments which would certainly not create damages until the boiling water reactor experiment (BORAX) was constructed at the National Reactor Testing Station. However, many nondestructive experiments were conducted in research reactors and valuable information on safety characteristics of different reactors was obtained.

Kinetic Experiments in LITR

The Low Intensity Testing Reactor (LITR) is a heterogeneous light water-cooled, light water-moderator, beryllium-reflected, enriched-uranium reactor. It was originally built from left-over parts from the Materials Testing Reactor as a mock-up assembly for measuring the water flow and the mechanical characteristics and has become since a more or less permanent reactor at the Oak Ridge National Laboratory. Standard MTR fuel elements are used.

A series of experiments were conducted in the LITR to determine kinetic behavior when sufficient reactivity was introduced to cause local boiling on the fuel elements, and in these tests it was demonstrated that power excursions are limited in magnitude by steam formation and that the reactor power oscillated after the initial power burst. The extent of the power excursions which could be safely conducted at the X-10 site were limited and further exploration of this phenomenon was left to the Argonne National Laboratory with their facility remotely located at the National Reactor Testing Station as described below.

Boiling Water Reactor Experiments

The first series of experiments conducted by the Argonne National Laboratory in the Borax facility constructed at the National Reactor Testing Station are described by J. R. Dietrich in Vol. 13, *Proc. 1st United Nations Int. Conf. on the Peaceful Uses of Atomic Energy*, United Nations, New York.

The reactor tank for the first experiments was contained in a large shield tank of 10 ft diameter which was sunk part-way in the ground and had earth piled around it for additional shielding. Adjacent to the shield tank was a pit with concrete walls in which was installed equipment for filling and emptying the reactor and shield tanks and for preheating the water in the reactor tank. The reactor tank, 4 ft in diameter and about 13 ft high, contained the reactor core, which consisted of an adjustable number of plate-type fuel elements held at the bottom by a supporting grid and at the top by a removable cover. In operation the reactor tank was filled with water to a height of 3 to $4\frac{1}{2}$ ft above the top of the core; this water constituted the reflector, moderator and coolant. The shield tank was filled with water only when the reactor was shut down.

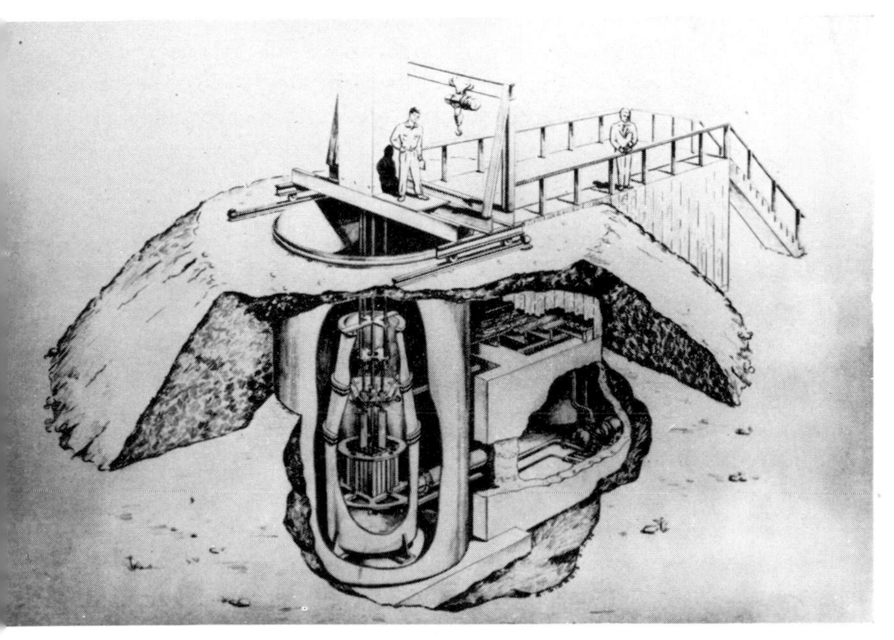

FIG. 1. Borax 1. (Argonne National Laboratory photograph.)

The reactor contained five cadmium control rods operated by drive mechanisms located in a rectangular housing above the shield tank. The connection from the mechanism to the rods was through spring-loaded magnetic couplings. These couplings could be released in unison or individually, allowing the rods to drop freely downward under the acceleration of the springs plus gravity. When released, the central control rod dropped out of the reactor core to apply the excess reactivity used for the experiments. The other four rods when released dropped into the reactor core to terminate the experiments. Each rod traveled the length of the core in about 0.2 sec.

The fuel elements were made of aluminum-clad, aluminum–uranium, alloy plates, of 60 mils total thickness, fastened into aluminum side plates to make boxes roughly 3 in. square.

Experiments were made in different reactors which were also used for investigation of the steady state characteristics of boiling reactors. The pertinent differences between the first two reactors lay in their power characteristics. These differences will be described, but differences in mechanical detail of the two reactors will be omitted. The two reactors will be designated I and II. Each reactor was loaded for any given experiment with the number of fuel elements which would give a convenient amount of reactivity.

Experiments were made by the following procedure. The reactor water temperature was adjusted to the desired value and the reactor was made critical at a low power (about 1 W) by appropriate positioning of the control rod. The central control rod was then dropped out of the reactor core. Initial power was sufficiently low and the speed of the rod ejection was sufficiently high so that in almost all cases the rod was completely out of the core and the reactor reached its stable value before the reactor power had risen high enough to provide significant thermal effects. The power was allowed to continue to rise until the formation of steam in the reactor core reduced the reactivity below criticality and caused the power to fall to a low value. After it was evident that the power had been safely limited by the formation of steam, the remaining four rods were dropped into the reactor to terminate the experiment, referred to as a power excursion. By proper adjustment of the number of fuel elements in the reactor core and of the position of the four outer control rods, the reactor could be made critical with the central control rod inserted to any desired degree in the core. The magnitude of excess reactivity applied by ejection of the central rod could thus be adjusted at will.

Figure 2 is a reproduction of a typical chart from the multi-channel magnetic oscillograph which recorded the data on the experiments. In this case, the applied excess reactivity was 1.4 per cent k_{eff} and reactor I was used. The neutron flux (proportional to reactor power) was recorded over about three decades by three different neutron-sensitive ion chambers working through logarithmic amplifiers. The stable reactor periods as indicated by the three ion chamber records were 0.0096, 0.0107, and 0.0109 sec, respectively. The temperature of one of the fuel plates, which was situated at roughly the highest flux position in the core, was recorded by two fast-response thermocouples. Both of the couples were located near the position of maximum flux. There is little difference between the two temperatures, because of the high thermal conductivity of the thin plate.

The ion chambers, which were calibrated in terms of absolute power by thermal methods, indicated that the reactor power reached the maximum value of 330 megawatts before formation of steam checked the rise. Further generation of steam reduced the activity below the critical value and caused the power to decrease very rapidly to a value of about 0.2 megawatts.

Once the initial power excursion had been checked by boiling in the reactor, the specific power variation depended both quantitatively and qualitatively

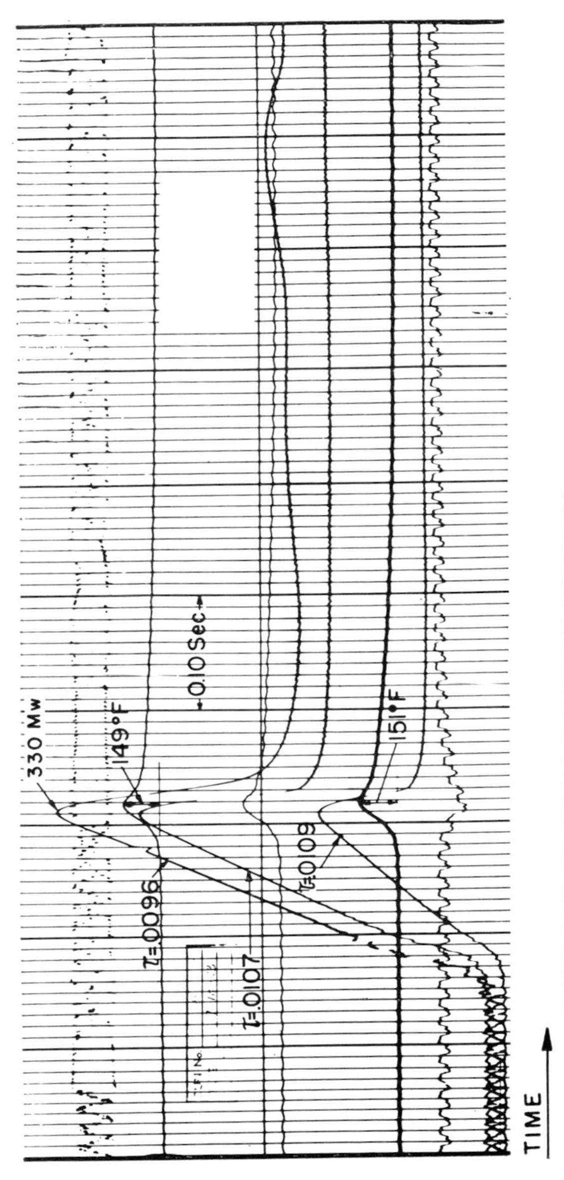

REACTOR WATER WAS INITIALLY AT SATURATION TEMPERATURE.
THE MEASURE PERIODS (τ) ARE MARKED ON THE THREE (LOGARITHMIC) ION CHAMBER RECORDS.
THE TEMPERATURE RISE OF 149F IS AT THE SURFACE OF FUEL PLATE 11; THE RISE OF 151F IS AT THE CENTER OF THE SAME FUEL PLATE.
THE NOISY TRACE NEAR THE TOP OF THE RECORD IS FROM A MICROPHONE AT THE REACTOR.
THE BOTTOM-MOST TRACE IS FROM A PRESSURE TRANSDUCER IN THE REACTOR CORE.

FIG. 2. Typical record of power excursion. (Argonne National Laboratory photograph.)

upon the amount of excess reactivity to which the reactor was initially sub
jected and upon the bulk temperature of the reactor water. When the exces
reactivity applied was low, corresponding to a reactor period of about 0.03 se
or longer, the reactor power after the initial surge settled down to a relativel
steady value in the neighborhood of half a megawatt. For this type of excur
sion the self-regulating characteristics of the reactor operate rapidly enoug
to stabilize power at a steady value characteristic of the amount of applie
excess reactivity. After the power has reached this steady value, furthe
reactivity could, of course, be applied, and the reactor would continue t
operate stably in steady boiling at a higher power.

If the excess reactivity which was applied by ejection of the control ro
exceeded that corresponding to a period of 0.02 or 0.03 sec, the initial powe
excursion was followed by a series of qualitatively similar excursions of smalle
amplitude, which occur at intervals of about 1 sec. The amplitudes of th
successive excursions, although they varied in an irregular manner, had sus
tained tendency to increase or decrease for this condition, simulating a ste
increase in reactivity. This type of operation is hereafter referred to a
"chugging".

When the applied excess reactivity was greater than that corresponding t
about 0.01 sec, the chugging was no longer observed, and the power afte
the first surge remained at a low value. This permanent shutdown was n
doubt the result of expulsion of sufficient water from reactor tank to partiall
uncover the reactor core. The occurrence of this behavior in the reacto
would, of course, depend upon the specific design of the reactor in question.

When the applied excess reactivity was increased to about 2 per cent k_{eff} t
give periods in the 0.005 sec range, the qualitative behavior of the reacto
power remained the same, but the fuel plate temperatures did not dro
immediately after the power surge. The fuel plate temperatures remaine
high for almost a second after the power surge and decreased by small jumps
as though the plate had been blanketed by steam for some time after th
power excursion.

Both the total nuclear energy liberation and the maximum fuel plate tem
peratures reached during the power excursions depended upon the amount o
excess reactivity involved in the excursion. For both the cases of the saturatio
temperature and room temperature water, the total energy liberated durin
the power excursion was nearly proportional to the product of the maximun
power and the reactor period.

The transient experiments were continued with reactor II to investigat
power transients at reactor pressures as high as 300 psi. The pressurizatio
of the reactor was by the vapor from the reactor water. Consequently, onl
the saturated condition could be investigated. The power excursions wer
run with the reactor tank completely closed. In no case did the pressure in th
steam space above the reactor rise by more than about 5 psi as a result of a

FIG. 3. Typical Borax experiment. (Argonne National Laboratory photograph.)

excursion. The effect of pressurization was to decrease both the energy released in an excursion of given period and the maximum temperature rise of the fuel plates.

In the short period experiments with Reactor I at atmospheric pressure, the steam pressure which built up in forcing the water rapidly from the reactor resulted in permanent deformation of the fuel plates. Because of this effect, it was not possible to extend the experiments to periods shorter than about 0.005 sec without damage to the reactor to the point where it became unuseable. Despite this mechanical damage the maximum temperatures reached by the fuel plates did not approach the melting temperature. It was decided that the reactor, which by this time had fulfilled its other purposes,

FIG. 4. Typical Borax 1 experiment. (Argonne National Laboratory photograph.)

would be sacrificed in an experiment which was violent enough to melt the fuel plates. For this purpose a control rod worth 4 per cent k_{eff} was completely ejected from the reactor core. To increase the severity of the experiment it was run with the reactor water at room temperature. Although the ejection of the rod only required about 0.2 sec, the rod was only about 80 per cent out of the core when the reactor reached its peak value. The minimum period resulting from this ejection was 0.0026 sec. The power excursion melted most of the fuel plates. The pressure resulting from the molten metal in contact with the reactor water burst the reactor tank and ejected most of the contents of the shield tank into the air. The sound of the explosion at the control station, half a mile away, was comparable to that resulting from the explosion of 1 to 2 lb of 40 per cent dynamite on the bare ground at the same distance. Figures 5 to 13, taken from a motion picture record of the experiment show stages of the explosion.

FIG. 5. Final Borax 1 experiment. (Argonne National Laboratory photograph.)

FIG. 6. Final Borax 1 experiment. (Argonne National Laboratory photograph.)

FIG. 7. Final Borax 1 experiment. (Argonne National Laboratory photograph.)

FIG. 8. Final Borax 1 experiment. (Argonne National Laboratory photograph.)

FIG. 9. Final Borax 1 experiment. (Argonne National Laboratory photograph.)

FIG. 10. Final Borax 1 experiment. (Argonne National Laboratory photograph.)

FIG. 11. Final Borax 1 experiment. (Argonne National Laboratory photograph.)

FIG. 12. Final Borax 1 experiment. (Argonne National Laboratory photograph.)

FIG. 13. Final Borax 1 experiment. (Argonne National Laboratory photograph.)

The total energy release during the excursion was estimated to be 135 MW sec. Analysis of the mechanical damage to some components indicated the peak pressures were at least 6,000 psi, and were probably as high as 10,000 psi. It was evident from examination of the reactor debris that many of the fuel plates had practically completely melted. Others, evidentally those at the edge of the core, had only partly melted and portions of them remained fastened to the side plates of the fuel elements (Figs. 14–15). Some of the fragments which had evidently been molten appeared as spongy metallic globules. Other fragments appeared to have been molten inside while the outside clad remained solid.

Most of the heavy debris fell to the ground near the shield pit. The control rod drive mechanism, which weighed about a ton, fell to the side of the earth shield after having been thrown about 30 ft in the air. Recognizable fuel plate fragments were thrown as far as 200 ft from the reactor site. Surveys of the total fission-product radioactivity of all the debris indicated that practically all the fuel originally in the reactor could be accounted for in a radius of 350 ft around the original reactor location and no large section of the reactor core material left the site in the form of airborne material. At the time of the experiment the wind velocity was 8 mph at ground level, and 20 mph at 250 ft above the ground. Fifteen minutes after the experiment the

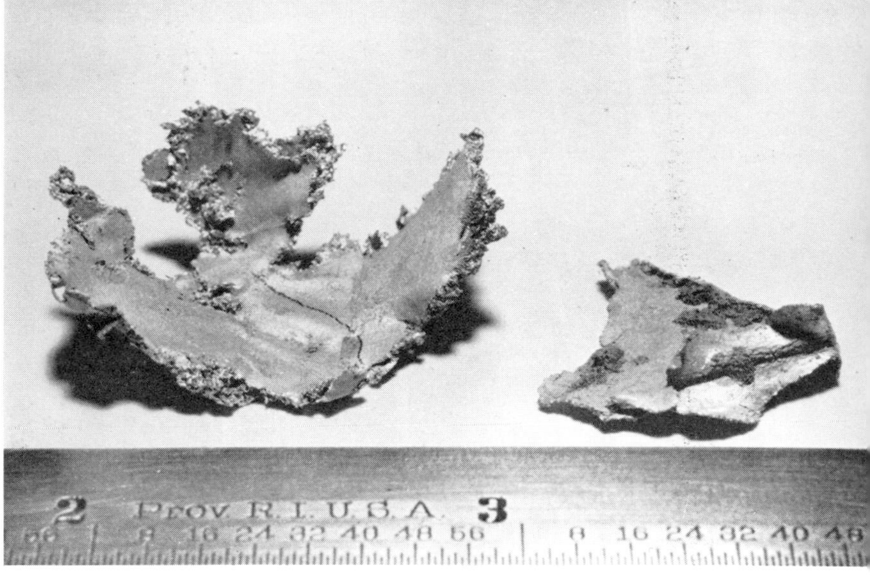

FIG. 14. Remains of fuel element from final Borax 1 experiment. (Argonne National Laboratory photograph.)

FIG. 15. Remains of fuel plate from final Borax 1 experiment (Argonne National Laboratory photograph.)

total beta plus gamma activity level, 3 ft above the ground, at a point 0.8 miles directly downwind of the reactor, was 5 mr/hr. At all points further from the reactor the effects of fall-out were less than this value. Momentarily during the explosion, a gamma dose rate in excess of 400 mr/hr was indicated on the survey meter $\frac{1}{2}$ mile from the reactor. This indication decayed rapidly and the total dose received at a half-mile point "crosswind" was less than 10 mr.

Both the observed radiation intensity and mechanical damage were roughly consistent with the measured nuclear energy release of 135 MW sec. Although

FIG. 16. Remains of fuel plate from final Borax 1 experiment. (Argonne National Laboratory photograph.)

the explosion was spectacular, its effects were comparable to those which could be caused by a moderate amount of chemical explosive. The destruction of the reactor tank was not surprising since it was constructed of relatively thin ($\frac{1}{2}$ in.) steel. Most of the equipment outside the shield tank was either undamaged or repairable, and much of it, including the control rod drive mechanism, was decontaminated, reconditioned and reused on Reactor II.

There was no evidence that the power-limiting process in the destructive experiment differed qualitatively from that which was effective in the earlier, non-destructive experiments. It was quite evident that the nuclear power release was terminated at an early stage of the explosion; indeed, high-speed motion pictures recorded the white flash emitted by the reactor as it reached high power and showed that it was extinguished before any ejected materia appeared over the top of the shield tank. The flash lasted about 0.003 sec The energy stored in the fuel plates as sensible heat and latent heat of fusior during the relatively short nuclear power burst was, of course, released during the much longer explosion process.

These experiments with boiling reactors proved that the reactors investi gated possessed a high degree of inherent safety, and indicated that it i possible to design practical reactors of this type which are safe against any

FIG. 17. Spert-I reactor building, National Reactor Testing Station. (AEC Idaho Operations Office photograph.)

reactivity accident which might be expected to occur in practice. However, results of these experiments indicated that an extended experimental program would be required to adequately explore all of the areas of interest to the reactor safety problem. The permanent facility was, therefore, constructed at the National Reactor Testing Station and a continuing program of investigation was initiated designated SPERT for special power excursion reactor tests. This facility is operated for the Atomic Energy Commission by the Phillips Petroleum Company. Three reactors have been constructed for this program. SPERT-I is a heterogeneous, water-moderated, enriched-fuel, research reactor constructed for the purpose of conducting transient experiments on unpressurized systems. The reactor consists of a plate-type, uranium–aluminum core immersed in an open vessel. There are five blade-type control rods, of which the outer four serve as shim-safety rods. The central rod is used for the initiation of power excursions. The control center is located approximately 0.5 miles from the reactor. A limited adjustment of water temperature in circulation is obtained by means of immersion heaters and a power-stirrer. The hydrostatic head of the water can be adjusted within limits. The reactor construction is simple and readily modified and the core is easily accessible for static measurements and core modifications. The SPERT-II reactor was constructed for tests extending the range of pressure

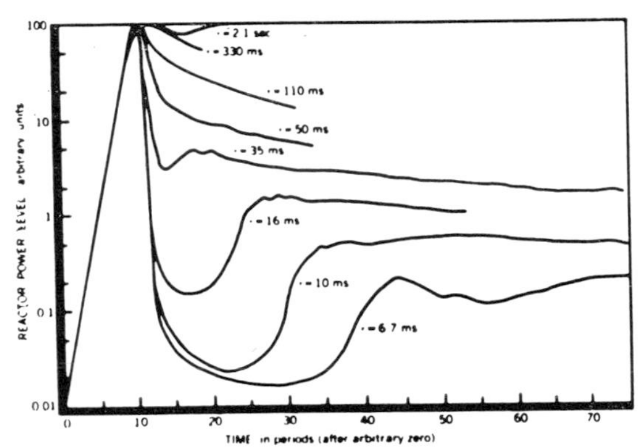

Normalized Power vs. Time
As the transient period is decreased, power behavior changes from damped to oscillatory and the ratios of both peak power to the power minimum, and peak power to the equilibrium power, increased.

FIG. 18. Spert-I transient tests. (B. J. Garrick, (Editor), *Research and Development in Reactor Safety*, U.S. Government Printing Office.)

and temperature beyond those of SPERT-I, and for tests involving either upward or downward flow of water in the reactor core. Initially the primary purpose of this facility is for nonflow tests of D_2O moderated and reflected reactor cores for comparison with tests of H_2O moderated and reflected cores. Subsequently the facility is to provide for tests with reflective materials other than water and heavy water. The SPERT-III reactor was constructed before SPERT-II with the objective of permitting tests over a wide range of flow, temperature and pressure. This reactor is water cooled and moderated and has a two-loop primary system for heat removal at power levels up to 60 megawatts of approximately 30 min duration. The primary system is capable of sustaining pressures up to 2500 psi and temperatures up as high as 670°F. In addition to other experiments it will be possible to study the cold water accident in SPERT-III.

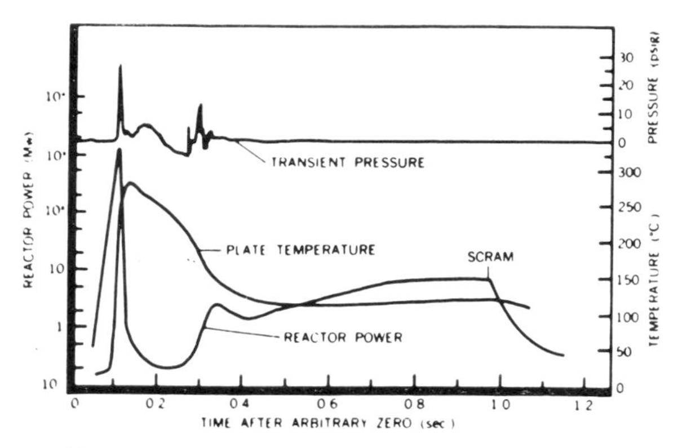

Temperature, Transient Pressure, and Power vs. Time

For an instantaneous reactivity addition of 1.4%, the maximum power level reached is 1300 megawatts and the fuel plate surface temperature in the center of the core was approximately 150 C in excess of the saturation temperature. Transient pressures in the fuel assembly end box in the core center reached about 25 psig.

FIG. 19. Spert-I transient tests. (B. J. Garrick, (Editor), *Research and Development in Reactor Safety*, U.S. Government Printing Office.)

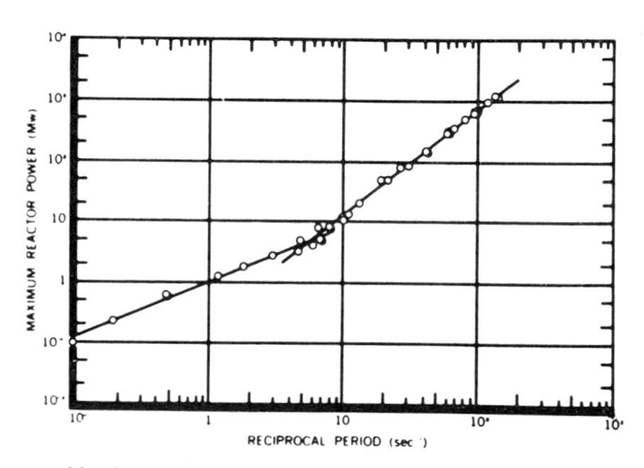

Maximum Power vs. Reciprocal Period

The point of intersection occurs at a reciprocal period corresponding to a Δk just below prompt critical.

FIG. 20. Spert-I transient tests. (B. J. Garrick, (Editor), *Research and Development in Reactor Safety*, U.S. Government Printing Office.)

The initial effort in experimenting with the SPERT-I studies was directed towards repeating transient tests made by other experimenters and extending the work to the area between the destructive and the non-destructive tests. In experiments with rapid insertion of reactivity, the resulting kinetic behavior was observed and recorded for a particular set of reactor conditions, such as temperature, initial power and height of water above the core. The step transients covered reactor periods from 10 sec to 5 msec. Ramp rate tests were conducted in which reactivity was added linearly to the critical system by withdrawal of the control rod bank at a fixed rate. Rapid transient explorations were conducted for a wide range of starting powers and reactivity insertion rates. Some of the data obtained from SPERT-I experiments are illustrated in Figs. 19 through 23. It has been found that an increase of

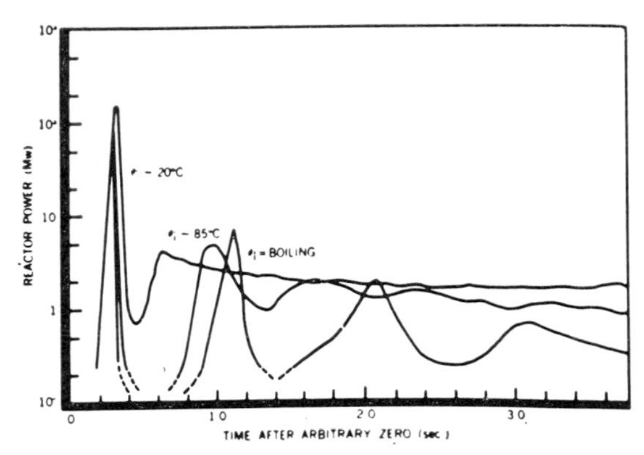

Power vs. Time

For an instantaneous reactivity addition of approximately 0.95%, a change in initial reactor temperature from 20 C to boiling decreased the maximum peak power from approximately 160 megawatts to 50 megawatts. Change in initial temperature also affected the damping characteristics.

FIG. 21. Spert-I transient tests. (B. J. Garrick, (Editor), *Research and Development in Reactor Safety*, U.S. Government Printing Office.)

water head or water temperature decreased stability. Also the void worth varied significantly with the void size and location. It was found that detailed static measurements must be made to interpret void worth during a transient. These observations apply only to the SPERT-I reactor.

Some conclusions having general application to all reactor systems are as follows:

1. The reactivity compensation required to terminate an excursion is less than the injected reactivity for an important class of tests. The immediate practical importance of this result is that for incidents involving reactivity

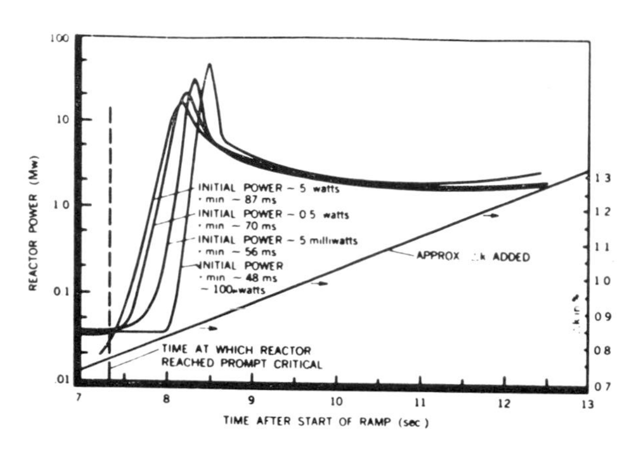

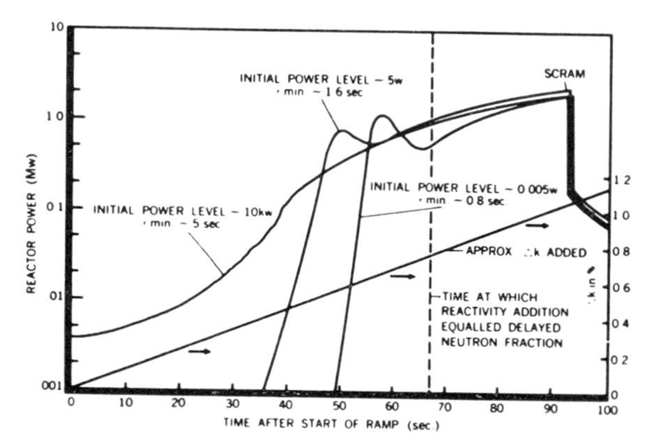

Power Behavior vs. Time As A Function of Initial Power Level

These data show the comparative insensitiveness of peak power and minimum period to different starting power levels, and the strong dependence of these same parameters on changes in ramp rate.

FIG. 22. Spert-I transient tests. (B. J. Garrick, (Editor), *Research and Development in Reactor Safety*, U.S. Government Printing Office.)

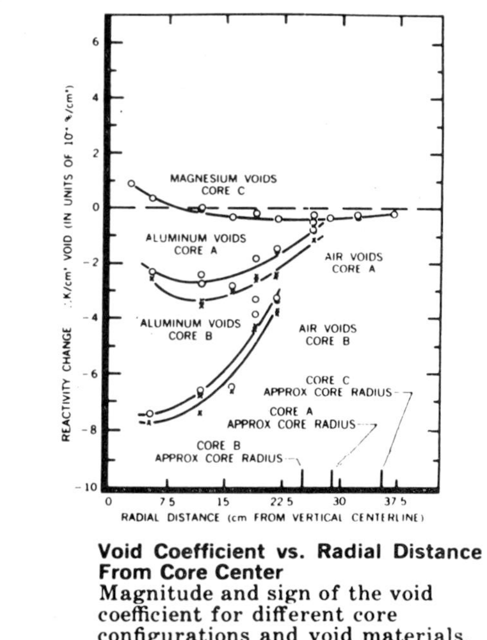

Temperature Coefficient vs. Temperature
Sign and magnitude of the temperature coefficient for different core configurations and fuel plate spacing.

Void Coefficient vs. Radial Distance From Core Center
Magnitude and sign of the void coefficient for different core configurations and void materials.

FIG. 23. Spert-I static tests. (B. J. Garrick, (Editor), *Research and Development in Reactor Safety*, U.S. Government Printing Office.)

N

injections up to the region of control by prompt neutrons only, the emergenc
control of reactor may not require a fast-acting, strong control. A fast
acting, small control will reduce the needed delay to permit the insertion of
slower strong control. This simplified view does not adequately describ
reactors which can be autocatalytic for part of an excursion.

2. The severity of a ramp-produced excursion is strongly dependent o
ramp rate, and weakly dependent on initial power over a very wide range.

3. It is possible for the small initial reactivity disturbance to lead to larg
self-induced reactivity changes associated with runaway.

4. Static measurements of reactor properties are required in detail fo
interpretation of theories on transient behavior and for safety evaluation.

Under some conditions, extremely unstable operation was found i
SPERT-I. Power versus time after start of a ramp insertion is shown i
Fig. 24. The occurrence of this condition depended in part on the hydro

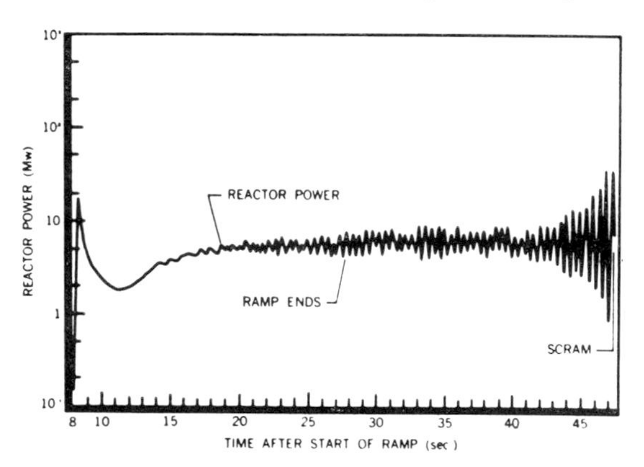

Power vs. Time After Start of Ramp
These data show power behavior for a
ramp test indicating instability.

FIG. 24. Spert-I transient tests. (B. J. Garrick, (Editor), *Research and Development in Reactor Safety*, U.S. Government Printing Office.)

dynamics of the system since in some cases a reactor would be stable wit
1.5 per cent reactivity held in voids and a 2 ft head of water over the cor
whereas if the head of water were increased to 4 ft divergent oscillations i
reactor power occurred. Ramp insertions of 1.5 per cent reactivity cause
instability for both boiling and subcooled tests.

From experience with SPERT-I and other boiling reactors, it appears tha
stable generation of steam can be obtained for sufficiently low power level
and that increasing of power levels beyond some limiting point may produc
power oscillations. The nature of these oscillations will depend upon th
operating conditions of the system and in some cases will be limited while i
other cases they may diverge. In the operating conditions of greatest interes

he oscillations are usually the order of 100 per cent of full power and have a requency of the order of seconds with a non-sinusoidal wave form indicating he non-linear characteristic of the oscillations. The point at which oscillations ppear has been found to be sharply influenced by pressure with higher pecific power being obtained as the pressure is increased over the experi-nental range of atmospheric pressure to 300 lb/in^2. Increased pressure xhibits a strong stabilizing effect on a boiling water reactor. A reactor vhich exhibited oscillatory tendencies with less than 1 per cent reactivity in oids at atmospheric pressure was found to operate without oscillations with .5 per cent reactivity in voids at 300 psi. Oscillations occurred in the Experi-nental Boiling Water reactor at low pressure when the power was 1 MW vhile at 600 psi pressure no power oscillations were detected at a power level of 20 MW.

Safety of a small non-circulating homogeneous reactor of the type used or research has been systematically investigated by Remley and others at Atomics International in the KEWB program (Kinetic Experiments on Water Boilers). From this work it appears that the formation of gas bubbles by adiolysis of the water is the principle shutdown mechanism.[4]

The results of boiling reactor experiments have been correlated by Corbin.[7, 8]

WATER-COOLED AND MODERATED RESEARCH REACTORS

The aluminum–uranium alloy fuel elements, of the type developed for the Materials Testing Reactor, have been used for the construction of a number of research reactors ranging in power from a few kilowatts to the megawatt ange. The first facility of this type was the Bulk Shielding Facility at the Oak Ridge National Laboratory. A tank of water 20 ft wide, 40 ft long and 20 ft deep, filled with water, provided cooling, moderation and shielding. The size of the tank resulted in the designation “Swimming Pool Reactor”. The fuel elements were assembled in a grid which hangs from a movable bridge so that the active lattice can be moved to different positions as required by the experiments. Where higher power operation is required provisions may be made to circulate water through the core or the core may be enclosed n a separate tank for adequate water circulation.

Water-moderated and cooled reactors of this type have special safety problems which require consideration in addition to those normally associ-ated with water reactor as follows:

(a) The control mechanism is open and available for manual manipulations.

(b) The fuel elements are inserted manually with the aid of long tongs and the manual insertion of a fuel element could greatly change the reactivity.

(c) The lattice grid usually contains more spaces than the number of fuel assemblies required and a better moderator than water may be inserted in the empty positions. Such moderators might include graphite o beryllium and the addition of such materials would increase reactivity

(d) For the cases where the core is movable, its motion away from experi ments which poison the reactor may increase reactivity. Also, if th core is moved closer to a moderator such as graphite, reactivity woul be increased. Usually there are provisions which positively preven motion of the reactor when the control rods are withdrawn.

(e) Since a research reactor is built to make it possible to perform experi ments usually of a non-routine nature, it can be expected that man types of experiments will be constructed which change the reactivit by the presence of poison materials or the addition of fuel near th core. These assemblies may be capable of fairly rapid motion manuall or by collapsing of a structure or an assembly may fall.

(f) Voids may be necessary near the active lattice for beam holes or shield ing experiments and these voids decrease reactivity. The sudde flooding or collapse of a void could increase reactivity at a rapid rate.

In addition to the safety features of the control and regulating systen additional safety is assured in such research reactors by limitations on th amount of fuel in the reactor and the excess reactivity that may be availabl in the control rods or in experiments including beam holes.

ZIRCONIUM–HYDRIDE, WATER MODERATED REACTORS

A series of reactors have been developed with fuel elements made of ar alloy of uranium 20 per cent enriched in U^{235} and zirconium hydrided tc approximately one hydrogen atom for each zirconium atom. A typical fue element is a right circular cylinder about 1.5 in. in diameter and 14 in. lon with 4 in. graphite slugs at each end of the cylinder to act as top and botton reflectors. Between the graphite and fuel-moderator material are insertec aluminum disks containing burnable poison. The fuel elements are clad witl aluminum 0.030 in. thick. The reactor core is located in a tank of wate which serves as moderator, radiation shield and coolant (Fig. 25). The cor is surrounded by a graphite reflector 12 in. thick.

The transient testing of this reactor type, known as TRIGA, has beer conducted by General Dynamics Corporation's General Atomic Div. tc determine the safety characteristics, and is described in Reference 9. It wa found in initial experiments that when the fuel and water are heated by th same amount, there is only a slight change in reactivity varying from 0.1 cents/°C to −0.1 cents/°C. However, if the reactivity is operated at a sufficien power level to raise the temperature of the fuel elements above that of th core water, there is a large decrease in reactivity of approximatel −1.5 cents/°C average fuel-temperature rise. The average core water tem perature did not change significantly compared with that of the fuel element during the experiments that were conducted. As the uranium and th hydrogen in the fuel-moderator elements are intimately mixed, there is no time delay. The large negative temperature coefficient is attributed to the bound hydrogen in the zirconium hydride. The moderating properties of chemically bound hydrogen in zirconium hydride are much like those of free hydrogen above 0.13 eV, however, they are greatly inhibited below this energy and are then not effective as a moderator. During a transient the hydride temperature and the fuel temperature rise simultaneously and the neutron spectrum

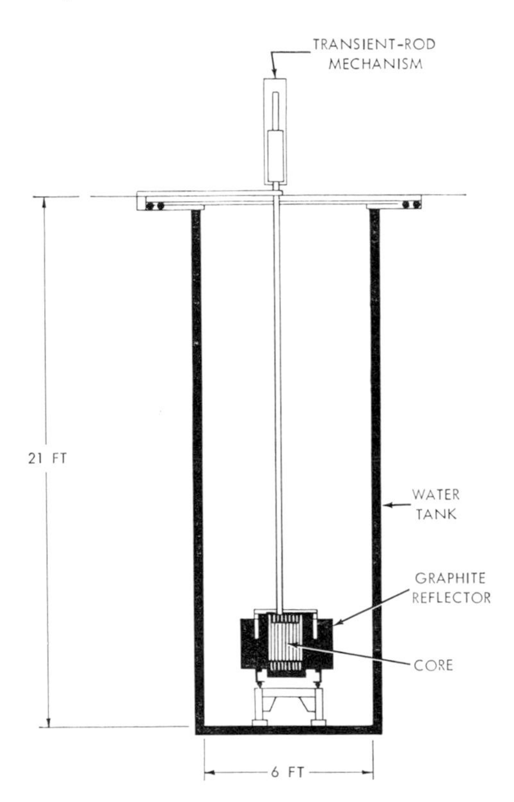

FIG. 25. Triga reactor.

is hardened promptly. This results in a decrease of fission probability and in an increase in the fraction of neutrons lost because of leakage from the core and parasitic capture in the water and in the control rods. The water temperature does not increase significantly over the short time interval of a prompt radiation burst. In the transient experiments there was no evidence of system instability, boiling of water in the core, or disturbance of the water level in the tank. The major portion of the transient energy was stored as thermal energy in the fuel elements during the transient and subsequently diffused slowly into the cooling water.

A series of transient experiments have been conducted including the step-insertion of excess reactivity which resulted in a power burst and an increase in fuel temperature. Tests including thermal cycling indicate that high transient temperatures can be obtained without significant damage to the fuel elements. On the basis of the negative-temperature-coefficient data obtained, it is expected that a heavy neutron pulse can be permitted without damages.[9]

REFERENCES

1. DIETRICH, J. R., Experimental determination of the self-regulation and safety of operating water-moderated reactors, *Proc. 1st United Nations Int. Conf. on the Peaceful Uses of Atomic Energy*, Vol. 13, 88–101, United Nations, New York.
2. DIETRICH, J. R., Experimental investigation of the self-limitation of power during reactivity transients in a subcooled, water-moderated reactor, Report AECD-3668.
3. ULRICH, A. J., Results of recent analysis of Borax II transient experiments, Report ANL-5532, April 1956.
4. GARRICK, B. J., Editor, *Research and Development in Reactor Safety*, U.S. Government Printing Office.
5. MONTGOMERY, C. R., *et al.*, Summary of the Spert I, II and III reactor facilities, IDO 16418, Nov. 1957.
6. NYER, W. E., and FORBES, S. G., Spert program review, IDO-16415, 1957.
7. HORNING, W. A., and CORBEN, H. C., Theory of power transients in Spert I reactor, IDO-16446, Aug. 1957.
8. CORBIN, H. C., and HORNING, W. A., Theory of power transients in the Spert I reactor, IDO-16434, Jan. 1958.
9. STONE, R. S., SLEEPER, H. P., STAHL, R. H., and WEST, G., Transient behavior of TRIGA, a zirconium–hydride, water-moderated reactor, *Nucl. Sci. and Engng.*, 6,255-259 (1959).

CHAPTER 7

DISPERSION OF RADIOACTIVE MATERIAL

INTRODUCTION

THE problem of estimating the possible hazards to people in the area surrounding a nuclear installation can be divided into estimates of the quantity and conditions or manner of release of radioactive material and estimates of the dispersion of the hazardous material in the environs. The many uncertainties in any estimates of the release process have been emphasized in the preceding chapters. When these uncertainties are compounded by variations in the weather possible at the time of release, it is found that calculated exposures varying over several orders of magnitude can be predicted depending on the assumptions and the viewpoint of the author. A set of calculations giving assurance for the safety of the public can be presented or a basis for an alarming estimate of possible off site exposures can be found for most reactor installations.

In the case of a release of material over a long period of time, the conditions can be expected to approach average values which are usually well known and, therefore, accurate estimates can be made of exposures from routine operations not subject to the possible variations in meteorological conditions at the time of an accidental release of material.

PROBLEMS IN ROUTINE OPERATION

Air cooled reactors such as the X-10 and the Brookhaven Research Reactor during operation continuously discharge radioactive material into the atmosphere due to neutron activation of the air. This release of radioactive material into the air is in addition to any release of fission products that may result from a fuel element failure and particularly the release of gaseous fission products. Since the cooling air is filtered to remove dust before entering and after passing through the reactor, particulate materials are not a major problem although during the initial period of operation of the X-10 reactor, the particulate material problem due to ruptured slugs and dust caused serious site contamination. This experience led to the installation of filters and other equipment for the removal of small particles of radioactive material. Since the Brookhaven air cooled reactor is located near populated areas, there is a need to forecast the expected exposure from activity, principally, argon 41, in the discharged air based on predicted weather conditions and records of actual off site exposures so that the accumulated exposures are not excessive. For this purpose elaborate meteorological studies have been made and specialized procedures have been developed for predicting exposures with accuracy. Only a very unusual prolonged period of unfavorable weather conditions would require a reduction in reactor operating power for reasons of accumulated off site exposure at this site.

The single pass air and water cooled reactors which continuously discharge radioactive material to the environs might be considered as an unusual case and usually essentially no activity is discharged to the environs. Contaminated liquids of any significance are concentrated and shipped to some disposal area or dumped at sea in concrete-weighted drums. In most reactors with closed primary coolant loops, the amount of gaseous radioactive wastes is very small. For small homogeneous research reactors the contaminated gases can be stored in cylinders for decay and eventual disposal. Thus, there is no appreciable radioactive discharge to the atmosphere or surface water during normal conditions for most reactors designed for operation near populated areas.

The event that is most feared in reactor operation is an uncontrolled increase in temperature leading to melting and vaporization of core materials followed by disruption of the core structure and perhaps accompanied by burning or other chemical reactions of core materials. The hazard to the surrounding area is from the fraction of the fission products released in a sudden burst or a release of material over a period of time from a leaking enclosure or a burning mass of material. The distribution of material from a puff or plume with deposition of radioactive gaseous and particulate material is usually the major problem to be considered in evaluating the possible exposure to people in the area.

Atmospheric Diffusion

The calculation of the rate of diffusion and deposition of radioactive material from a source is an essential step in hazards evaluation and it is, therefore, necessary to consider some of the factors which govern atmospheric diffusion. This information is also useful as a basis for considering elementary meteorological problems in site selection. The following discussion of meteorology and atmospheric diffusion has been taken from various reports prepared by the Scientific Services Division, U.S. Weather Bureau for the U.S. Atomic Energy Commission. The principal report on this subject is AECU 3066.[1]

The following general discussion of atmospheric diffusion is intended only to give some understanding of the factors involved and an appreciation for the complexity of the problem. It is usually advisable to obtain the help of a meteorologist before proceeding very far with the interpretation of local weather data or in making complex diffusion calculations. Fortunately such assistance can usually be obtained from the U.S. Weather Bureau.

Temperature Profiles

The temperature of the air under "normal" atmospheric conditions decreases with altitude and for a standard atmosphere the rate of change of temperature with altitude, or lapse rate, is $-3.5°F/1000$ ft. The temperature change of the air with altitude corresponding to an adiabatic expansion process from conditions at the surface to the pressure at the altitude selected is termed the dry adiabatic lapse rate which has a value of $-5.4°F/1000$ ft. Over a limited altitude range the normal variation of temperature with altitude may be reversed and the temperature then increases with altitude and this condition is designated an inversion. The altitude over which this condition prevails is termed the height of the inversion. A plot of different temperature profiles is shown in Fig. 1.

If a volume of air is displaced vertically in an atmosphere having exactly an adiabatic lapse rate, the density of the volume of air always changes with altitude so as to be identical in density with the atmosphere at that altitude. This is termed a neutral condition. If the lapse rate is a larger negative value than the adiabatic rate, an initial displacement of a volume of air will be accelerated and the air is said to be unstable. Also, if the lapse rate is a smaller negative value than the dry adiabatic condition, or even a positive value as for an inversion, the displacement of a volume of air will result in a change of density which tends to restore the volume of air to its original altitude, and the atmosphere is said to be stable. A neutral condition is not the most common condition since normally stable conditions prevail at night and unstable conditions are most frequent during the day. The dry adiabatic lapse rate is observed during windy weather with good mixing under cloud cover which limits radiation.

Since the adiabatic lapse rate is greater than the lapse rate for a standard atmosphere, at some altitudes there usually occurs a very stable condition which is termed a turbulent inversion. An inversion may be of other origins. A wind shift may bring in an air mass of different temperature. If the new mass is warmer than the air already present, the warm air over-rides the cooler air because of the density differences, thus creating an inversion. If colder air is entering an area, the wind front wedges under the lighter warm air, and the effect is similar. Another type of inversion is caused by nocturnal terrestrial radiation, which occurs when there is little or no wind and cloud cover. An inversion of this type begins as soon as the sun sets and even before sunset, if a location is in shadow. The inversion is caused by the cooling of air in contact with surface objects which become cooled by radiation. If there is considerable cloud coverage, thermal radiation is reflected to the earth's surface, and the total cooling may be very little. Also, if there is an appreciable wind, there is surface turbulence and mixing which destroys stratification.[(3)]

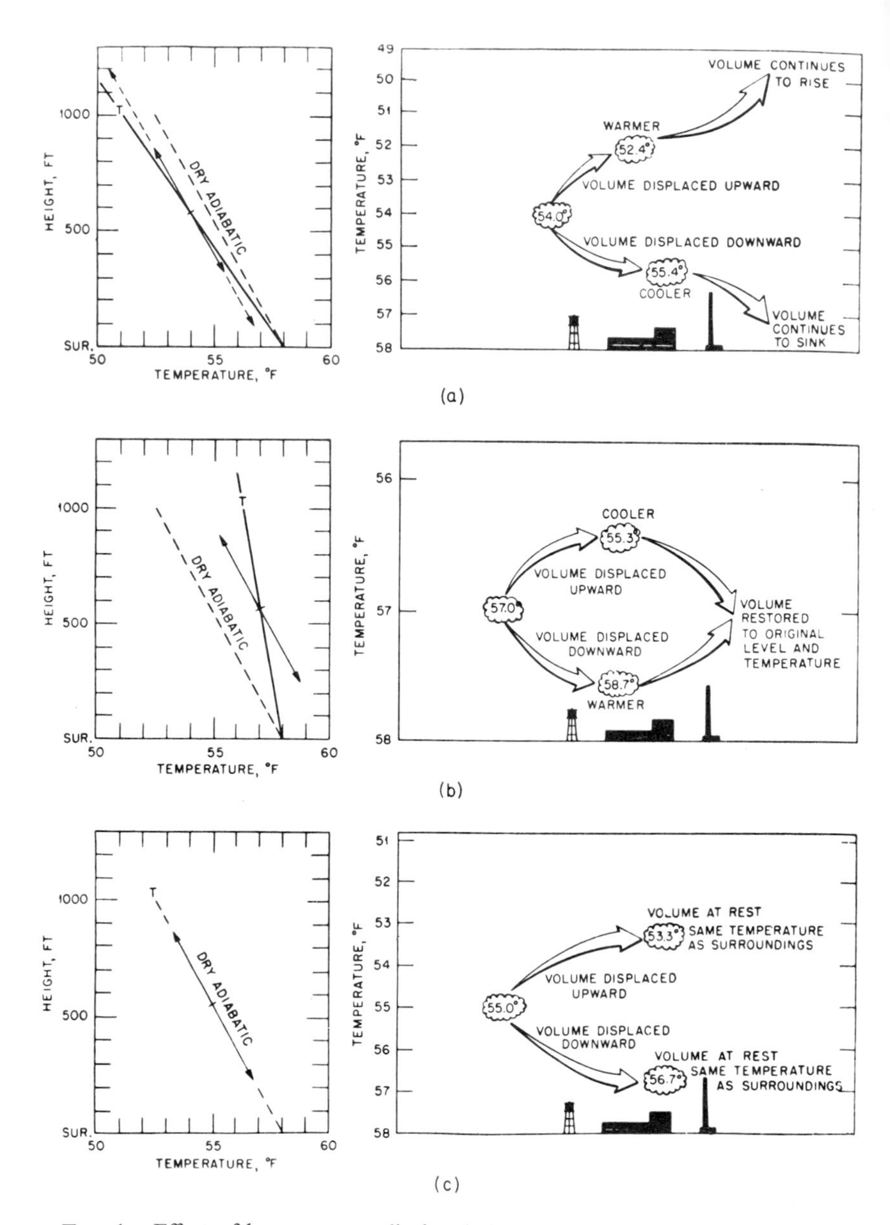

FIG. 1. Effect of lapse rate on displaced air volumes. (a) Unstable lapse rate. (b) Stable lapse rate. (c) Neutral lapse rate. (Reproduced from "Meteorology and Atomic Energy," AECU 3066, U.S. Weather Bureau.)

The pattern of wind and temperature gradient will be very much altered y the location of an area in relation to large bodies of water. The mid-ontinent area will have much larger variations in daily lapse rates, as well s larger seasonal variations. Certain combinations of hemispheric wind, cean currents and ocean–continent relationships may result in persistent mperature inversions. The west coast of North America from southern alifornia southward to central Mexico, the Peruvian and Chilean coasts of outh America, and the African coast from Capetown to near Sao Paulo and om Dakar to near Gibraltar are all subject to these inversions and as a esult have entirely different diffusion regimes than the mid-United States.[1] hese persistent inversions in the Los Angeles basin make this area somewhat ndesirable for the location of large nuclear reactors because any radioactive aterial discharged into the atmosphere might remain in the area with very ttle dilution from mixing.

Behavior of Effluents

It has been found that the appearance of effluent plumes from stacks is egulated largely by the vertical gradient of air density (as measured by the erical gradient of air temperature) in the neighborhood of a stack. Since here are a limited number of configurations of the vertical temperature radients that are likely to occur in nature, the kinds of plume behavior likely o be observed from any particular stack are also limited. Although some ariations may be found, it is believed that there are five major types of plume ehavior, and these are shown schematically in Fig. 2, along with the eneral form of vertical temperature distribution causing the behavior. These lassifications are mostly applicable to single and more or less isolated stacks.

A description of smoke behavior and accompanying meteorological onditions for each type is given below:

lume types

(1) *Looping*. Looping occurs with super adiabatic (very unstable) temper-ture lapse rates. A stack effluent, if visible, appears to loop because of hermal eddies in the wind flow. Gases diffuse rapidly, but sporadic puffs aving strong concentrations are occasionally brought to the ground near the ase of the stack for a few seconds during light winds. This is usually a fair veather daytime condition since strong solar heating of the ground is required. ooping is not favored by cloud, snow cover or strong winds.

(2) *Coning*. This type of plume occurs with a temperature gradient between lry adiabatic and isothermal and may occur periodically (between thermals) vith a superadiabatic lapse rate. The effluent plume is shaped like a cone with he axis horizontal. The distance from the stack at which the effluent first comes o the ground is greater than with looping conditions because thermal turbu-ence, and hence vertical motion, is less. It usually is favored by cloudy and

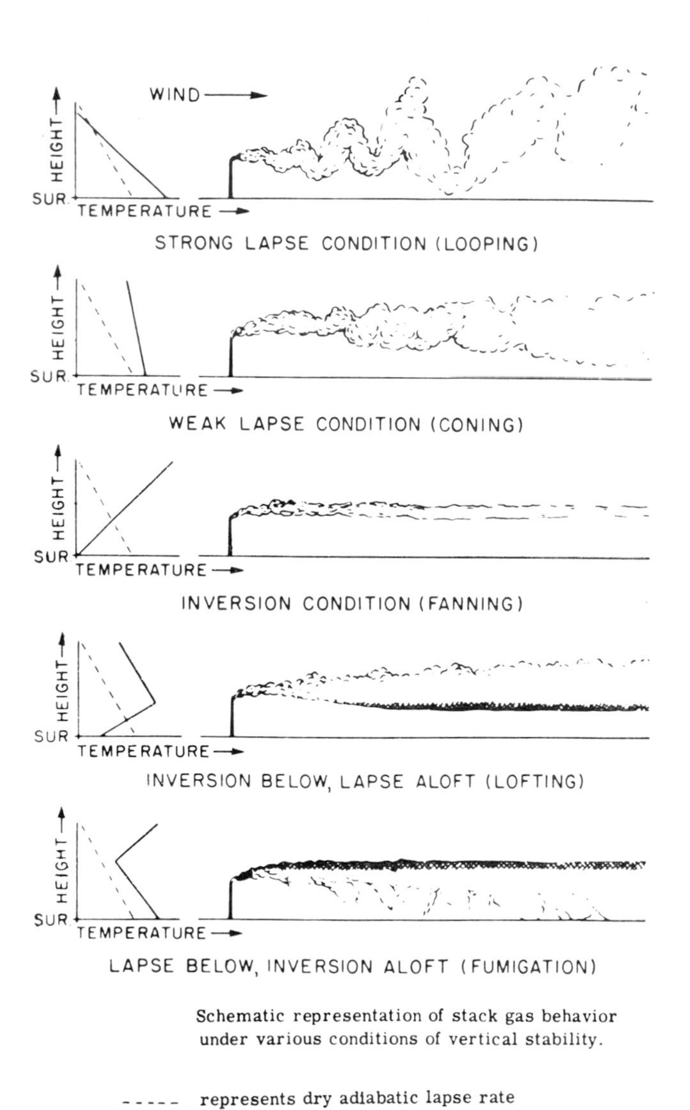

FIG. 2. Stack gas behavior. (Reproduced from "Meteorology and Atomic Energy," AECU–3066, U.S. Weather Bureau.)

windy conditions and may occur day or night. In dry climates it may occur infrequently, and conversely, in cloudy climates it may be the most frequent type observed. Much of the development of diffusion formulae has been for stability conditions similar to this type of lapse rate.

(3) *Fanning*. Fanning occurs with temperature inversions or near-isothermal lapse rates. This type of plume with its very smooth, almost laminar, flow also depends somewhat on wind speed and the ruggedness of the terrain. The stack effluent diffuses practically not at all in the vertical. The effluent trail may resemble a meandering river, widening gradually with distance from the stack. Depending on the duration of the stable period and wind speed at the stack level, the effluent may travel for many miles with very little dilution. The concentration within the stream varies inversely as the wind speed, for the distance that it travels is directly proportional to the wind speed, the variation in wind direction, and the duration of this type of flow pattern. Since fanning is usually associated with surface inversions it is mostly a night-time condition. It is favored by light winds, clear skies, and snow cover. The condition can persist for several consecutive days in winter in some climates, especially at high altitude. In most cases the fanning behavior of plumes is not considered an unfavorable condition for stack release, even though the effluent undergoes little dilution after leaving the stack. The important feature is that an effluent of neutral bouyancy tends not to spread to the ground during inversion conditions. Fanning might be unfavorable, however, in the following circumstances:

(a) Where the stack is short with respect to surrounding buildings or other objects needing freedom from pollution.
(b) Where the effluent plume contains radioactive wastes which may radiate harmfully to the ground.
(c) Where there is a group of stacks of various heights giving an extensive cloud of effluent. This is especially bad when long periods of calm occur.
(d) When lateral spread and variability of plume direction are restricted (as by a deep narrow valley) so that the waste passes repeatedly over the same places.

The likelihood that an effluent would accumulate in the neighborhood of a stack during inversion situations is smaller than one might expect. Wind speeds so light that they cannot be detected by ordinary instruments (1 m.p.h. or less) will remove effluent sufficiently fast to prevent accumulation for most stacks regardless of discharge rate. Also, since the tendency is far greater for the plume to spread horizontally than vertically, there is little likelihood that such an accumulation would cause a plume to spread downward to the ground as long as the inversion persists.

The theoretical treatment of diffusion is on the most uncertain grounds under these conditions. The assumption of isotropy, or the same rate of diffusion in the vertical and horizontal direction, is probably in error since the horizontal diffusion, although much less than for unstable regions, is larger than the vertical spread. A ratio of horizontal to vertical diffusion coefficients C_y/C_z, of 3 may be used for calculations.

(4) *Lofting*. This type of plume occurs with the transition from lapse to inversion and should thus be observed most often near sunset. Depending upon the height of the stack and the rate of deepening of the inversion layer, the lofting condition may be very transitory or may persist several hours. It has infrequently been found to persist throughout the night for the 250 ft stacks at the National Reactor Testing Station (NRTS), although its usual duration is 1–3 hr. The zone of strong effluent concentration, shown by shading in Fig. 2, is caused by trapping by the inversion of an effluent carried into the stable layer by turbulent eddies that penetrate the layer for a short distance. Except when the inversion is very shallow, the lofting condition may be considered as a most favorable diffusion situation. The inversion prevents effluent from reaching the ground, and at the same time the effluent may be rapidly diluted in the lapse layer above the inversion. If sufficient meteorological instrumentation is available to detect lofting conditions, it may be used for "dumping" an effluent that would cause trouble during other meteorological conditions. Obviously the duration of lofting conditions depends on the rate of increase of depth of the nocturnal inversion and the height of the stack. Since this type of lapse rate will prevent effluent from reaching the ground, diffusion computations would often not be required. However, if the air concentration above the inversion is of interest, an approximation could be obtained by assuming the inversion top to be an impermeable surface and, choosing the stability parameter appropriate to the layer above the inversion, computing concentrations at the inversion top. The height in this instance would be the distance from the effective stack height extended above the inversion top. Since the inversion would not be completely impermeable, these values would be extremely crude.

(5) *Fumigation*. This condition occurs at the time when the nocturnal inversion is being dissipated by heat from the morning sun. The lapse layer usually begins at the ground and works its way upward less rapidly in winter than in summer. At some time, the inversion is still present just above the top of the stack and acts as a lid, for convective eddies mix the effluent plume within the shallow lapse layer and near the ground. This condition may also develop in sea breeze circulation during late morning or early evening. Large concentrations are brought to the ground along the entire effluent stream (which may be quite long owing to the previous presence of fanning conditions) by thermal eddies in the lapse layer. The zone of strong concentration shown by shading in the Fig. 2 is that portion of the plume that has not yet been mixed downward. The fumigation deserves special attention in stack

planning since it (1) provides a method whereby strong effluent concentrations can be brought to the ground, at least briefly, at great distance from the stack and (2) gives stronger sustained ground concentrations in the neighborhood of a stack than loop, coning, etc. During smoke experiments at Brookhaven using a 355 ft stack, it was found that fumigation concentration averaged 20 times the maximum concentration computed by the Sutton equation over a period of about 15 min. This figure has been used at Brookhaven to compute exposures for the reactor stack effluent. Concentrations thus computed allow the severity of the fumigation to vary with stack height and wind speed and thus should be sufficiently applicable to other stacks. The duration of fumigation conditions may vary widely from place to place. It will depend upon the rate of deepening of the lapse layer as the nocturnal inversion dissipates and upon the height of the stack. In some places inversion "lids" are shown to persist for several days. Fumigation conditions may persist for prolonged periods in deep layers of radiation fog.

It should be pointed out that the nocturnal inversion does not always break gradually, in the manner that results in the fumigation. It is likely that in some climates fumigation conditions occur almost daily, whereas in others they would occur only rarely.

For stack planning it is well to determine, if possible, the frequency, duration, and severity of fumigation as well as wind direction at the time they may occur. This, of course, requires suitable instrumentation, the most important component of which is continuous measurement of temperature at two or more levels above the ground.

Effect of Terrain on Wind

In evaluating a reactor site and estimating possible hazards to populated areas, the local conditions and effects of terrain must be carefully considered, in addition to the general problem of atmospheric diffusion since the local conditions can be the controlling factors in the distribution of hazardous materials. The following discussion is intended to point out certain typical effects of terrain as examples of important local factors. These examples emphasize the need in site evaluation for actual meteorological data for the particular area under consideration both on the surface and at several altitudes. For flat country without nearby hills or bodies of water, meteorological observations from another site within several miles might be suitable for complete site evaluation. In many situations it may be desirable, after a preliminary site survey and selection of an area for detailed considerations, to obtain data at the proposed location to determine the local effects. For large installations where atmospheric contamination is considered a major problem, meteorological equipment including a tower for determining wind and temperature at several altitudes may be required.

Local winds caused by differential heating of the terrain are often the dominant circulation over a site. Although valley-slope winds may occur over gently sloping terrain, they are most pronounced and frequent in mountainous areas. During the day the air over a slope may be warmer than the air at the same height over the valley. The rising of this warmer air creates a well defined wind up the slope. The reverse is true at night and the cooler air over the slope flows downward into the valley.

In the case of the sea (or lake) breeze, the strong heating of air over land during the daytime leads to an inflow of air at the surface from water to land which, in turn, causes a return circulation aloft from land to water. At night a reverse flow occurs. Principal among the factors which favor a sea breeze is a relatively high intensity of solar radiation; however, a weak general pressure gradient is required to permit formation of a true sea breeze. Thus in tropical regions one might expect the sea breeze to occur throughout the year, whereas in temperate latitudes the sea breeze is most frequent during spring and summer. Ordinarily the sea breeze begins early in the day with relatively low wind speeds and attains its maximum about 2:00 p.m. Its depth in temperate latitudes is usually less than 2000 ft. As the sea breeze penetrates inland, its speed and depth are reduced considerably. Speeds measured 10 to 15 miles inland may be about one-half the value at the coast line, and the depth is correspondingly reduced. The land breeze, the return flow from land to sea, begins about 2 hr after sunset and continues until about 2 or 3 hr after sunrise. The land breeze is not so well marked a wind phenomenon as the sea breeze, and the depth is only a few hundred feet.

There are other types of local winds caused by differential heating of the terrain. The heating of one side of a valley while the other side is relatively shady may produce overturning of the air in the valley.

A coastal location may be subjected to a combination of sea breeze and up-slope wind. The effects which produce local winds also combine with the effects of a large scale pressure system to increase or decrease wind speed and to alter wind direction.

Channeling may occur when the general wind pattern is across a valley but not at right angle to the valley. For example the region in which the Oak Ridge National Laboratory is located consists of a series of mountain ranges and valleys running northeast and southwest. Consequently, almost half of the time the winds are either northeasterly or southwesterly.

As the air mass moves over the terrain, there will be changes in the flow velocity due to changes in shape of the air mass. If the volume expands vertically in passing across a valley, the velocity is reduced at the floor of the valley. However, in passing over a hill the air mass is compressed and the wind speeds will be greater. A mass of cold air pouring through a mountain pass can cause strong winds due to compression of the volume. The "Santa Ana" wind of Southern California blowing from the desert into the Los Angeles rea, and the Columbia Gorge winds are examples of this phenomenon in he United States.

Wind Roses

Long periods of wind records may be summarized in a wind rose, such as he one shown in Fig. 3. The wind rose may be used to estimate the probaility that material may move in a particular direction and at a certain range

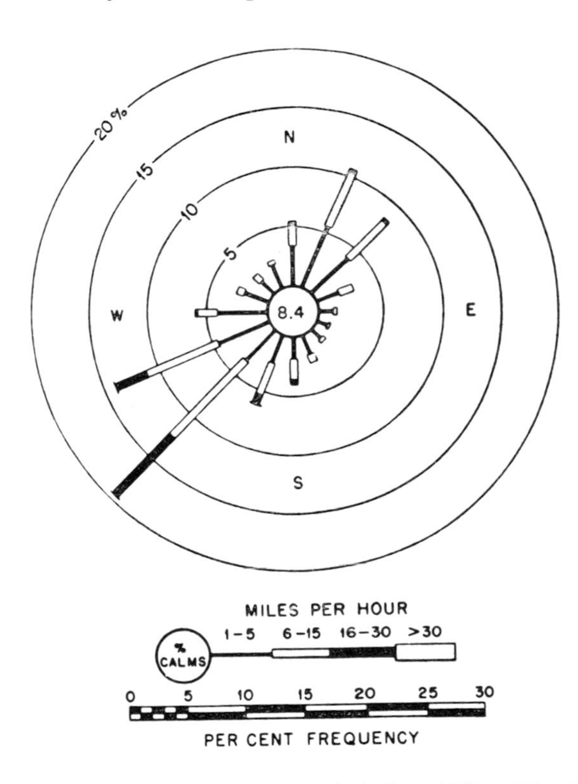

Fig. 3. Annual wind rose for two-year period, June 1950 to May 1952. Wind at 20 ft level. Radial lines show direction from which wind blows (NRTS). (Reproduced from "Meteorology and Atomic Energy," AECU–3066, U.S. Weather Bureau.)

of speed. These summaries may be constructed for specific elevations, seasons, time of day or meteorological conditions to provide data for specific problems.

The configuration of a wind rose integrates the results of air currents on a planetary scale, the passing of high- and low-pressure systems, and local winds. In standard practice, winds are shown as the direction from which they occur and are measured clockwise from north with north as 360° (or 0°). Thus a southwest wind is at the 225° position on the circle.

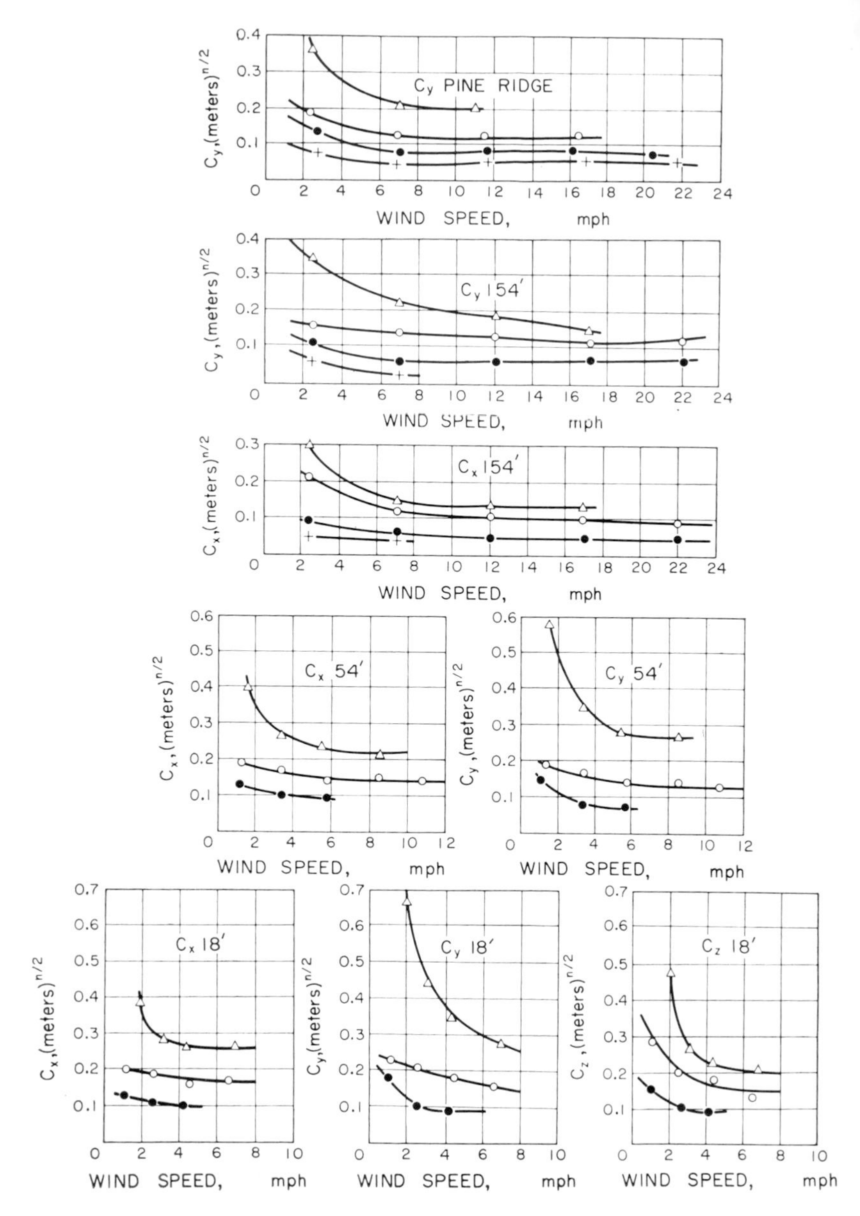

FIG. 4. Observed C_x, C_y and C_z vs. wind speed and stability (Holland). △, lapse ($\leqslant -1$°F/100 ft, n=0.15 to 0.20). 0, neutral (-0.5 to 0°F/100 ft, n=0.25) ●, moderate inversion ($\leqslant 2$°F/100 ft, n=0.30 to 0.40). +, large inversion ($>$ 2°F/100ft n=0.50). (Reproduced from "Meteorology and Atomic Energy," AECU–3066 U.S. Weather Bureau.)

DIFFUSION THEORY

There are several theories for predicting atmospheric diffusion and each theory has certain advantages and areas of application. The equations developed by Sutton are usually applied to reactor hazards problems. The experimental basis for this theory is limited to observations extending only a few miles. The equations developed by Roberts are applied to very long distances of the order of hundreds of miles. The developments in the following sections are based on O. G. Sutton's theory of atmospheric diffusion as described in References 1 and 4.

The concentration of radioactive material, χ, expressed as grams per cubic meter or curies per cubic meter, is to be calculated at any point x, y, z, downwind, crosswind, and vertically measured from a ground point beneath a continuous source and from the center of the moving cloud in the instantaneous case. Distance is expressed as meters. The source strength Q is expressed as grams or curies for the instantaneous source and as grams per second or curies per second for a continuous source. Sutton's theory defines coefficients C_x, C_y, and C_z, $(\text{meters})^{n/2}$ in the x, y, and z planes, respectively. The generalized diffusion coefficient C applies for isotropic turbulence, i.e. $C = C_x = C_y = C_z$. Time, t, is in sec; $\bar{u}$ is the mean wind speed in meters per sec; and h is the height of the source, the height of the plume, or the stack height.

Using a nondimensional parameter, n, associated with stability, the diffusion equation for a point source takes the form

$$\chi(x, y, z, t) = \frac{Q}{\pi^{3/2} C_x C_y C_z (\bar{u}t)^{3(2-n)/2}} \times \exp\left[-(\bar{u}t)^{n-2}\left(\frac{x^2}{C_x^2} + \frac{y^2}{C_y^2} + \frac{z^2}{C_z^2}\right)\right].$$

No deposition on the ground from the cloud is assumed in the derivation of this diffusion function.

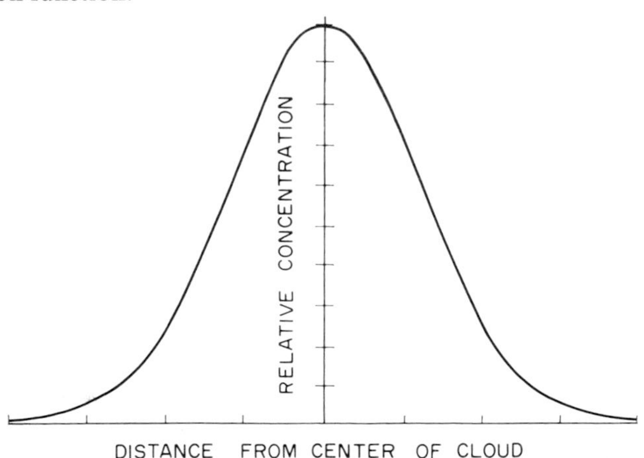

FIG. 5. Concentration across cloud width. (Reproduced from "Meteorology and Atomic Energy," AECU–3066, U.S. Weather Bureau.)

The diffusion coefficients, C_x, C_y, and C_z are a function of the air viscosity wind speed, and the stability parameter n. For a large lapse rate, the value of n may be as low as 0.15, whereas under a large inversion the value of n may be 0.50. Observed values of the diffusion coefficient as a function of wind speed and stability are shown in Fig. 4.

From the form of the diffusion equation it can be seen that the concentration of material varies across the width of a cloud or plume as shown in Fig. 5. The width of a cloud is usually expressed as the distance between points where the concentration is 0.1 of the value at the center of the cloud or plume.

By the process of integration with respect to one or more of the coordinate directions and time, other diffusion formulae for more complex sources may be built up. The sources most often encountered in practical applications are shown in Table I.

Table I

Source type	Application
Continuous point	Smokestack or vent
Continuous line	Array of stacks or vents
Instantaneous volume	Explosion

Another class of useful formulae involves certain geometrical properties of any of these results, namely the maximum concentration and its location and the plume width or height. Finally, a group of special results and extensions involving corrections for radioactive decay, ground surface deposition, diffusion in very unstable atmospheres, and many other important modifications can be deduced.

Fall-out, Wash-out and Rain-out

Fall-out is used to designate deposition during nonprecipitating weather and includes the effects of both gravitational settling and impaction. Rain-out is the removal due to association of the particulate or gaseous matter with precipitation prior to descent as in the case of debris clouds from nuclear weapons. The wash-out process is of greater interest and is applicable to airborne matter entirely below the bases of precipitating clouds.

It is noted in Reference 1 that with reasonable assumptions of rainfall rates, drop sizes, and collection efficiencies, high ground concentration of radioactivity may result from wash-out. It is almost always found that a maximum hazard can be assumed if sufficient of the radioactive material can be washed out of the traveling cloud and deposited on the ground. A cloud of fission products entering a heavy local shower could produce extreme ground concentrations by wash-out.

Radiation from Airborne Material

The exclusion distance formula discussed in Chapter 1 was derived from the assumption that the spread of the cloud was one seventh the travel with uniform distribution of material within the cloud. It was further assumed that the cloud was large as compared with the range of gamma radiation so that within the cloud, radiation was absorbed per unit volume at the same rate as emitted. It has been seen in the above section that the distribution of material inside the cloud is Gaussian and that the cloud size is a complex function of several parameters. Beta radiation has a range in air of only a few meters and therefore the external beta dose rate is proportional to the local concentration of beta emitting material. However, gamma radiation has a mean free path in air of the order of 300 m and in many problems of interest the size of the cloud may not be large compared with the range of gamma radiation. Also there are effects of absorption in air, multiple scattering and radioactive decay that may be important. The mathematical solution of the exposure problem by multiple integration of the relations over the range of variable involved becomes quite complex. Solutions to these difficult problems for typical situations have been arrived at by the Scientific Services Division, U.S. Weather Bureau and are presented in both analytical and graphical form in References 1, 2 and 3 and these references should be consulted for the detailed development of these complex problems.

Reliability of Diffusion Predictions

From several comparisons of calculated concentration values with experimental data it appears that over relatively short ranges of the order of 1600 ft, it can be expected from Sutton's formula that 68 per cent of the observed concentration values will lie within ± 5 times the values predicted by Sutton's theory and 95 per cent lie within an order of magnitude. From experience with release of radioactivity it appears that for very long trajectories errors greater than an order of magnitude may be expected.

Because of the range of values that the variables and parameters assume, numerical evaluations of the diffusion equations are fairly tedious. Furthermore, since in most cases the use of tables of logarithms cannot be avoided, direct numerical calculations will ordinarily produce a degree of accuracy far greater than that warranted by the diffusion theories in themselves. Therefore, graphical solution of the diffusion equations is particularly convenient and sufficiently accurate for most purposes.

NOMOGRAMS FOR GRAPHICAL SOLUTIONS*

Diffusion Equations

Because of the uncertainties always present in the values of the parameters involved and the complication of the calculations, a graphical solution of the diffusion equations is particularly useful. A nomogram for the solution of the Sutton equations is presented in Fig. 6.

* From Meterorology and Atomic Energy, U.S. Weather Bureau. This arrangment of the material is reproduced from Reference 5.

An example of the use of the nomogram for the solution of a typical problem follows.

Problem: Find the relative dilution χ/Q, at a point at ground level beneath the axis of a plume, 1000 m downwind from a continuously emitting stack of effective height 100 m.

The parameters have the values $n = 0.5$, $C^2 = 0.2$ and $U = 2$ m/sec.

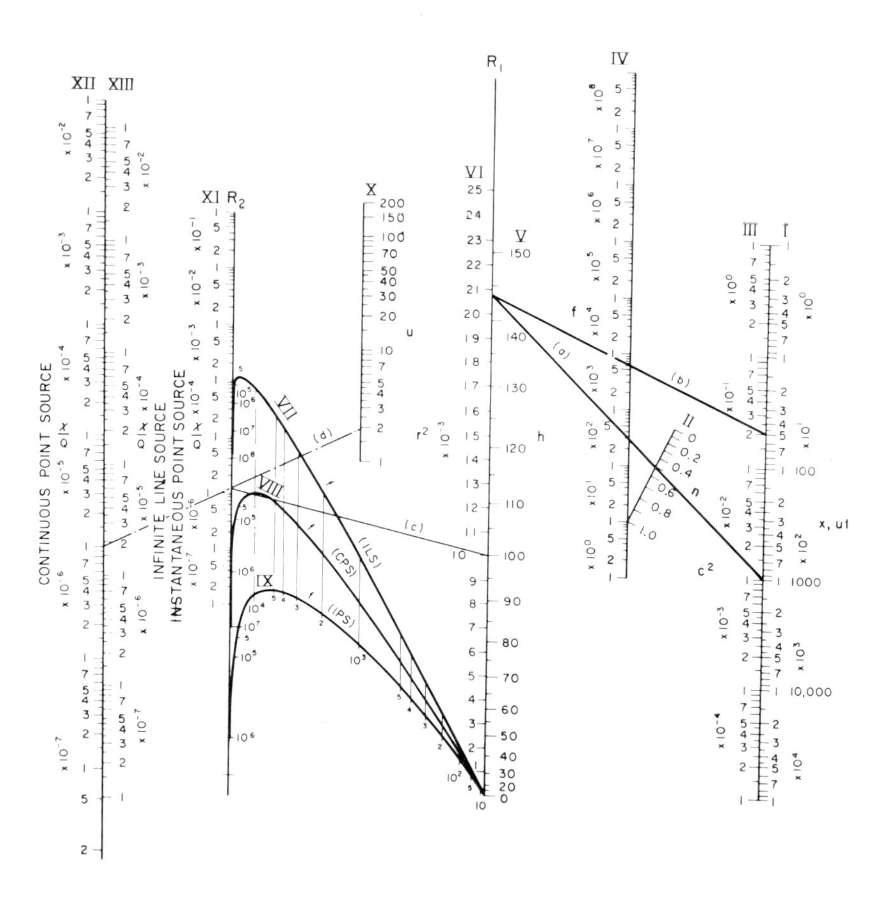

FIG. 6. Nomogram for atmospheric diffusion calculations. (Reproduced from "Meteorology and Atomic Energy," U.S. Weather Bureau.)

Solution using Fig. 6.

1. Align $X = 1000$ m, scale I, with n, scale II, locating a point on reference scale R_1 (light line (a)).
2. Align this point with C^2, scale III, locating a value f on scale IV (line (b), $f = 6 \times 10^3$).

Table II

Program for Solving Meteorological Diffusion Problems on Fig. 6.

Source type	Quantity sought	Symbol	Necessary alignments (symbols refer to scales on Fig. 7.6)	Remarks
Instantaneous point Source at the ground	Ground concentration	χ	R_1-II-I; R_1-IV-III; XI-IX-VI	Use $r^2=x^2+y^2$ scale VI*
	Concentration at height h	χ	R_1-II-I; R_1-IV-III; XI-IX-VI	Use $r^2=x^2+y^2+h^2$, scale VI*
Instantaneous point source at height h	Ground concentration	χ	R_1-II-I; R_1-IV-III; XI-IX-VI	Use $r^2=x^2+y^2+h^2$, scale VI*
	Distance to maximum ground concentration	d_{max}	R_2-(t) IX-V; R_1-IV-III; R_1-II-I	**
	Maximum ground concentration	χ_{max}	XI-(t) IX-V	**
	Integrated ground concentration	TID	R_1-II-I; R_1-IV-III; R_2-VIII-VI; XII-R_2-X	
Continuous point source at the ground	Downwind concentration at height h	χ	R_1-II-I; R_1-IV-III; R_2-VIII-V; XII-R_2-X	*
	Ground concentration at lateral distance y from plume axis	χ	R_1-II-I; R_1-IV-III; R_2-VIII-V; XII-R_2-X	Use $h=y$, scale V*
Continuous point source at height h	Maximum ground concentration	χ_{max}	R_2-(t) VIII-V; XII-R_2-X	**
	Distance to maximum concentration	d_{max}	R_2-(t) VIII-V; R_1-IV-III; R_1-II-I	**
	Downwind ground concentration	χ	R_1-II-I; R_1-IV-III; R_2-VIII-V; XII-R_2-X	*, †
	Ground concentration at a lateral distance y from plume axis	χ	R_1-II-I; R_1-IV-III; R_2-VIII-VI; XII-R_2-X	Use $r^2=y^2+h^2$, scale VI*
Continuous infinite crosswind line source at height h	Downwind ground concentration	χ	R_1-II-I; R_1-IV-III; R_2-VII-V; XIII-R_2-X	*
	Maximum ground concentration	χ_{max}	R_2-(t) VII-V; XIII-R_2-X	**
	Distance to maximum concentration	d_{max}	R_2-(t) VII-V; R_1-IV-III; R_1-II-I	**

* Other problems than the ones listed here may be solved on the nomogram. See Reference 1 for a more nearly complete list. In particular, the source strength Q can be found in all problems marked with an asterisk if the concentrations are known, since the quantity found from the nomogram is χ/Q.

** The symbol (t) means "find a tangent to."

† A problem of this type is given as a text example.

(Reproduced from Meteorology and Atomic Energy, U.S. Weather Bureau.)

3. Align the stack height h, scale V with the value of f obtained in step 2, on scale VIII, locating a point on reference scale R_2, (line (c)).
4. Align this point with U, scale X and find $\chi/Q = 1 \times 10^{-5}$, scale XII (line (d)).

If the lateral ground level concentration at a given distance y from the point below the cloud axis had been required, the steps would have been the same, except that instead of using h, scale V, the value $r^2 = y^2 + h^2$ on scale VI would have been used.

To calculate the maximum concentration χ_{max}, find a line through h, scale V, tangential to scale VIII, locating a point on R_2. Then align this point and U, scale X and read χ_{max}/Q from scale XII. The product of this and the emission rate is the desired maximum ground level concentration.

Table II lists the alignments, in the proper order, by means of which the most commonly occurring diffusion problems may be solved. At each step there are aligned points on three scales, two of which are known, either from the conditions of the problem or from a previous step. The unknown quantity in the final step is in each case the quantity desired.

The nomogram may also be used as follows to find the eddy peak ground level concentration for looping conditions. Obtain the value of f in the same way as above but use $(X^2 + h^2)^{1/2}$ instead of X (if these differ appreciably) on scale I. Then follow steps 3 and 4 but use $h = 0$ on scale V.

To solve the instantaneous volume source case use $X = X_0 + Ut$ in the instantaneous point source case.

Gamma Dose Nomogram—Power Excursion

A nomogram for obtaining the gamma ray dose at ground level is given in Fig. 7. Each particle of the cloud is assumed to travel with the average wind velocity and the distribution of particles in the cloud is assumed to obey the above concentration equation for isotropic diffusion. The nomogram is designed to give dosage estimates accurate to ± 20 per cent.

(a) *Nomogram Instructions, Point Source*

To calculate the dosage at ground level resulting from the release of all fission products of a power excursion of integrated energy of Q (MW-sec) at a distance D (m) from the origin given the height of rise h (m), the wind speed U (m/sec) and the Sutton diffusion parameters C (m)$^{n/2}$ and n, the procedure is as follows:

1. Place a straightedge through D on the $D + X_0$ scale and n. Mark the point of intersection on the $D^{(2-n)/2}$ scale.
2. Place the straightedge through this point and C. Mark the intersection on the σ scale.

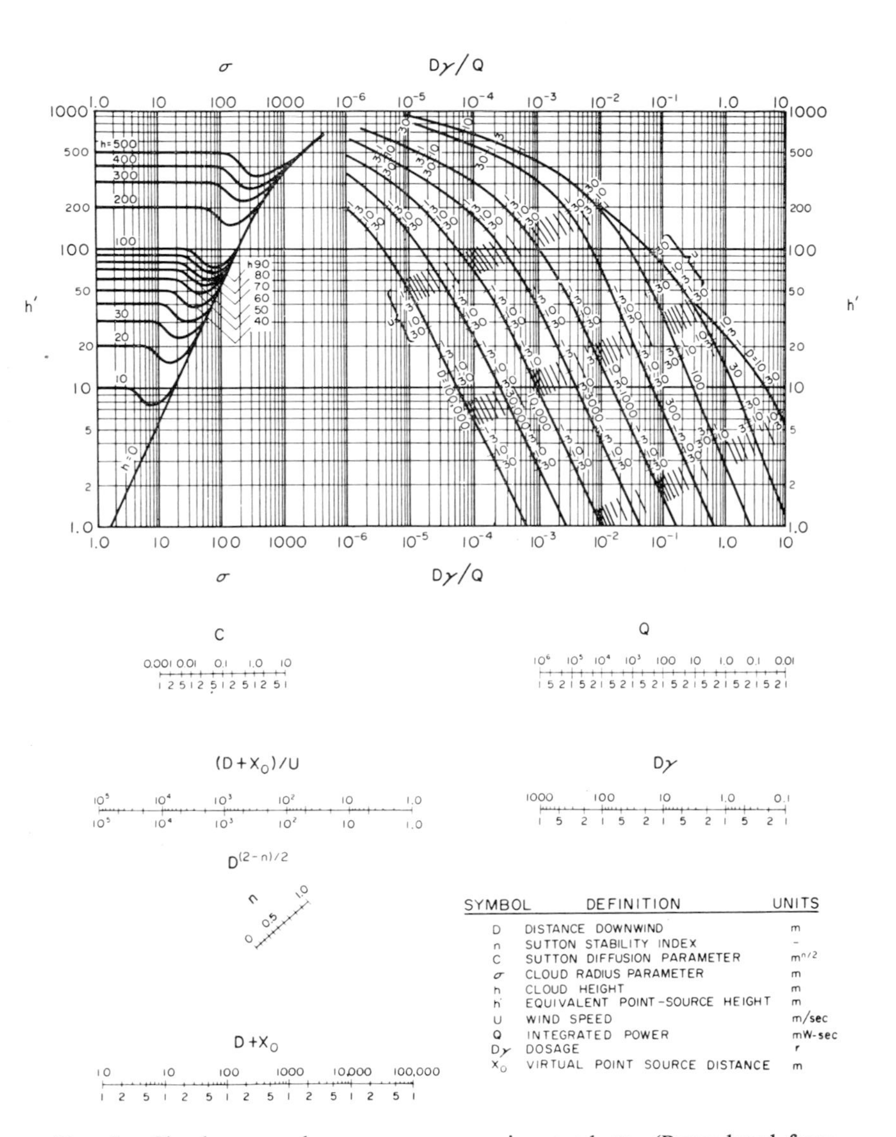

FIG. 7. Cloud gamma dosage, power excursion products. (Reproduced from "Meteorology and Atomic Energy," AECU–3066, U.S. Weather Bureau.)

3. Align the straightedge parallel to the vertical axis through this point on the σ scale and mark the intersection on the curve representing h (or a suitably interpolated point). The ordinate is h'.
4. Align the straightedge horizontally through this h' value. Mark the intersection with the curve representing D or a suitably interpolated point. (The abscissa is the dosage for a 1 MW-sec source and a 1 m/sec wind speed.)
5. Displace the point found in step 4 along the line of constant D, a distance equal to the distance between the 1 m/sec mark and that representing the wind speed U. (It will be noted that there are several differing wind speed corrections for each D curve. The wind correction nearest to the point found in step 4 should be used, interpolating if necessary.) Lay the straightedge parallel to the vertical axis through this point to obtain D_γ/Q.
6. Lay the straightedge through D_γ/Q from step 5 and Q on the Q scale; the intersection on the D_γ scale is the gamma dosage in roentgens.

Note that the assumption is made that all the fission products escape to the air. The result may be scaled to conform to the postulated percentage release.

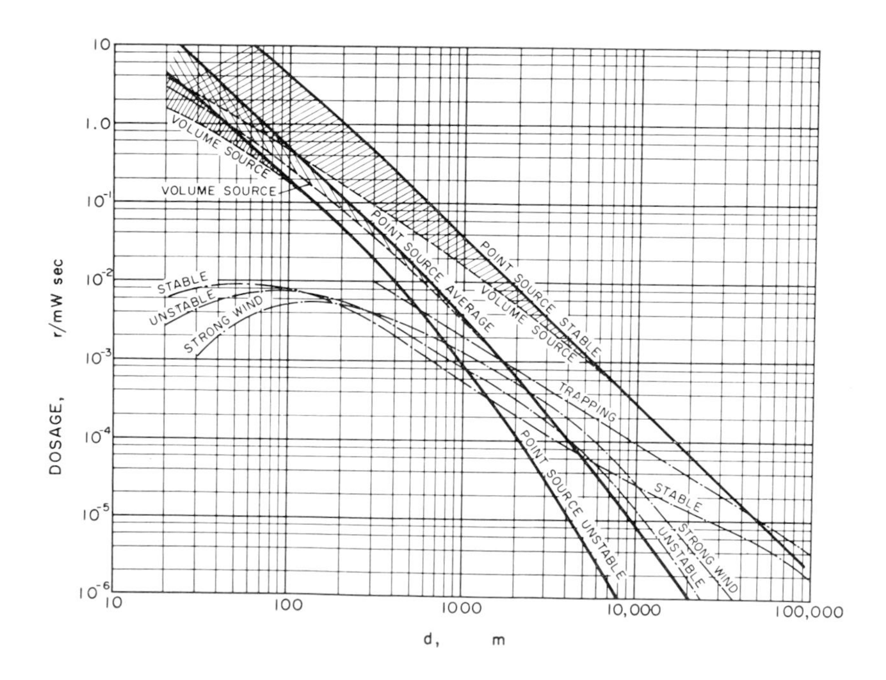

FIG. 8. Gamma dosage variability, power excursion. —, $h=0$ (point source). —, —, $h=200$ m. (point source).

b) *Nomogram Instructions, Virtual Source*

Since about 90 per cent of the cloud is contained within a sphere of liameter 4.3σ, the virtual source correction X_0 can readily be determined. ay the straightedge through the point on the σ scale representing $\frac{1}{4}$ the nitial cloud diameter and follow steps 1 and 2 in reverse to find X_0 on the) + X_0 scale or X_0/U on the $(D + X_0)/U$ scale, respectively. The value of X_0 thus attained can then be added to the value of D for which the dosages re required. The addition is made only at step 1. The value of d used in tep 4 must be measured from the actual origin.

An idea of the values to be expected under a plausible variety of meteoro- ogical conditions can be obtained from Fig. 8. These values were obtained rom the nomogram, Fig. 7 using the meteorological parameters in Table III.

Table III

Meteorological Parameters for Sample Cases

	n	C, $m^{n/2}$ $h = 0$	50	200	500	$\bar{u}$, m/sec $h = 0$	50	200	500
Average conditions	0.25	0.20	0.15	0.10	0.08	3	5	7	9
Stable conditions	0.50	0.05	0.03	0.02	0.01	1	3	6	9
Unstable conditions	0.20	0.50	0.20	0.15	0.10	7	10	12	13
Strong wind conditions	0.25	0.20	0.15	0.10	0.08	15	22	30	35
Trapping		$(\sigma = 0.75h)$					5	7	9

(a) *Nomogram Instructions, Steady Power Fission Products*

Given:

steady power P in kW
Distance d downward from origin in m
Height of rise of cloud h in m
Wind speed U in m/sec
Generalized diffusion coefficient C in $m^{n/2}$
Virtual point source distance X_0 in m
Stability parameter n

Procedure:

1. Same as for power excursion case above.
2. Same as for power excursion case above.
3. Same as for power excursion case above.
4. Same as for power excursion case above.
5. Lay the straightedge parallel to the vertical axis through the point on the D scale (step 4) and mark the intersection on the 1 m/sec wind speed line.

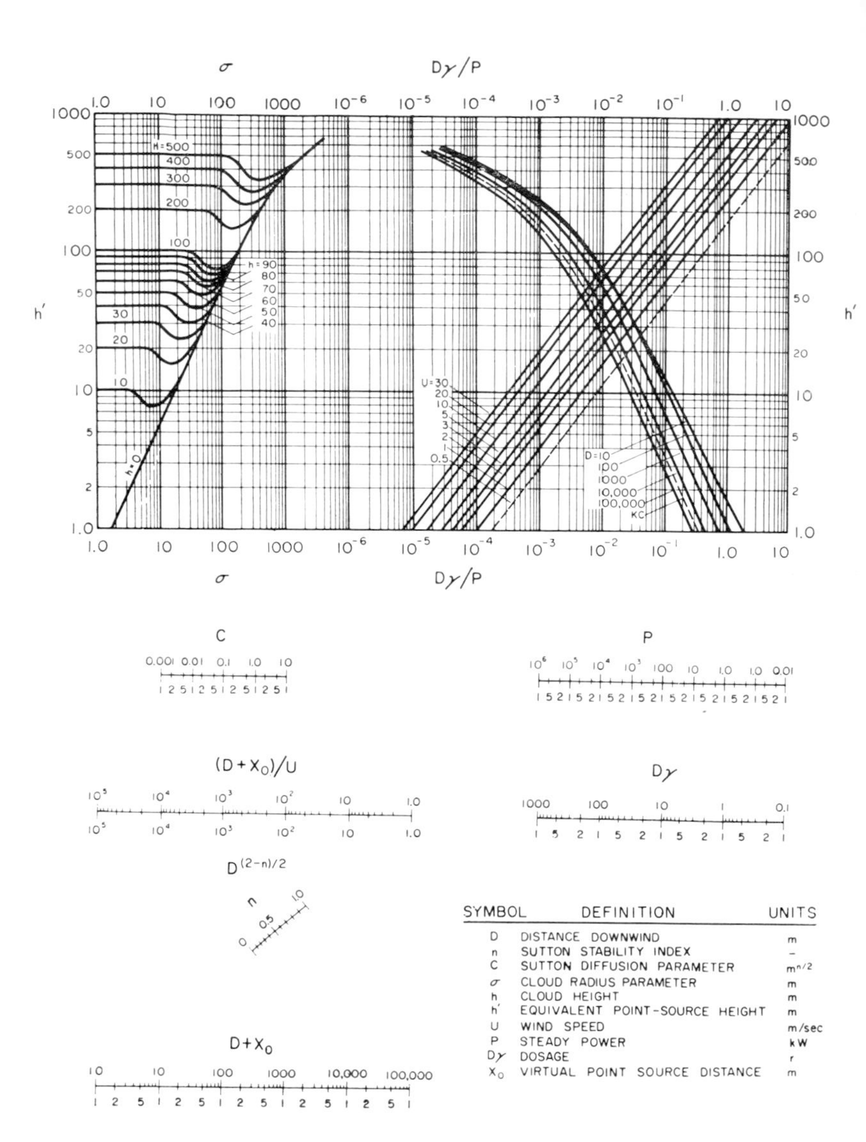

FIG. 9. Cloud gamma dosage, steady power products (Holland). (Reproduced from "Meteorology and Atomic Energy," AECU–3066, U.S. Weather Bureau.)

6. Lay the straightedge horizontally through this point and mark the intersection with the line representing the wind speed U.
7. Lay the straightedge vertically through this point and mark the intersection with the D_γ/P axis.
8. Lay the straightedge through D_γ/P from step 7 and P; the intersection on the D_γ scale is the gamma dose in roentgens.

(b) *Nomogram Instructions, Specific Radioisotopes*

The dashed curve marked KC in Fig. 9 is included to permit computation of the gamma dosage of a cloud of any specific radioisotope whose gamma

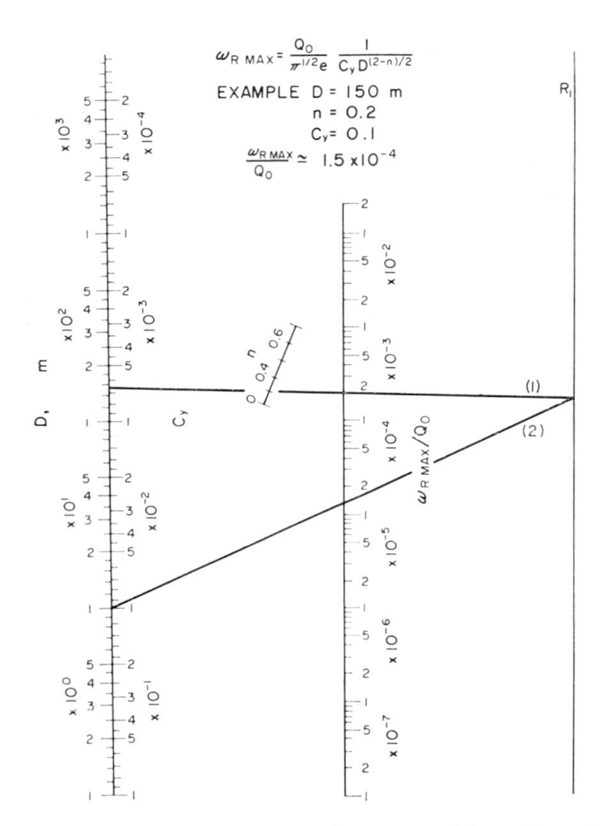

FIG. 10. Nomogram for computing maximum deposition. (Reproduced from Meteorology and Atomic Energy, AECU–3066, U.S. Weather Bureau.)

energy is of the order of 0.5 to 1.0 MeV and whose half-life is large compared to the cloud passage time (e.g. 1 hr or more). In this case follow the same procedure as that for the steady power case up to step 4 then use the dashed curve marked KC to obtain the dosage per kilocurie of release for $E = 0.7$

MeV and $U = 1$ m/sec. To obtain the gamma dosage in roentgens, this should be multiplied by

$$\frac{Q_i E \exp\left(\frac{-0.693D}{UT}\right)}{0.7U}$$

where

Q_i = initial source strength in kcur
E = gamma ray energy in MeV
T = half-life in sec
D = distance of the receptor in m
U = wind speed in m/sec

(c) *Nomogram Instructions, Continuous Release*

The *KC* curve of Fig. 9 may be used in this same manner for continuous source problems (except for the continuous emission of fission products as they are formed in which case Fig. 7 is applicable). For these problems Q is the rate of emission in kc/hr and D_γ is the gamma dose rate in roentgens/hr. Q may require a further correction for decay if there is an appreciable holdup time prior to release to the air.

(d) *Nomogram Instructions, Dosage Contours*

If it is desired to compute dosages for locations other than directly under the axis of the cloud ($y = 0$), these values can be obtained for any distance y from the x-axis by substituting $\sqrt{(y^2 + h^2)}$ for h on the nomogram. Computation of a sufficient number of these values for various distances will permit the ready construction of isodose lines.

Check List of Meteorological Considerations Pertinent to Reactor Hazards Analysis

The check list below is intended to indicate those things of a meteorological nature to which attention should be given in a hazards analysis report. Not all hazards analyses would, or should, contain all the elements listed below. The type of reactor, its design power level, the site locations, etc., must be considered in fixing the scope of a hazards analysis. Only a first presentation of a high power reactor would require treatment of all items. Thus, analyses for reactors at a site for which previous reports have been prepared may well refer to these reports for climatology, site description, etc., and deal with only items or hazards unique to the device under consideration.

The items listed in the outline should not be considered separately, but rather should be integrated to whatever extent is necessary into the entire hazards analysis. Obviously, a health physicist and reactor engineer will have to supply much of the required information.

For safety the most pessimistic accident is given most attention.

(1) What will be the nature of the radioactive cloud before the diffusion begins?

(a) From where will the cloud enter the atmosphere?
  (i) A stack?
  (ii) Building openings or outlets?
  (iii) A demolished building or enclosure?
(b) What will be the type of source?
  (i) Instantaneous?
  (ii) Continuous?
(c) What (and how much) radioactivity may be released? (If the source is not instantaneous, give rate of emission.)
  (i) Fission products (mainly short- or long-lived)?
  (ii) Irradiated material?
(d) What will be the physical composition of the cloud, especially with regard to particulates which will fall out of it?
(e) What will be the initial temperature of the cloud?
(f) What will be the initial volume of the cloud when it reaches ambient pressure and temperature?
(g) What will be the approximate shape of the cloud just after the disaster? How will its shape change with time? What is the initial shape assumed for the purpose of making diffusion estimates?
(h) What distribution of material is assumed in the cloud prior to diffusion (uniform, Gaussian, etc.)?
(i) How high will the cloud rise in the atmosphere? (What is the "effective height" at which diffusion is assumed to begin?)

(2) What are the meteorological conditions chosen for estimating the effects of diffusion?

(a) Wind speeds?
(b) Diffusion parameters?

(3) How will the cloud diffuse in the atmosphere?

(a) In the case of the instantaneous source.
  (i) What will be the approximate dimensions of the cloud at various distances? What will be the changes in concentrations near the ground as it moves outward from the source?
  (ii) Assuming no radioactive decay, what will be the maximum ground concentrations along the line traversed by the center of the cloud?
  (iii) Considering wind speed, how will radioactive decay affect dosages at the ground?

(b) In the case of the continuous point source the treatment may be similar to the instantaneous point source except that exposure might be figured on the basis of the duration of the emission rather than on the time required for cloud passage.

(4) How significant is fallout and the scavenging action of precipitation?

(a) What will be the hazard from particulates at various distances as the radioactive cloud passes?

(b) How will fall-out affect concentrations and dosages near the ground?

(c) How will the scavenging action of precipitation affect concentrations and dosages near the ground?

(d) What will be the ground deposition?

(i) Due to fall-out?

(ii) Due to the scavenging action of precipitation?

(5) How representative are available meteorological data?

(a) What is the period of record? (At least 5 years of data should be considered, if available.)

(b) If the observing station is not on the site, how does its location differ from that of the site?

(c) What significant differences, if any, exist between meteorological conditions at the observing station and the site?

(6) What are the significant meteorological conditions?

(a) Wind

(i) What is the surface wind direction and speed distribution annually (annual wind rose)?

(ii) What are the significant diurnal differences? (Are there local winds, such as sea breezes, mountain winds, etc.)

(iii) What are the significant seasonal differences?

(iv) What is the speed and direction distribution during precipitation?

(v) What is the speed and direction distribution during stable and unstable conditions?

(vi) What are the significant differences between surface winds and those at the height a radioactive cloud may occur?

(vii) Will a cloud trajectory away from the site rise, descent, curve, or follow a straight line because of topography?

(b) Stability

(i) What is the frequency of unstable and stable conditions?

(α) Annually?

(β) Seasonally?

(ii) How is stability affected by local conditions (i.e. valley effects, West Coast subsidence inversion)?

(iii) What is the average and maximum duration of inversion conditions?
(iv) What significant facts are known about the depth and intensity of inversions?

(c) Precipitation
(i) What are the normal monthly and annual totals, and what extremes have occurred?
(ii) Are there significant variations?
(α) Seasonally?
(β) Diurnally?
(iii) What is the monthly and annual snowfall?
(iv) What are the greatest amounts of precipitation recorded for various internals, such as one hr, one day, etc.?
(v) What is the average duration of precipitation?
(vi) What is average duration of the intervals between occurrences of precipitation?

(d) Severe weather
(i) What are the possibilities of floods?
(ii) What are the possibilities of wind storms (i.e., severe thunderstorms, gusts, tornadoes, and hurricanes)?

(e) Miscellaneous (where significant for special designs)
(i) What natural atmospheric dustiness conditions prevail?
(ii) Do rapid temperature changes occur which might make automatic control difficult?

7) Are meteorological conditions taken into consideration in the scheduling f operations? Are adverse meteorological conditions being avoided by cheduling with respect to seasonal condition, time of day, etc.? Will schedulng be done on the basis of climatology, forecasts, or the actual weather onditions as observed at the time of the operation?

REFERENCES

. Meteorology and atomic energy, Report AECU 3066, U.S. Department of Commerce Weather Bureau for the U.S. Atomic Energy Commission.
. HOLLAND, J. Z., Radiation from clouds of reactor debris, *Proc. 1st United Nations Int. Conf. on the Peaceful Uses of Atomic Energy*, Vol. 13, United Nations, New York.
. WEXLER, H., MACHTA, L., PACK, D. H., and WHITE, F. D., Atomic energy and meteorology, *Proc. 1st United Nations Int. Conf. on the Peaceful Uses of Atomic Energy*, Vol. 13, United Nations, New York.
. SUTTON, O. G., *Micrometeorology*, McGraw-Hill, New York, 1953.
. SOUDAK, H., Hazards to surrounding areas—meteorological considerations, Chapter 9, AEC Safety Monograph.

CHAPTER 8

EXTENT OF POSSIBLE DAMAGES

INTRODUCTION

THE possible damages that could result from a malfunction have been th subject of study and speculation since the construction of the first nucle reactor was planned. This subject has been highly controversial because, i the limit of the worst possible assumptions, the damages that most nucle installations could cause are very large when expressed in terms of possib casualties, damage to the installation itself, loss of crops and loss of use c land and buildings. Estimates of the probable extent of damages that wou be expected to result from the operation of a large number of nuclear installa tions for a period of say 20 years are difficult to make and justify because c the limited experience with reactor operation and uncertainty as to whethe our favorable experience to date is typical or the result of extreme caution i the design and operation. Many studies and records of discussions of thes subjects can be found in the literature.

MAXIMUM PERMISSIBLE DOSE

Several organizations have issued recommendations on the maximur permissible radiation exposures to man. Those of the International Commis sion on Radiation Protection (ICRP) are followed closely by the Nationa Committee on Radiation Protection and Measurements (NCRP), of th United States National Bureau of Standards. Such recommendations of th NCRP are usually published as handbooks of which the National Bureau c Standards Handbook 69 "Maximum Permissible Body Burdens and Max mum Permissible Concentrations of Radionuclides in Air and Water fo Occupational Exposure" is of interest. In the United States, the Atomi Energy Commission establishes standards for protection against radiatio hazards arising out of activities under license issued by the AEC. The AE regulations are published as Title 10 of the Code of Federal Regulation The International Atomic Energy Agency (IAEA) is empowered to provid for the application of standards of safety for protection against radiation t its own operations and to operations making use of assistance provided by or with which it is directly associated. The IAEA publication "Safe Hand ling of Radioisotopes" follows the recommendations of the Internationa

Commission for Radiological Protection. The International Labor Organization (ILO) also has issued standards on protection against ionizing radiation including a manual of Industrial Radiation Protection.

The exposure limits given in the following paragraphs have been taken from the publication of the National Bureau of Standards as representing typical values. It should be recognized that these limits may be different from those in the U.S. Atomic Energy Commission radiation standards. Also radiation standards are being continuously reviewed and are occasionally revised, usually downward.

The NCRP has promulgated four basic rules and recommendations concerning exposure to ionizing radiation as outlined in NBS Handbook 69:

a. *The Maximum Permissible Dose* (MPD) to the most critical organs, accumulated at any age, shall not exceed 5 rems multiplied by the number of years beyond 18, and the dose in any thirteen consecutive weeks shall not exceed 3 rems. This exposure limit applies to radiation workers and not to the general public.

b. *The Maximum Permissible Dose* to the skin of the whole body shall not exceed 10 rems multiplied by the number of years beyond the age 18, and the dose in any thirteen consecutive weeks shall not exceed 6 rems. This rule applies to radiation of low penetrating power. The dose to the lens of the eye shall be limited by the dose to the head and trunk. The MPD for hands and forearms, feet and ankles, shall not exceed 75 rems/yr, and the dose in any thirteen consecutive weeks shall not exceed 25 rems.

c. The permissible levels from internal emitters will be consistent as far as possible with the age-proration and dose principles given above. Control of the internal dose will be achieved by limiting the body burden of radioisotopes. In the table are given selected values of *maximum permissible body burdens* and *maximum permissible concentrations* of radionuclides in air and water for occupational exposure.

An accidental or emergency dose of 25 rems to the whole body or a major portion thereof occurring only once in the lifetime of the person, need not be included in the determination of the radiation exposure status of that person. Radiation exposure resulting from necessary medical and dental procedures need not be included in the determination of the radiation exposure status of the person concerned.

The radiation of radioactive material outside a controlled area, attributed to normal operations within the controlled area, shall be such that it is improbable that any individual will receive a dose of more than 0.5 rem in any one year from external exposure. The maximum permissible average body burden of radionuclides in persons outside the controlled area and attributed to the operations within the controlled area shall not exceed one-tenth of that for radiation workers.

ACUTE NUCLEAR EXPOSURE†

It has long been known that excessive exposure to nuclear radiations, suc as X-rays, alpha and beta particles, gamma rays, and neutrons, which ar capable of producing ionization either directly or indirectly, can cause injur to living organisms. As has been mentioned above, after the discovery c X-rays and radioactivity, towards the end of the 19th century, serious an sometimes fatal exposure to radiation was sustained by radiologists befor the dangers were realized. Before the development of nuclear weapons radiation injury was a rare occurrence and relatively little was known of th phenomena associated with radiation sickness.

The harmful effects of radiation appear to be due to ionization (and excita tion) produced in the cells composing living tissue. As a result of ionizatior some of the constituents which are essential to their normal functioning, ar damaged or destroyed. In addition, the products formed may act as ce poisons. Among the observed consequences of the action of nuclear (o ionizing) radiations on cells is breaking of the chromosomes, swelling of th nucleus and of the entire cell, destruction of cells, increase in viscosity of ce fluid and increased permeability of the cell membrane. In addition, th process of cell division is delayed by exposure to radiation. Frequently, th cells are unable to undergo division, so that the normal cell replacemen occurring in the living organism is inhibited.

It has been established that all radiations capable of producing ionizatio (or excitation) directly, e.g., alpha and beta particles, or indirectly, e.g. X-rays, gamma rays and neutrons, can cause radiation injury of the sam general type. However, although the effects are qualitatively similar, th various radiations differ in the depths to which they penetrate the body an in the degree of injury corresponding to a specified amount of radiatio absorption. This difference is (partly) expressed by means of the *relativ biological effectiveness* (or RBE). For beta particles, the RBE is close t unity; this means for the same amount of energy absorbed in living tissue beta particles produce about the same extent of injury within the body as d X-rays or gamma rays. The RBE of alpha particles from radioactive source is variously reported to be from 10 to 20, but this is believed to be too larg in most of the cases. Fast neutrons have an RBE for acute radiation injur of about 1.7, but this value is appreciably larger where the formation o opacities of the lens of the eye (cataracts) are concerned. In other words neutrons are more effective than other nuclear radiation in causing cataracts

The effects of nuclear radiation on living organisms depend not only on th total dose, that is, on the amount absorbed, but also on the rate of absorption i.e., on whether it is acute or chronic, and on the region and extent of the bod

† Material for this section has been taken from *The Effects of Atomic Weapons*, SAMUE GLASSTONE, Editor, U.S. Atomic Energy Commission.

exposed. A few radiation phenomena such as genetic effects, apparently depend only upon the total dose received and are independent of the rate of delivery. In other words, the injury caused by radiation to the gene cells is cumulative. In the majority of instances, however the biological effect in a given dose of radiation decreases as the rate of exposure decreases. Thus, to cite an extreme case, 700 roentgens in a single dose would probably be fatal, if the whole body were exposed, but it would not cause death or have a noticeable external effect, if applied more or less evenly over a period of thirty years. A skin exposure dose of 700 roentgens of X-rays will cause a certain degree of reddening, if administered locally to a small area over a period of one hour. However, the same exposure of 700 roentgens over the entire body, would probably be fatal. Different portions of the body show different sensitivities to radiation although there are undoubtedly variations of degree among individuals. In general, the most radiosensitive parts include the lymphoid tissue, bone marrow, spleen, organs of reproduction, and gastrointestinal tract. Of intermediate sensitivity are the skin, lungs, kidney and

Table I

Expected Effects of Acute Whole-Body Radiation Doses

Acute dose (roentgens)	Probable effect
0 to 50	No obvious effect, except possibly minor blood changes.
80 to 120	Vomiting and nausea for about 1 day in 5–10 per cent of exposed personnel. Fatigue but no serious disability.
130 to 170	Vomiting and nausea for about 1 day, followed by other symptoms of radiation sickness in about 25 per cent of personnel. No deaths anticipated.
180 to 220	Vomiting and nausea for about 1 day, followed by other symptoms of radiation sickness in about 50 per cent of personnel. No deaths anticipated.
270 to 330	Vomiting and nausea in nearly all personnel on first day, followed by other symptoms of radiation sickness. About 20 per cent deaths within 2 to 6 weeks after exposure; survivors convalescent for about 3 months.
400 to 500	Vomiting and nausea in all personnel on first day, followed by other symptoms of radiation sickness. About 50 per cent deaths within 1 month; survivors convalescent for about 6 months.
550 to 750	Vomiting and nausea in all personnel within 4 hours from exposure, followed by other symptoms of radiation sickness. Up to 100 per cent deaths; few survivors convalescent for about 6 months.
1000	Vomiting and nausea in all personnel within 1–2 hr. Probably no survivors from radiation sickness.
5000	Incapacitation almost immediately. All personnel will be fatalities within 1 week.

liver, whereas muscle and full-grown bones are the least sensitive. From experiments with animals, as well as from observations made from the effect of atomic weapons on individuals, the expected effects of acute whole-body radiation doses have been estimated as shown in Table I.

There is considerable variation among individuals in the effects of acute whole-body radiation. This difference in response to radiation by individuals is brought about by the fact that not all members of a group of human beings react in the same manner. For example, only 20 per cent of those exposed would be expected to succumb to an acute dose in the vicinity of 300 rems. The other 80 per cent will suffer from radiation sickness but will probably recover. It appears from recent studies that there is probably no single value for the median lethal dose that applies under all conditions. However, at present the value of 450 rems has been adopted as a reasonable average value of the median lethal dose. It is also usually assumed that for exposures below 25 rems no effects can be observed clinically.

INTERNAL SOURCES OF RADIATION†

Whenever radioactive material is released into the lower atmosphere there is a possibility that such material will enter the body through the digestive tract due to consumption of food and water contaminated with fission products, through the lungs by breathing air containing particulate material or through wounds or abrasions. A very small amount of radioactive material present in the body can cause considerable injury since radiation exposure of various organs and tissues from internal sources is continuous and further the body tissues in which injury may occur are near the source of radiation and not shielded from it by intervening materials. This is of particular importance with alpha and beta particles which cannot reach sensitive organs, except the outer layers of the skin, if originating outside the body. But, if the sources, e.g. plutonium (alpha-particle emitter) or fission products (beta-particle emitters) are internal, the particles can dissipate their entire energy within a small, possibly sensitive, volume of body tissue, thus causing considerable damage. This situation is sometimes aggravated by the fact that certain chemical components tend to concentrate in specific cells or tissues, some of which are highly sensitive to nuclear radiation. The fate of a given isotope which has entered the blood stream will depend upon its chemical nature. Radioisotopes of an element which is a normal constituent of the body will follow the same metabolic processes as normally occurring inactive isotope of the same element. This is the case, e.g. with iodine which tends to concentrate in the thyroid gland. An element not usually found in the body, except

† Material for this section has been taken from *The Effects of Atomic Weapons*, SAMUEL GLASSTONE, Editor, U.S. Atomic Energy Commission and *Selected Materials on Employee Radiation Hazards and Workmen's Compensation*, Joint Committee on Atomic Energy, Congress of the United States, February 1959.

perhaps in minute traces, will behave like one with similar chemical properties that is normally present. Thus, among the fission products, strontium and barium, which are similar chemically to calcium are largely deposited in the calcifying tissue of the bone. The radioisotopes of the rare earth elements, e.g. cerium, which constitute a considerable portion of the fission products, and plutonium, are also "bone-seekers". All bone-seekers, are potentially very hazardous because they can injure the sensitive bone marrow where many blood cells are produced.

In order to constitute an internal radiation source, the active materials must gain access to the circulating blood, from which they can be deposited in the bones, liver, etc. While the radioactive substances are in the lungs, stomach, and intestines, they are, for all practical purposes, an external, rather than an internal, source of radiation. The extent to which fall-out contamination can get into the blood stream will depend upon two main factors: the size of the particles and the solubility in the body fluids. Elements which do not tend to concentrate in any particular part of the body are eliminated rapidly by natural processes.

If other things, e.g. particle size and solubility are equal, a greater proportion of the material entering the body by breathing will find its way into the blood than entering the digestive tract. The amount of radioactive material absorbed by inhalation, however, appears to be relatively small. The reason is that the nose can filter out almost all particles over 10 microns in diameter and 95 per cent of those exceeding 5 microns. Consequently, only a small portion of the large particles present in the air will usually succeed in reaching the lung.

The isotopes representing the greatest potential internal hazard are those with short radioactive half-lives and comparative long biological half-lives (defined as time taken for the amount of a particular element in the body to decrease to half its initial value due to the elimination by natural (biological) processes). A certain mass of an isotope of short radioactive half-life will emit particles at a greater rate than the same mass of another isotope, possibly of the same element, having a longer half-life. Further, a long biological half-life means that the active material will not be readily eliminated from the body by natural processes. For an example, the element iodine has a biological half-life of about 180 days, because it is quickly taken up by the thyroid gland from which it is eliminated slowly. The radioisotope iodine-131, a fairly common fission product, has a radioactive half-life of only 8 days. Consequently, if a sufficient quantity of this isotope enters the body blood stream, it is capable of causing serious damage to the thyroid gland. In addition to radioiodine, the most important potentially hazardous fission products, assuming sufficient amounts get into the body, fall into two groups. The first, and most significant, contain strontium-89, strontium-90, and barium-140, whereas the second consists of a group of rare earths and related elements, particularly cerium-144 and the chemically similar yttrium-91. These elements are readily deposited and held in various parts of the bone where the emitted beta and gamma radiations can injure blood forming tissues and may also cause tumor formation. Another potentially hazardous element, which may be present, is plutonium in the form of the alpha-particle emitting isotope plutonium-239. In addition to concentrating in skeletal tissue, strontium, barium and plutonium are found to accumulate to some extent in both the liver and spleen. The rare earths also deposit in the liver and to the lesser extent in the spleen. However, many radioisotopes are readily eliminated from the liver. It is of interest to note that in spite of the large amount of radioactive material that may pass through the kidneys, in the process of elimination, these organs ordinarily are not greatly affected. By contrast, uranium causes damage to the kidneys, but as a chemical poison rather than because of its radioactivity.

The genetic effects of radiation are of a long term character which produce no visible injury in the exposed individual but may have noticeable consequences in future generations.

Experiments with various types of animals have shown that the increased frequency of the occurrence of gene mutations, as a result of exposure to radiation, is approximately proportional to the total amount of radiation, absorbed by the gonads (from the beginning of their development to the time of conception). There is apparently no amount of radiation, however small, that does not cause some increase in normal mutation frequency. The dose rate of the radiation exposure or its duration have little influence; it is the total accumulated dose to the gonads that is the important quantity. It should be pointed out however that a large dose of radiation does not mean that the resulting mutations will be more harmful than for a small dose. With a large dose the mutations will be of the same general type as for a small dose, or as those which occur spontaneously, but there will be more of them in proportion to the dose.

The possibility of genetic damage has been known for some 25 years but knowledge of this was very qualitative until a few years ago. Even now most knowledge is based on animal experience but the burst of effort in studying genetic effects of radiation over the past ten to fifteen years has substantially enhanced man's knowledge of the subject. Whatever the effects of radiation on genetic status may be, its principal impact will not be upon the generation of individuals exposed. Rather will it be distributed over future generations up to perhaps fifty in number. Whether the effects on future generations are to be good or bad—and on the average the prediction is bad—the control of that future lies in the hands of those living today.

In general, the risks from radiation use are not very clear and are particularly difficult to define on a cause and effect basis. Except in relatively few instances, there is almost no knowledge relating death, shortening of life expectancy or genetic abnormalities in man to a specific dose or particular radiation exposure. It is unfortunate, for analytical purposes, that practically any radiation effect may also be caused by other means, known or unknown. Most radiation effects can be detected only by statistical studies on relatively large population samples.

The establishment of permissible levels of radiation exposure is not basically a scientific problem. Indeed, it is more a matter of philosophy, of morality, and of wisdom. There is today little or no direct, positive proof that there does or does not, exist some level of exposure to radiation below which harm will not result. Therefore, today the term "tolerance dose" is not used since it implies that there was some degree of radiation that was wholly without harm. In its place is used a term "permissible dose," which, while not completely unobjectionable, does not carry the connotation of absolute safety. Since it is well established that atomic energy has many benefits to man and working with atomic energy involves risks of exposure to ionizing radiation, the establishment of permissible doses depends upon a weighing of the benefits against the risks of possible damage. The levels presently recommended by the International Commission on Radiation Protection (ICRP) reflect the best decision on this delicate question that can be reached today. As has been mentioned previously, the latest levels as recommended by the ICRP average about 5 rems/yr for occupational exposure to adults. From 1950 to 1956, the average was about 15 rems/yr; from 1934 to 1950 the average was 60 rems/yr (in the United States the National Committee on Radiation Protection in 1935 recommended 30 rems/yr); prior to 1934 the levels were of the order of 100 rems/yr. Thus, as additional information has accumulated on the effects of exposure to ionizing radiations, the estimates of permissible exposures have been reduced and it does appear that any unnecessary exposure to ionizing radiation should be avoided.

EXPOSURE CALCULATIONS†

A convenient unit for expressing exposure to a cloud of fission products is curie-seconds per cubic meter (C-sec/m^3). The time of passage of the cloud certainly will take more than a few second so that the subject will take enough breaths of air into his lungs to come into equilibrium with the surroundings. The time will be, however, short enough so that the dose can be considered as being delivered in one single exposure. It is clear that within these limits it makes no difference whether a dose of 100 C-sec/m^3 results from 100 sec′ exposure to a cloud concentration of 1 C/m^3 or 10,000 sec′ exposure to a cloud concentration of 10 m C/m^3. In calculations using the unit C-sec/m^3 the assumption is made that the energy of the radiation remains constant over the time of the exposure. Although this is usually approximately true

† Material in this section has been taken from "Theoretical Possibilities and Consequences of Major Accidents in Large Nuclear Power Stations," Wash-740.

Table II

Maximum Permissible Body Burdens and Maximum Permissible Concentrations of Radionuclides in Air and in Water for Occupational Exposure

Radionuclide and type of decay		Organ of reference (critical organ in boldface)	Maximum permissible burden in total body q(μc)	Maximum permissible concentrations			
				For 40 hr week		For 168 hr week**	
				$(MPC)_w$ $\mu c/cm^3$	$(MPC)_a$ $\mu c/cm^3$	$(MPC)_w$ $\mu c/cm^3$	$(MPC)_a$ $\mu c/cm^3$
$_1H^3(\beta^-)$	(Sol)	Body tissue	10^3	0.1	2×10^{-5}	0.03	5×10^{-6}
		Total body	2×10^3	0.2	2×10^{-5}	0.05	7×10^{-6}
(H_2^3)	(Immersion)	Skin			2×10^{-3}		4×10^{-4}
$_6C^{14}(CO_2)(\beta^-)$	(Sol)	Fat	300	0.02	4×10^{-6}	8×10^{-3}	10^{-6}
		Total body	400	0.03	5×10^{-6}	0.01	2×10^{-6}
		Bone	400	0.04	6×10^{-6}	0.01	2×10^{-6}
$_{11}Na^{24}(\beta^-, \gamma)$	(Sol)	GI (SI)*		6×10^{-3}	10^{-6}	2×10^{-3}	4×10^{-7}
		Total body	7	0.01	2×10^{-6}	4×10^{-3}	6×10^{-7}
	(Insol)	GI (LLI)		8×10^{-4}	10^{-7}	3×10^{-4}	5×10^{-8}
		Lung			8×10^{-7}		3×10^{-7}
$_{15}P^{32}(\beta^-)$	(Sol)	Bone	6	5×10^{-4}	7×10^{-8}	2×10^{-4}	2×10^{-8}
		Total body	30	3×10^{-3}	4×10^{-7}	9×10^{-4}	10^{-7}
		GI (LLI)		3×10^{-3}	6×10^{-7}	9×10^{-4}	2×10^{-7}
		Liver	50	5×10^{-3}	6×10^{-7}	2×10^{-3}	2×10^{-7}
		Brain	300	0.02	3×10^{-6}	8×10^{-3}	10^{-6}
	(Insol)	Lung			8×10^{-8}		3×10^{-8}
		GI (LLI)		7×10^{-4}	10^{-7}	2×10^{-4}	4×10^{-8}
$_{18}A^{37}(\varepsilon)$	(Immersion)	Skin			6×10^{-3}		10^{-3}

Nuclide	Form	Organ					
$_{18}A^{41}(\beta^-, \gamma)$	(Immersion)	Total body			2×10^{-6}		4×10^{-7}
$_{27}Co^{60}(\beta^-, \gamma)$	(Sol)	GI (LLI)		10^{-3}	3×10^{-7}	5×10^{-4}	10^{-7}
		Total body	10	4×10^{-3}	4×10^{-7}	10^{-3}	10^{-7}
		Pancreas	70	0.02	2×10^{-6}	7×10^{-3}	6×10^{-7}
		Liver	90	0.03	10^{-6}	9×10^{-3}	5×10^{-7}
		Spleen	200	0.05	4×10^{-6}	0.02	2×10^{-6}
		Kidney	200	0.07	6×10^{-6}	0.03	2×10^{-6}
$_{36}Kr^{85}m(\beta^-, \gamma)$	(Immersion)	Total body			6×10^{-6}		10^{-6}
$_{36}Kr^{85}(\beta^-)$	(Immersion)	Total body			10^{-5}		3×10^{-6}
$_{36}Kr^{87}(\beta^-, \gamma)$	(Immersion)	Total body			10^{-6}		2×10^{-7}
$_{38}Sr^{89}(\beta^-)$	(Sol)	Bone	4	3×10^{-4}	3×10^{-8}	10^{-4}	10^{-8}
		GI (LLI)		10^{-3}	3×10^{-7}	4×10^{-4}	9×10^{-8}
		Total body	40	2×10^{-3}	2×10^{-7}	7×10^{-4}	6×10^{-8}
	(Insol)	Lung			4×10^{-8}		10^{-8}
		GI (LLI)		8×10^{-4}	10^{-7}	3×10^{-4}	5×10^{-8}
$_{38}Sr^{90}(\beta^-)$	(Sol)	Bone	2	4×10^{-6}	3×10^{-10}	10^{-6}	10^{-10}
		Total body	20	10^{-5}	9×10^{-10}	4×10^{-6}	3×10^{-10}
		GI (LLI)		10^{-3}	3×10^{-7}	5×10^{-4}	10^{-7}
	(Insol)	Lung			5×10^{-9}		2×10^{-9}
		GI (LLI)		10^{-3}	2×10^{-7}	4×10^{-4}	6×10^{-8}
$_{39}Y^{90}(\beta^-)$	(Sol)	GI (LLI)		6×10^{-4}	10^{-7}	2×10^{-4}	4×10^{-8}
		Bone	3	10	5×10^{-7}	4	2×10^{-7}
		Total body	20	80	3×10^{-6}	30	10^{-6}
$_{44}Ru^{103}(\beta^-, \gamma, e^-)$	(Sol)	GI (LLI)		2×10^{-3}	5×10^{-7}	8×10^{-4}	2×10^{-7}
		Kidney	20	0.08	10^{-6}	0.03	3×10^{-7}
		Total body	50	0.2	3×10^{-6}	0.08	9×10^{-7}
		Bone	100	0.6	7×10^{-6}	0.2	2×10^{-6}

Table II—cont.

Radionuclide and type of decay	Organ of reference (critical organ in boldface)	Maximum permissible burden in total body q(μc)	Maximum permissible concentrations			
			For 40 hr week		For 168 hr week**	
			$(MPC)_w$ $\mu c/cm^3$	$(MPC)_a$ $\mu c/cm^3$	$(MPC)_w$ $\mu c/cm^3$	$(MPC)_a$ $\mu c/cm^3$
$_{44}Ru^{106}(\beta^-, \gamma)$	GI (LLI)		4×10^{-4}	8×10^{-8}	10^{-4}	3×10^{-8}
(Sol)	Kidney	3	0.01	10^{-7}	4×10^{-3}	5×10^{-8}
	Bone	10	0.04	5×10^{-7}	0.01	2×10^{-7}
	Total body	10	0.06	7×10^{-7}	0.02	3×10^{-7}
(Insol)	Lung			6×10^{-9}		2×10^{-9}
	GI (LLI)		3×10^{-4}	6×10^{-8}	10^{-4}	2×10^{-8}
$_{51}Sb^{125}(\beta^-, \gamma, e^-)$	GI (LLI)		3×10^{-3}	6×10^{-7}	10^{-3}	2×10^{-7}
	Lung	40	0.04	5×10^{-7}	0.01	2×10^{-7}
(Sol)	Total body	60	0.05	6×10^{-7}	0.02	2×10^{-7}
	Bone	70	0.06	8×10^{-7}	0.02	2×10^{-7}
	Liver	3×10^{3}	3	3×10^{-5}	0.9	10^{-5}
	Thyroid	7×10^{4}	60	7×10^{-4}	20	2×10^{-4}
(Insol)	Lung			3×10^{-8}		9×10^{-9}
	GI (LLI)		3×10^{-3}	5×10^{-7}	10^{-3}	2×10^{-7}
$_{53}I^{126}(\beta^-, \varepsilon, \gamma)$	Thyroid	1	5×10^{-5}	8×10^{-9}	2×10^{-5}	3×10^{-9}
(Sol)	Total body	90	6×10^{-3}	9×10^{-7}	2×10^{-3}	3×10^{-7}
	GI (LLI)		0.05	10^{-5}	0.02	4×10^{-6}
(Insol)	Lung			3×10^{-7}		10^{-7}
	GI (LLI)		3×10^{-3}	5×10^{-7}	9×10^{-4}	2×10^{-7}
$_{53}I^{129}(\beta^-, \gamma, e^-)$	Thyroid	3	10^{-5}	2×10^{-9}	4×10^{-6}	6×10^{-10}
(Sol)	Total body	200	2×10^{-3}	2×10^{-7}	5×10^{-4}	7×10^{-8}
	GI (LLI)		0.1	3×10^{-5}	0.04	9×10^{-6}
(Insol)	Lung			7×10^{-8}		2×10^{-8}
	GI (LLI)		6×10^{-3}	10^{-6}	2×10^{-3}	4×10^{-7}

$_{53}I^{131}(\beta^-, \gamma, e^-)$	(Sol)	Thyroid	0.7	6×10^{-5}	9×10^{-9}	2×10^{-5}	3×10^{-9}
		Total body	50	5×10^{-3}	8×10^{-7}	2×10^{-3}	3×10^{-7}
		GI (LLI)		0.03	7×10^{-6}	0.01	2×10^{-6}
	(Insol)	GI (LLI)		2×10^{-3}	3×10^{-7}	6×10^{-4}	10^{-7}
		Lung			3×10^{-7}		10^{-7}
$_{53}I^{132}(\beta^-, \gamma, e^-)$	(Sol)	Thyroid	0.3	2×10^{-3}	2×10^{-7}	6×10^{-4}	8×10^{-8}
		GI (SI)		0.01	3×10^{-6}	4×10^{-3}	9×10^{-7}
		Total body	10	0.1	2×10^{-5}	0.04	6×10^{-6}
		GI (ULI)		5×10^{-3}	9×10^{-7}	2×10^{-3}	3×10^{-7}
		Lung			7×10^{-6}		2×10^{-6}
$_{53}I^{133}(\beta^-, \gamma, e^-)$	(Sol)	Thyroid	0.3	2×10^{-4}	3×10^{-8}	7×10^{-5}	10^{-8}
		GI (SI)		0.02	4×10^{-6}	6×10^{-3}	10^{-6}
		Total body	20	0.02	4×10^{-6}	9×10^{-3}	10^{-6}
	(Insol)	GI (LLI)		10^{-3}	2×10^{-7}	4×10^{-4}	7×10^{-8}
		Lung			10^{-6}		4×10^{-7}
$_{53}I^{134}(\beta^-, \gamma)$	(Sol)	Thyroid	0.2	4×0^{-3}	5×10^{-7}	10^{-3}	2×10^{-7}
		GI (S)		0.02	4×10^{-6}	6×10^{-3}	10^{-6}
		Total body	10	0.3	5×10^{-5}	0.1	2×10^{-5}
	(Insol)	GI (S)		0.02	3×10^{-6}	6×10^{-3}	10^{-6}
		Lung			2×10^{-5}		7×10^{-6}
$_{53}I^{135}(\beta^-, \gamma, e^-)$	(Sol)	Thyroid	0.3	7×10^{-4}	10^{-7}	2×10^{-4}	4×10^{-8}
		GI (SI)		0.01	3×10^{-6}	5×10^{-3}	10^{-6}
		Total body	20	0.05	7×10^{-6}	0.02	3×10^{-6}
	(Insol)	GI (LLI)		2×10^{-3}	4×10^{-7}	7×10^{-4}	10^{-7}
		Lung			3×10^{-6}		10^{-6}
$_{54}Xe^{131}(\gamma, e^-)$	(Immersion)	Total body			2×10^{-5}		4×10^{-6}
$_{54}Xe^{133}(\gamma, e^-)$	(Immersion)	Total body			10^{-5}		3×10^{-6}
$_{54}Xe^{135}(\beta^-, \gamma)$	(Immersion)	Total body			4×10^{-6}		10^{-6}

Table II—cont.

Radionuclide and type of decay		Organ of reference (critical organ in boldface)	Maximum permissible burden in total body q(μc)	Maximum permissible concentrations			
				For 40 hr week		For 168 hr week**	
				$(MPC)_w$ $\mu c/cm^3$	$(MPC)_a$ $\mu c/cm^3$	$(MPC)_w$ $\mu c/cm^3$	$(MPC)_a$ $\mu c/cm^3$
$_{55}Cs^{137}(\beta^-, \gamma, e^-)$	(Sol)	Total body	30	4×10^{-4}	6×10^{-8}	2×10^{-4}	2×10^{-8}
	(Sol)	Liver	40	5×10^{-4}	8×10^{-8}	2×10^{-4}	3×10^{-8}
	(Sol)	Spleen	50	6×10^{-4}	9×10^{-8}	2×10^{-4}	3×10^{-8}
	(Sol)	Muscle	50	7×10^{-4}	10^{-7}	2×10^{-4}	4×10^{-8}
	(Sol)	Bone	100	10^{-3}	2×10^{-7}	5×10^{-4}	7×10^{-8}
	(Sol)	Kidney	100	10^{-3}	2×10^{-7}	5×10^{-4}	8×10^{-8}
	(Sol)	Lung	300	5×10^{-3}	6×10^{-7}	2×10^{-3}	2×10^{-7}
	(Sol)	GI (SI)		0.02	5×10^{-6}	8×10^{-3}	2×10^{-6}
	(Insol)	Lung			10^{-8}		5×10^{-9}
	(Insol)	GI (LLI)		10^{-3}	2×10^{-7}	4×10^{-4}	8×10^{-8}
$_{56}Ba^{140}(\beta^-, \gamma)$	(Sol)	GI (LLI)		8×10^{-4}	2×10^{-7}	3×10^{-4}	6×10^{-8}
	(Sol)	Bone	4	6×10^{-3}	10^{-7}	2×10^{-3}	4×10^{-8}
	(Sol)	Total body	9	0.01	3×10^{-7}	5×10^{-3}	10^{-7}
	(Sol)	Liver	10^3	2	5×10^{-5}	0.9	2×10^{-5}
	(Sol)	Lung	3×10^3	4	9×10^{-5}	2	3×10^{-5}
	(Sol)	Muscle	3×10^3	5	10^{-4}	2	4×10^{-5}
	(Sol)	Spleen	4×10^3	6	10^{-4}	2	4×10^{-5}
	(Sol)	Kidney	4×10^3	8	2×10^{-4}	3	5×10^{-5}
	(Insol)	Lung			4×10^{-8}		10^{-8}
	(Insol)	GI (LLI)		7×10^{-4}	10^{-7}	2×10^{-4}	4×10^{-8}
$_{58}Ce^{144}(\alpha, \beta^-, \gamma)$	(Sol)	GI (LLI)		3×10^{-4}	8×10^{-8}	10^{-4}	3×10^{-8}
	(Sol)	Bone	5	0.2	10^{-8}	0.08	3×10^{-9}
	(Sol)	Liver	6	0.3	10^{-8}	0.1	4×10^{-9}
	(Sol)	Kidney	10	0.5	2×10^{-8}	0.2	7×10^{-9}
	(Sol)	Total body	20	0.7	3×10^{-8}	0.3	10^{-8}
	(Insol)	Lung			6×10^{-9}		2×10^{-9}
	(Insol)	GI (LLI)		3.10^{-4}	6×10^{-8}	10^{-4}	2×10^{-8}

$_{84}Po^{210}(\alpha)$	(Sol)	Spleen	0.03	2×10^{-5}	5×10^{-10}	7×10^{-6}	2×10^{-10}
		Kidney	0.04	2×10^{-5}	5×10^{-10}	8×10^{-6}	2×10^{-10}
		Liver	0.1	7×10^{-5}	2×10^{-9}	3×10^{-5}	6×10^{-10}
		Total body	0.4	2×10^{-4}	5×10^{-9}	8×10^{-5}	2×10^{-9}
		Bone	0.5	3×10^{-4}	7×10^{-9}	10^{-4}	2×10^{-9}
		GI (LLI)		9×10^{-4}	2×10^{-7}	3×10^{-4}	7×10^{-8}
	(Insol)	Lung			2×10^{-10}		7×10^{-11}
		GI (LLI)		8×10^{-4}	2×10^{-7}	3×10^{-4}	5×10^{-8}
$_{90}Th\text{-}Nat(\alpha, \beta^-, \gamma, e^-)$	(Sol)	Bone	0.01	3×10^{-5}	(2×10^{-12})	10^{-5}	(6×10^{-13})‡
		Kidney	0.07	10^{-4}	(4×10^{-12})	4×10^{-5}	(2×10^{-12})
		Total body	0.07	2×10^{-4}	(9×10^{-12})	7×10^{-5}	(3×10^{-12})
		GI (LLI)		3×10^{-4}	6×10^{-8}	10^{-4}	2×10^{-8}
		Liver	0.3	5×10^{-4}	(2×10^{-11})	2×10^{-4}	(8×10^{-12})
	(Insol)	Lung			(4×10^{-12})		(10^{-12})
		GI (LLI)		3×10^{-4}	5×10^{-8}	10^{-4}	2×10^{-8}
$_{92}U^{233}(\alpha, \gamma)$	(Sol)	GI (LLI)		9×10^{-4}	2×10^{-7}	3×10^{-4}	7×10^{-8}
		Bone	0.05	0.01	5×10^{-10}	4×10^{-3}	2×10^{-10}
		Kidney	0.08	0.03	10^{-9}	0.01	4×10^{-10}
		Total body	0.4	0.04	2×10^{-9}	0.01	5×10^{-10}
	(Insol)	Lung			10^{-10}		4×10^{-11}
		GI (LLI)		9×10^{-4}	2×10^{-7}	3×10^{-4}	6×10^{-8}
$_{92}U^{234}(\alpha, \gamma)$	(Sol)	GI (LLI)		9×10^{-4}	2×10^{-7}	3×10^{-4}	7×10^{-8}
		Bone	0.05	0.01	6×10^{-10}	4×10^{-3}	2×10^{-10}
		Kidney	0.08	0.03	10^{-9}	0.01	4×10^{-10}
		Total body	0.4	0.04	2×10^{-9}	0.01	6×10^{-10}
	(Insol)	Lung			10^{-10}		4×10^{-11}
		GI (LLI)		9×10^{-4}	2×10^{-7}	3×10^{-4}	6×10^{-8}
$_{92}U^{235}(\alpha, \beta^-, \gamma)$	(Sol)	GI (LLI)		8×10^{-4}	2×10^{-7}	3×10^{-4}	6×10^{-8}
		Kidney	0.03	0.01	5×10^{-10}	4×10^{-3}	2×10^{-10}
		Bone	0.06	0.01	6×10^{-10}	5×10^{-3}	2×10^{-10}
		Total body	0.4	0.04	2×10^{-9}	0.01	6×10^{-10}
	(Insol)	Lung			10^{-10}		4×10^{-11}
		GI (LLI)		8×10^{-4}	10^{-7}	3×10^{-4}	5×10^{-8}

Table II—cont.

Radionuclide and type of decay		Organ of reference (critical organ in boldface)	Maximum permissible burden in total body q(μc)	Maximum permissible concentrations			
				For 40 hr week		For 168 hr week**	
				$(MPC)_w$ $\mu c/cm^3$	$(MPC)_a$ $\mu c/cm^3$	$(MPC)_w$ $\mu c/cm^3$	$(MPC)_a$ $\mu c/cm^3$
${}_{92}U^{236}(\alpha, \gamma)$	(Sol)	GI (LLI)		10^{-3}	2×10^{-7}	3×10^{-4}	7×10^{-8}
		Bone	0.06	0.01	6×10^{-10}	5×10^{-3}	2×10^{-10}
		Kidney	0.08	0.03	10^{-9}	0.01	4×10^{-10}
		Total body	0.4	0.04	2×10^{-9}	0.01	6×10^{-10}
	(Insol)	Lung			10^{-10}		4×10^{-11}
		GI (LLI)		10^{-3}	2×10^{-7}	3×10^{-4}	6×10^{-8}
${}_{92}U^{238}(\alpha, \gamma, e^-)$	(Sol)	GI (LLI)		10^{-3}	2×10^{-7}	4×10^{-4}	8×10^{-8}
		Kidney	5×10^{-3}	2×10^{-3}	7×10^{-11}	6×10^{-4}	3×10^{-11}
		Bone	0.06	0.01	6×10^{-10}	5×10^{-3}	2×10^{-10}
		Total body	0.5	0.04	2×10^{-9}	0.01	6×10^{-10}
	(Insol)	Lung			10^{-10}		5×10^{-11}
		GI (LLI)		10^{-3}	2×10^{-7}	4×10^{-4}	6×10^{-8}
${}_{92}$U-Nat$(\alpha, \beta^-, \gamma, e^-)$	(Sol)	GI (LLI)		5×10^{-4}	10^{-7}	2×10^{-4}	4×10^{-8}
		Kidney	5×10^{-3}	2×10^{-3}	7×10^{-11}	6×10^{-4}	3×10^{-11}
		Bone	0.03	6×10^{-3}	3×10^{-10}	2×10^{-3}	10^{-10}
		Total body	0.2	0.02	8×10^{-10}	7×10^{-3}	3×10^{-10}
	(Insol)	Lung			6×10^{-11}		2×10^{-11}
		GI (LLI)		5×10^{-4}	8×10^{-8}	2×10^{-4}	3×10^{-8}
${}_{94}Pu^{239}(\alpha, \gamma)$	(Sol)	Bone	0.04	10^{-4}	2×10^{-12}	5×10^{-5}	6×10^{-13}
		Liver	0.4	5×10^{-4}	7×10^{-12}	2×10^{-4}	2×10^{-12}
		Kidney	0.5	7×10^{-4}	9×10^{-12}	2×10^{-4}	3×10^{-12}
		GI (LLI)		8×10^{-4}	2×10^{-7}	3×10^{-4}	6×10^{-8}
		Total body	0.4	10^{-3}	10^{-11}	3×10^{-4}	5×10^{-12}
	(Insol)	Lung			4×10^{-11}		10^{-11}
		GI (LLI)		8×10^{-4}	2×10^{-7}	3×10^{-4}	5×10^{-8}

$_{94}Pu^{240}(\alpha, \gamma)$	(Sol)	Bone	0.04	10^{-4}	2×10^{-12}	5×10^{-5}	6×10^{-13}
		Liver	0.4	5×10^{-4}	7×10^{-12}	2×10^{-4}	2×10^{-12}
		Kidney	0.5	7×10^{-4}	9×10^{-12}	2×10^{-4}	3×10^{-12}
		GI (LLI)		8×10^{-4}	2×10^{-7}	3×10^{-4}	6×10^{-8}
		Total body	0.4	10^{-3}	10^{-11}	3×10^{-4}	5×10^{-12}
	(Insol)	Lung			4×10^{-11}		10^{-11}
		GI (LLI)		8×10^{-4}	2×10^{-7}	3×10^{-4}	5×10^{-8}
$_{94}Pu^{241}(\alpha, \beta^-, \gamma)$	(Sol)	Bone	0.9	7×10^{-3}	9×10^{-11}	2×10^{-3}	3×10^{-11}
		Kidney	5	0.04	5×10^{-10}	0.01	2×10^{-10}
		GI (LLI)		0.04	8×10^{-6}	0.01	3×10^{-6}
		Total body	9	0.06	8×10^{-10}	0.02	3×10^{-10}
		Liver	10	0.07	10^{-9}	0.03	3×10^{-10}
	(Insol)	Lung			4×10^{-8}		10^{-8}
		GI (LLI)		0.04	7×10^{-6}	0.01	2×10^{-6}
$_{95}Am^{241}(\alpha, \gamma)$	(Sol)	Kidney	0.1	10^{-4}	6×10^{-12}	4×10^{-5}	2×10^{-12}
		Bone	0.05	10^{-4}	6×10^{-12}	5×10^{-5}	2×10^{-12}
		Liver	0.4	2×10^{-4}	9×10^{-12}	7×10^{-5}	3×10^{-12}
		Total body	0.3	4×10^{-4}	2×10^{-11}	10^{-4}	5×10^{-12}
		GI (LLI)		8×10^{-4}	2×10^{-7}	3×10^{-4}	6×10^{-8}
	(Insol)	Lung			10^{-10}		4×10^{-11}
		GI (LLI)		8×10^{-4}	10^{-7}	2×10^{-4}	5×10^{-8}

* The abbreviations GI, S, SI, ULI, and LLI refer to gastro-intestinal tract, stomach, small intestines, upper large intestine, and lower large intestine, respectively.

** It will be noted that the MPC values for the 168 hr week are not always precisely the same multiples of the MPC for the 40 hr week. Part of this is caused by rounding off the calculated values to one digit, but in some instances it is due to technical differences discussed in the ICRP report. Because of the uncertainties present in much of the biological data and because of individual variations, the differences are not considered significant. The MPC values for the 40 hr week are to be considered as basic for occupational exposure, and the values for the 168 hr week are basic for continuous exposure as in the case of the population at large.

(Selected values taken from Handbook 69, National Bureau of Standards) for mixed fission products, in the case of certain volatile fission products tl energy changes over prolonged periods so that variations in fission produ energy must be used in calculating the dose rates. It has also been found th the total beta energy or the total gamma energy of the volatile fission produc can be represented by a $t^{-0.2}$ law over the time period of interest to react hazard evaluation. It is also customary to express the concentration fission products in the air at a fixed time after they are released and this tir

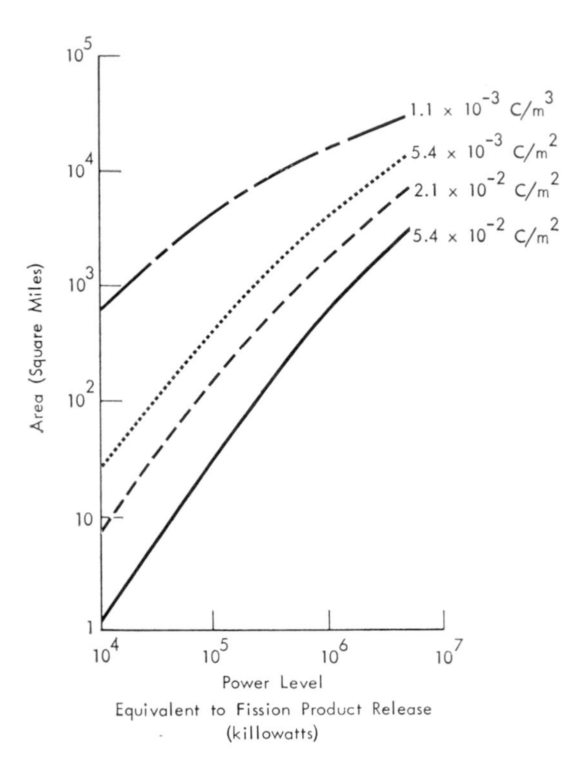

FIG. 1. (Taken from "Effects of a Major Reactor Disaster," by H. M. Parker and J. M. Healy, Vol. 13, *Proc. 1st United Nations Conf. on the Peaceful Uses of Atomic Energy*, United Nations, New York.)

is taken as 24 hr. Values of cloud concentration unless otherwise identifie are for 24 hr after release, although in most cases 2 hr is taken as the time actual exposure after the release.

An estimate of the whole-body gamma dose resulting from an exposure 1 C-sec/m^3 can be made by assuming the subject is immersed in a sen infinite cloud containing the radioactive material and neglecting bac scattering from the cloud. It will be noted that the dose rate is just half th obtained in an infinite medium since the cloud forms a hemisphere with t surface of the ground. An average gamma energy of 0.7 MeV is typical of fission products after prolonged reactor activation and 2 hr following reactor shutdown. An exposure of 1 C-sec/m^3 (24 hr value) to gross fission products, at 2 hr after release would give approximately a whole-body dose of 0.28 rad. Correspondingly an exposure to volatile fission products of 1 C-sec/m^3 would give a dose of 1.26 rad. It will be noted that the difference between these values results from the more rapid decay ($t^{-0.8}$ for volatile fission products as compared with $t^{-0.2}$ for average fission products) since the concentration of radioactive material is expressed on the basis of the 24 hr after release. These values are quite approximate since a number of assumptions have been made including that of an infinite cloud which assumption makes the results too high. On the other hand, neglect of radiation scattering from the ground and of gamma radiation received when not in the cloud at all make the results too low. The two effects have been assumed to cancel each other in the first approximation.

Whole body beta exposure from the passing cloud would be less than the gamma exposure and, since beta exposure affects principally the skin, this source should not contribute significantly to the total emergency dose. However it should be born in mind that beta dosage may become important, if the material is deposited on the body and not washed off promptly, and also exposure to the lens of the eye should be considered.

For the calculation of inhaled activity the standard man is assumed to have a respiratory rate of 20 l/min at work and 10 l/min during rest or light activity, and he is assumed to be at work 8 hr a day or one-third of the time. The average respiratory rate can be calculated to be 220 cm^3/sec, and an exposure of 1 C-sec/m^3 thus involves an inhalation of 220 microcuries. Only a fraction of an inhaled aerosol will be retained in the lungs. Gaseous activities and activities associated with very small particles will be exhaled and hence can contribute to the radiation dose only during the cloud passage. Activity associated with the larger particles will be removed in the upper respiratory tract and will never reach the lungs. Activity that is thus prevented from contributing to the lung dose is likely, however, to be swallowed and thus to present a possible hazard by ingestion. It is customary to estimate that 25 per cent of the inhaled activity will be retained in the lungs. If the 25 per cent figure is used, then for the 50 per cent fission product release, an exposure of 1 C-sec/m^3 would result in the retention of 55 microcuries as measured at 24 hr. For the volatile case, it is assumed that 20 per cent of the activity is in the form of mobile gases so that only 22 microcuries would be retained per C-sec/m^3.

The beta dose rate to the lungs, assuming a standard weight of 1 kg for the lungs can be calculated for the activity retained in the lungs, and assuming an average beta-particle energy of 0.4 MeV, 55 microcuries in the lungs would give a dose of 1.21 rad during the period from 2 hr to 24 hr following fission product release. The volatile fission products, assuming an average particl energy of 0.4 MeV and an equivalent of 22 microcuries in the lungs would giv a dose of 0.9 rad in the period from 2 to 24 hr following release.

It is necessary to consider the fate of the 55 microcuries of fission product which would be deposited in the lungs from an exposure of 1 C-sec/m^3. Th soluble fraction of these fission products probably remains in the lungs only a few hours before passing into the blood. Some of the insoluble material wil be removed and probably swallowed in the first few days; the rest will remain indefinitely. A convention sometimes used is to assume that half of th insoluble material is retained for 24 hr and the balance indefinitely. Materia caught in the upper respiratory tract will mostly be swallowed, althoug some will be removed by blowing the nose and coughing and expectora tion.

In considering the exposure to individual isotopes including radioactiv strontium, plutonium, thorium and the radioactive iodine, the maximun permissible burden in the total body is used as a unit of exposure, althoug little or no data exist on the additivity of partial body exposures. Howeve it does appear conservative simply to add up the partial body exposures as expressed as fractions of the maximum permissible burden in the total body for each isotope. It appears from such a summation that an exposure to 10 C-sec/m^3 for the case either of the release of 50 per cent of all of the fission products or an assumed release of only the volatile fission products will resul in an exposure of approximately equivalent to 1 maximum permissible bod burden of radioisotopes using the values for occupational exposure. An exposure of 10 C-sec/m^3 will thus be taken to be equivalent to 25 r of whole body gamma radiation in one exposure or 50 r in three months. Persons whos exposure falls below this level are assumed to receive no serious injury although it appears that exposures in this category might in a few cases hav undesirable consequences many years later.

The estimates of the combined exposures that might produce lethal effect are even more difficult and uncertain to make. Probably the exposure from bone-seekers such as strontium-89 and strontium-90 may be added togethe although the long time over which the strontium-90 delivers its dose make this assumption somewhat doubtful. Also the lung beta dose and the pluton ium dose may possibly be additive. An exposure to 400 C-sec/m^3 would give with these assumptions, a dose of 112 r to the whole-body, 540 rad to th lungs in the first day or 1620 in the first week, 942 rad to the bones over a few years and 224 rad to the intestinal tract. In addition there would be an undetermined amount of radiation to blood and other organs and 8000 rad to the thyroid, which would not be considered as contributing to morbidity It does not seem unlikely that such an exposure could cause death and there fore a value of 400 C-sec/m^3 would appear to be the lethal limit for fission product release.

If only the volatile fission products are released, the estimates of exposure are much less uncertain. The lung beta dose will result primarily from the radioactive iodine isotopes and probably only the first day of exposure need be considered. The thyroid dose can be neglected in estimates of fatal doses since complete destruction of this gland is not lethal. An exposure to 350 C-sec/m^3 would result in 450 r full-body exposure plus 325 rad to the lungs and 28 rad to the bones. An exposure to this limit would appear to be lethal in most cases.

If an exposure equivalent to 100 r is taken as the minimum that will produce illness it would appear, subject to the assumptions outlined above, that this exposure level would correspond to 90 C-sec/m^3 of total fission products or 78 C-sec/m^3 of the volatile fission products with both values being of course taken at 24 hr.

Comparatively little is known about the problems involved in living in an environment heavily contaminated by radioactive material. It is obvious that external whole-body gamma radiation would be significant at sufficiently high contamination levels and that even at low levels problems would exist in agriculture, particularly dairy farming. The extent to which dust deposited on the landscape would find its way into food and water supplies is not known.

From the publication *The Effects of Atomic Weapons*, (Samuel Glasstone, Editor), it appears that at 3 ft above a plane surface uniformly contaminated with 1 megacurie per square mile of material emitting 0.7 MeV gamma-rays, the radiation rate would be 4.2 r/hr. This value applies to a perfectly smooth surface and the build-up factor for gamma radiation in air has been ignored. The roughness of any ordinary ground surface would give some shielding and this might in general more than offset the build-up due to "sky shine". By conversion of units it is found that ground contaminated of 1 C/m^2 emitting 0.7 MeV gamma-rays would give a dose rate of 10.6 r/hr.

EVACUATION LIMITS

A dose rate such that 25 r could be received in the first 12 hr, would appear to call for urgent evacuation of a contaminated area. Such a rate would correspond to fission products surface contamination of about 0.2 C/m^2 of gross fission products or about 0.1 C/m^2 of the volatile fission products with both values being the 24 hr activity.

If the limit for evacuation from direct radiation be set at a dose of 50 r in 3 months, evacuation would be required for ground contamination of approximately 10^{-3} C/m^2 of gross fission products or 10^{-2} C/m^2 of the volatile fission products. In the volatile release the restrictions on occupancy would probably be temporary for a period not greater than three months, but for the full fission product release, some activities such as dairy farming might have to be prohibited indefinitely at this level of contamination.

The limiting levels of fission–product contamination for continued habitation has also been discussed by MARLEY and FRY in Vol. XIII, *Proceedings of the First United Nations Conference on the Peaceful Uses of Atomic Energy*, United Nations, New York. The limiting degree of hazard from possible ingestion was considered by these authors to be reached in normal living in a rural community at a contamination level of 10^{-3} C/m^2 measured at one day. The consumption of contaminated crops would contribute much of the hazard on the assumption that a man eats in one day leafy vegetables from about 0.05 m^2 of ground, that half the activity is removed by washing and that the consumption continues for 40 days, after which the leaves are cleaned as a result of new growth or continued washing by rain. At this level also the limiting level for a child, playing on the ground, would be reached on the assumption that a child actually ingests the contamination from 0.1 m^2 of ground before the activity is removed or fixed by natural processes.

The long-term ingestion hazard from the build-up of activity in the ground was estimated by Marley and Fry from the work of the University of California (Los Angeles) School of Medicine, from which it appears that the edible parts of plants take up strontium-90 from contaminated soil so that the content per gram dry weight is about the same as that in the ground. The build-up of activity would thus be about limiting from contamination at the level of 10^{-2} C/m^2 of mixed fission products measured at 1 day. At this level it also appears that open sources of drinking water would be contaminated to about limiting emergency ingestion levels, but in practice, there would be processes operating which would greatly clean up the drinking water. Temporary contamination of milk by strontium-89 may extend to zones where the fission product level is as low as 10^{-4} C/m^2 as a result of the wide area of herbage grazed each day by a cow. Supervision of the population would be needed down to this level in order to confirm that the basic tolerance was not being exceeded.

The results of these studies are tabulated in Table III taken from the publication *Theoretical Possibilities and Consequences of Major Accidents in Large Nuclear Power Plants* (WASH-740) and the above cited paper by Marley and Fry.

Three types of reactor accidents were considered in this study in order to indicate the range of public hazard which could result and to delineate the influence of the important variables. The three cases selected were:

(a) *The contained case*
(b) *The volatile release case*
(c) *The 50 per cent release case.*

In *the contained accident* it was assumed that all of the fission products from the 50 megawatt (thermal) reactor, after 180 days of operation, were released from the core and distributed uniformly throughout the interior of the

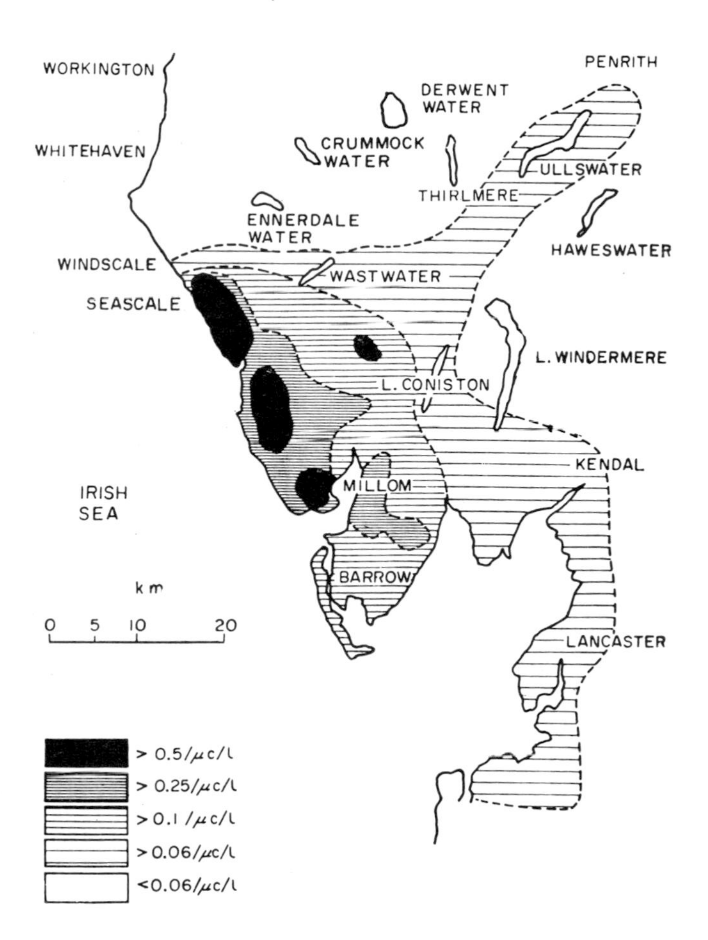

FIG. 2. I^{131} in milk, 13 October 1957. (Reproduced from Volume 18, *Proc. 2nd United Nations Int. Conf. on the Peaceful Uses of Atomic Energy*, United Nations, New York.)

Table III

Estimated Maximum Hazard Ranges From Reactor Fission Products

Type of hazard	Full fission product release (24 hr values)	Volatile fission products (24 hr values)	Dry weather values of B(km/MW$^{1/2}$)
Direct Radiation			
Lethal exposure	400 C-sec/m^3	350 C-sec/m^3	0.2
Illness likely	90 C-sec/m^3	80 C-sec/m^3	
No serious hazard	10 C-sec/m^3	10 C-sec/m^3	1.7
Ground Contamination			
Urgent evacuation within 12 hr	0.2 C/m^2	0.1 C/m^2	1.1
Evacuation necessary	10^{-2} C/m^2	10^{-1}	5.2
Crops, milk, water and lanu contaminated seriously			
Probable limit of temporary evacuation and restrictions on crops for first year	10^{-3} C/m^2		18
Probable limit of temporary milk contamination	10^{-4} C/m^2		(60)

containment building. None of the fission products were assumed to escape. Hazard to the public would arise from such an accident from the direct gamma radiation from the fission products dispersed inside the containment building. One inch of steel shielding for the walls of the building was assumed, with a site boundary at 2000 ft from the reactor.

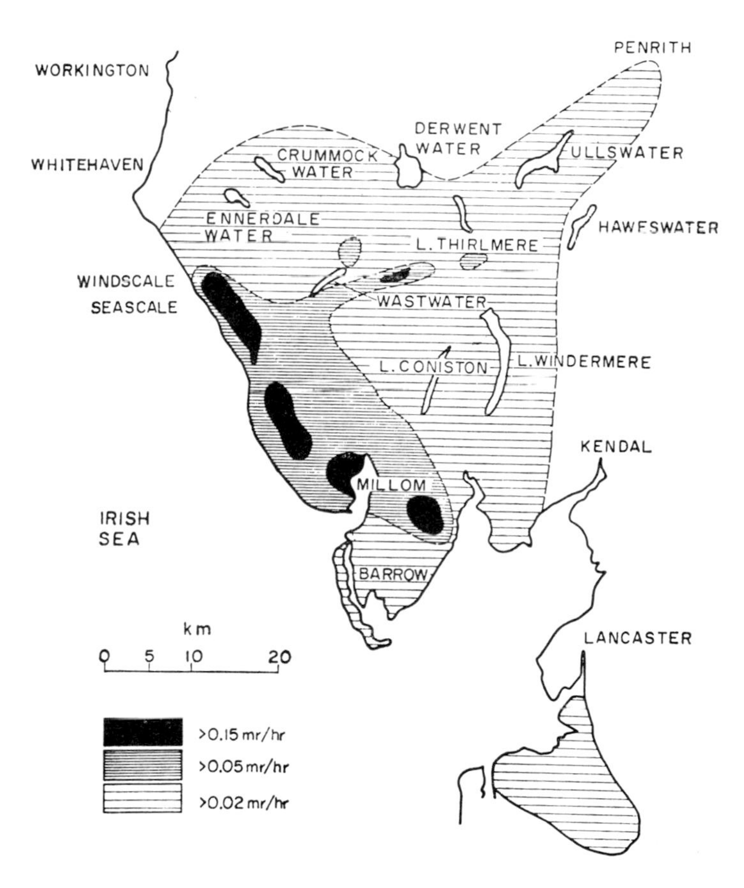

FIG. 3. Ground gamma radiation, 13 October 1957. (Reproduced from Vol. 18, *Proc. 2nd United Nations Int. Conf. on the Peaceful Uses of Atomic Energy*, United Nations, New York.)

In *the volatile release case* it is assumed that all of the volatile fission products in the reactor (500 MW thermal after 180 days), i.e., xenon, krypton, iodine, bromine and 1 per cent of the strontium were released from the containment building and subsequently disbursed into the atmosphere.

In *the 50 per cent release case* it was assumed that this fraction of all the fission products in the reactor were released from the containment.

DIFFUSION, DEPOSITION, AND RAIN-OUT OF THE ACTIVITY

The spread of radioactive material following a reactor incident with fall-out of particulate material and possibly rain-out depends on the conditions of release (hot or cold), precipitation, wind direction and atmospheric diffusion under the conditions of wind velocity, temperature profile and turbulence that may exist at the time of release and during the period of diffusion. There are the wide variations in each of these factors and the limits on combinations of conditions are difficult to predict. Marley and Fry[7] presented a useful analysis based on average conditions for diffusion from a ground level source with no thermal lift and for possible maximum deposition on the ground of activity under conditions of fair weather or rain. Their assumptions regarding particle size were made so as to give the maximum deposition at a few kilometers from the source since the contamination at this distance would be the maximum probable. The range downwind, R, at which various levels of ground contamination may be expected to extend was found to be roughly proportional to the square root of the amount of activity released and as indicated by the expression

$$R = BM^{1/2}$$

where M is the activity released expressed in megawatts, i.e. the total power of the reactor multiplied by the fractional release, and B is a constant having values for the various levels of contamination indicated in Table III.

Values of B as given in this table are those appropriate to ranges for the order of a few kilometers for lapsed conditions of turbulence in dry weather and wind speeds of 5 m/sec. The contamination zone would, of course, comprise only a narrow strip in the downwind direction, having an area

$$A(\text{km}^2) = R^2/14$$

The maximum deposition possible in rain at any given distance is about eight times that in dry weather and the maximum distance at which a given deposition may occur in the most unfavorable rain is thus about 2.8 times the distance in dry weather. The intensity of rain necessary to produce this contamination depends upon the distance downwind and the particle size and is unlikely to be reached in practice except for locations more than 5 km downwind. In inversion conditions the above simple formula no longer applies and greater ranges are possible.

The results of detailed calculations of dosage downwind from a cloud of fission products under various atmospheric conditions, rates of deposition of fission products on the ground as a function of particle size and atmospheric conditions, and rainout of fission products is presented in the publication WASH-740 from which Fig. 5–17 have been taken. From these figures can be seen the wide variation in radiation dosage and deposition that can be produced by meteorological factors.

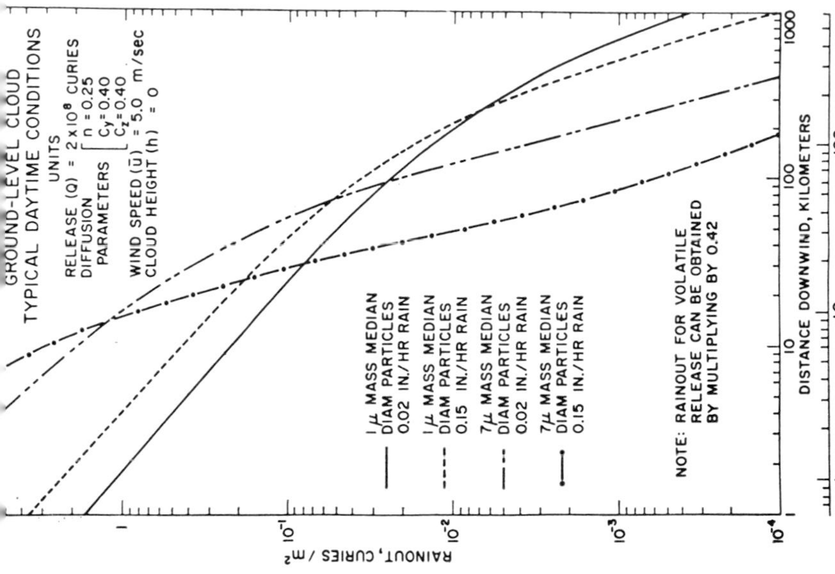

FIG. 4. (Reproduced from Report Wash–740 by permission of the United States Atomic Energy Commission.)

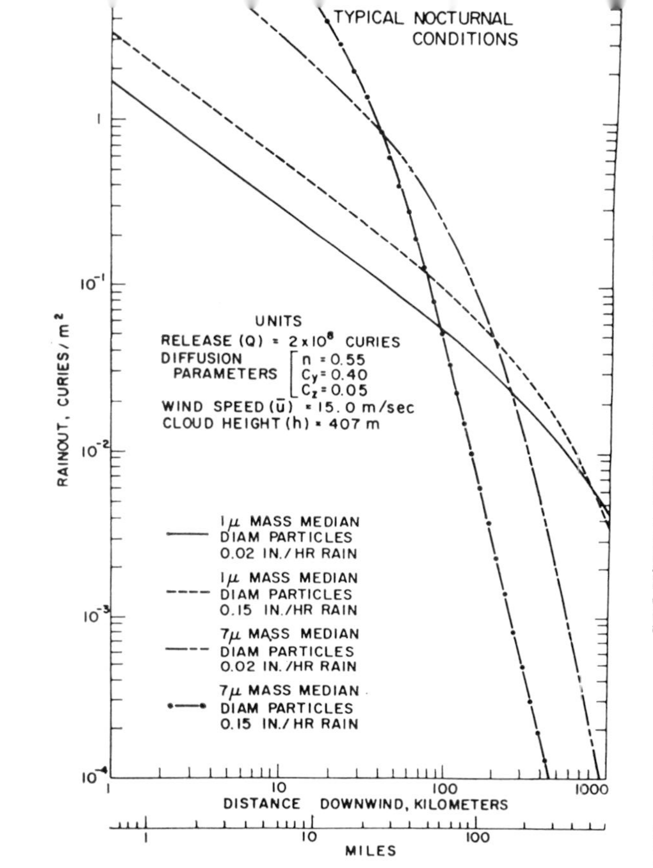

FIG. 5. (Reproduced from Report Wash–740 by permission of the United States Atomic Energy Commission.)

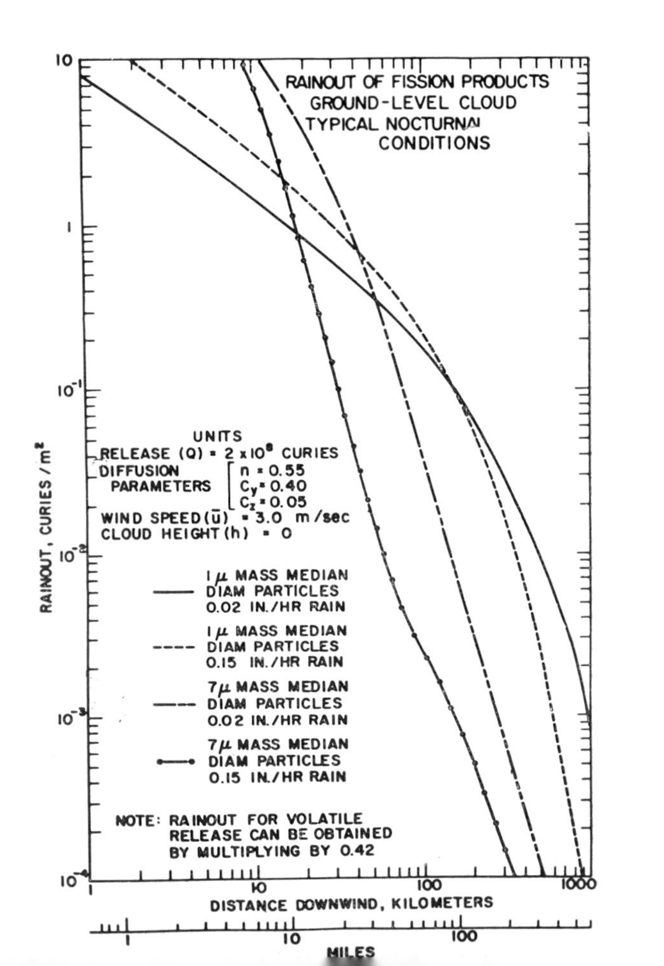

RAINOUT OF FISSION PRODUCTS
GROUND-LEVEL CLOUD
TYPICAL NOCTURNAL CONDITIONS
UNITS
RELEASE (Q) = 2 x 10^8 CURIES
DIFFUSION PARAMETERS n = 0.55 Cy = 0.40 Cz = 0.05
WIND SPEED (ū) = 3.0 m/sec
CLOUD HEIGHT (h) = 0
1 μ MASS MEDIAN DIAM PARTICLES 0.02 IN./HR RAIN
1 μ MASS MEDIAN DIAM PARTICLES 0.15 IN./HR RAIN
7 μ MASS MEDIAN DIAM PARTICLES 0.02 IN./HR RAIN
7 μ MASS MEDIAN DIAM PARTICLES 0.15 IN./HR RAIN
NOTE: RAINOUT FOR VOLATILE RELEASE CAN BE OBTAINED BY MULTIPLYING BY 0.42
RAINOUT, CURIES / m²
DISTANCE DOWNWIND, KILOMETERS
MILES

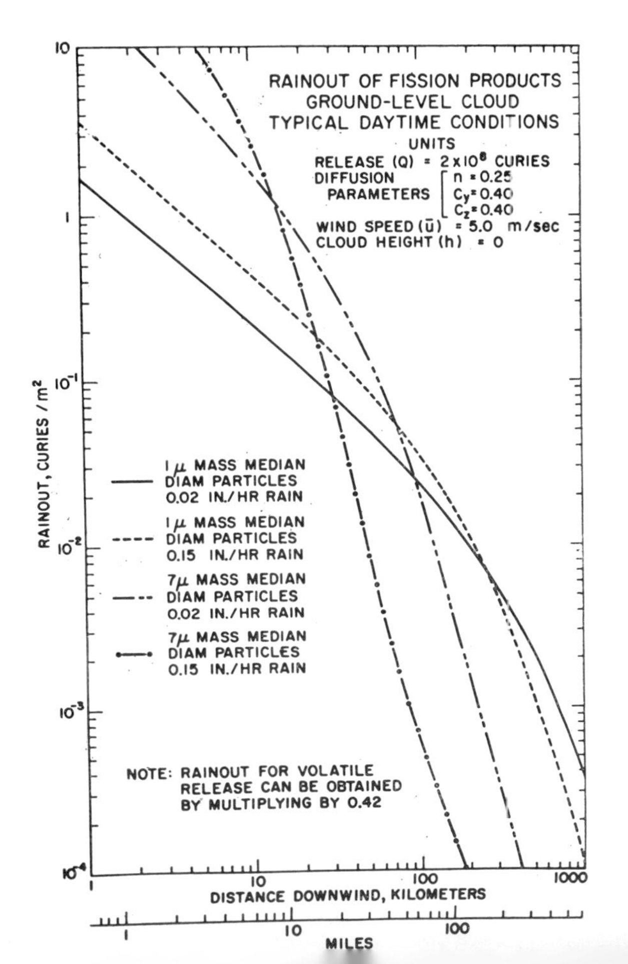

RAINOUT OF FISSION PRODUCTS
GROUND-LEVEL CLOUD
TYPICAL DAYTIME CONDITIONS
UNITS
RELEASE (Q) = 2 x 10^8 CURIES
DIFFUSION PARAMETERS n = 0.25 Cy = 0.40 Cz = 0.40
WIND SPEED (ū) = 5.0 m/sec
CLOUD HEIGHT (h) = 0
1 μ MASS MEDIAN DIAM PARTICLES 0.02 IN./HR RAIN
1 μ MASS MEDIAN DIAM PARTICLES 0.15 IN./HR RAIN
7 μ MASS MEDIAN DIAM PARTICLES 0.02 IN./HR RAIN
7 μ MASS MEDIAN DIAM PARTICLES 0.15 IN./HR RAIN
NOTE: RAINOUT FOR VOLATILE RELEASE CAN BE OBTAINED BY MULTIPLYING BY 0.42
RAINOUT, CURIES / m²
DISTANCE DOWNWIND, KILOMETERS
MILES

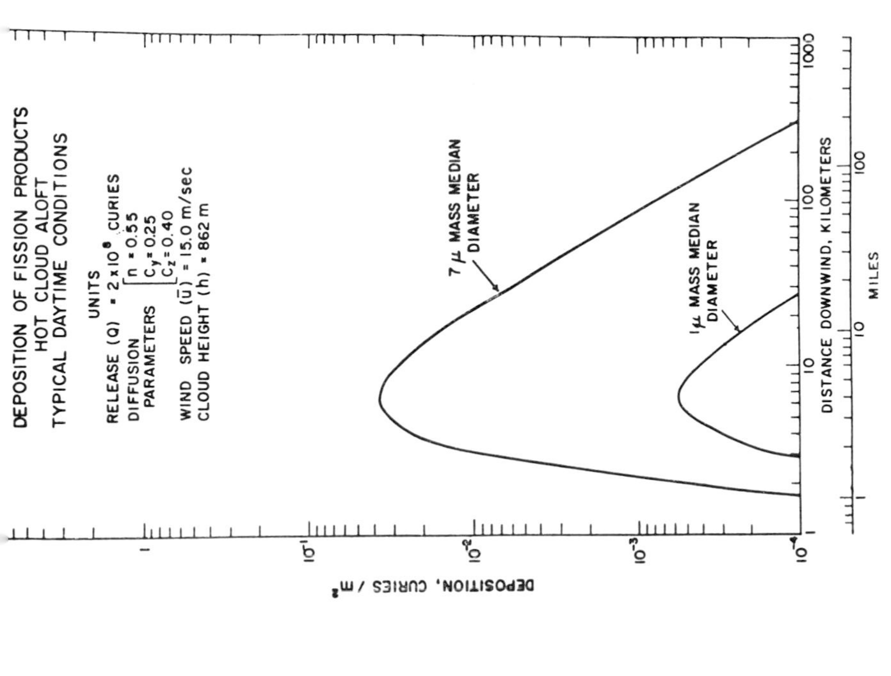

FIG. 9. (Reproduced from Report Wash-740 by permission of the United States Atomic Energy Commission)

FIG. 8. (Reproduced from Report Wash-740 by permission of the United States Atomic Energy Commission.)

It can be seen from Table III that the extent of personal injuries and property contamination from a nuclear incident is highly dependent on weather conditions at the time of the release. There are many uncertainties in such estimates of the extent of damages, however it is obvious from these results that the extent of possible damage under adverse conditions is extremely large.

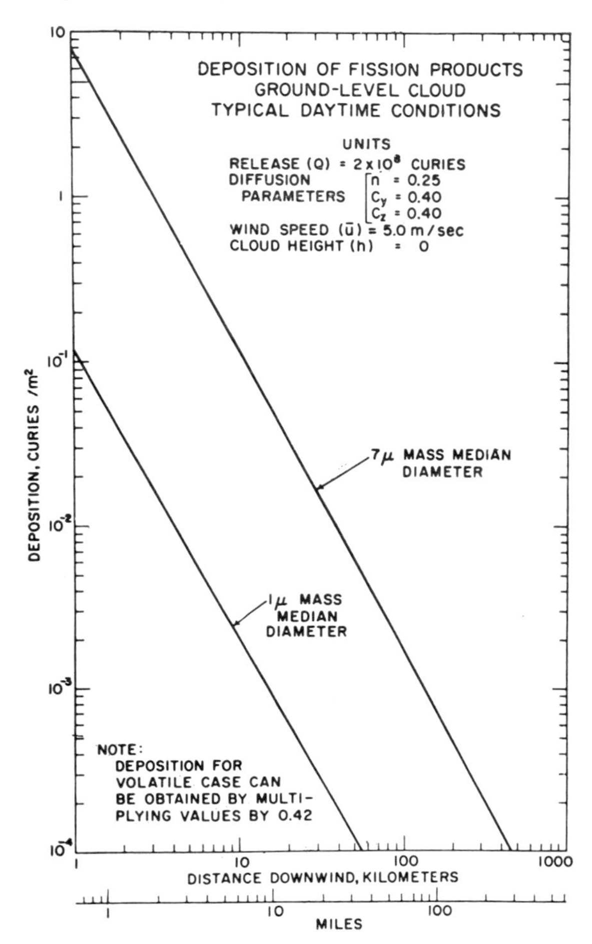

FIG. 10. (Reproduced from Report Wash–740 by permission of the United States Atomic Energy Commission.)

Other estimates of the extent of damages from a nuclear incident include the study by Parker and Healey, "Environmental effects of a major reactor disaster", *Proc. 1st United Nations Int. Conf. on the Peaceful Uses of Atomic Energy*, Vol. XI, United Nations, New York. The data in Fig. 1. have been taken from this study. Also Gomberg presented a paper "A quantitative approach to the evaluation of risk in locating a reactor at a given site", *Proc.*

Table IV

Integrated Gamma Doses (r) for Various Times After Shutdown from Fission Products Released Inside Container (Inferred from NDA–27–39).

Distance (ft)	Exposure times beginning at shutdown					
	1 hour	2 hours	1 day	1 week	1 month	100 days
1000	360	670	4700	24000	75000	120000
2000	18	34	220	1200	3400	5000
3000	1.5	2.8	16	89	270	340
4000	0.16	0.30	1.6	9.2	28	31
5000	0.020	0.037	0.19	1.1	3.3	3.4
6000	0.0027	0.0051	0.025	0.15	0.44	0.45

Reactor power = 500,000 kW   Operating time = ∞

Report Wash–740

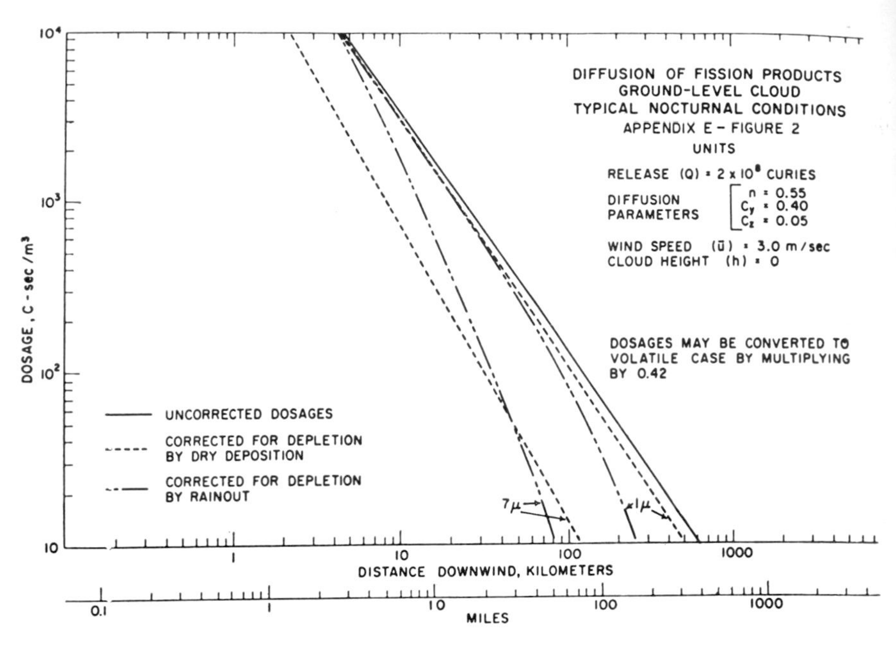

FIG. 11. (Reproduced from Report Wash–740 by permission of the United States Atomic Energy Commission.)

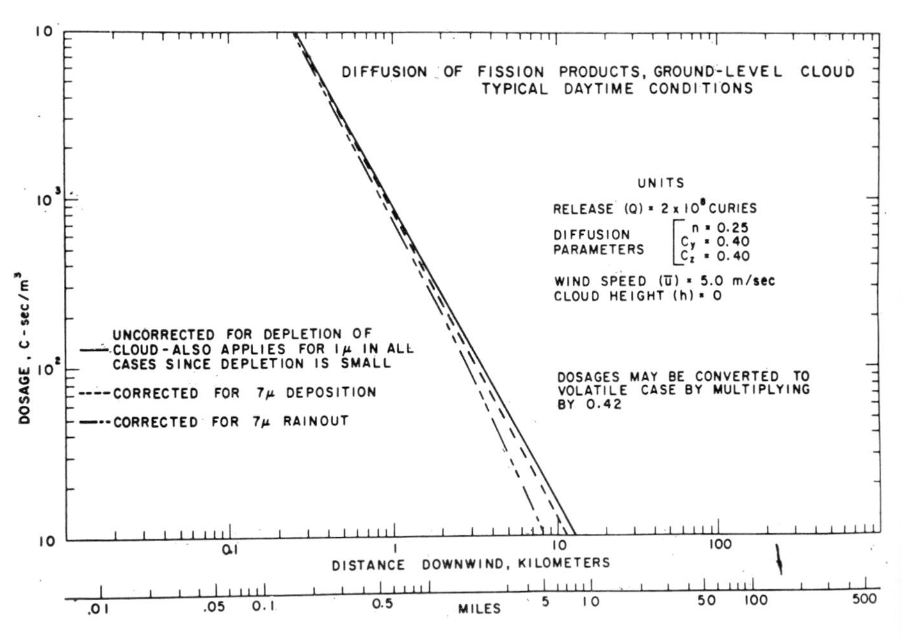

FIG. 12. (Reproduced from Report Wash–740 by permission of the United States Atomic Energy Commission.)

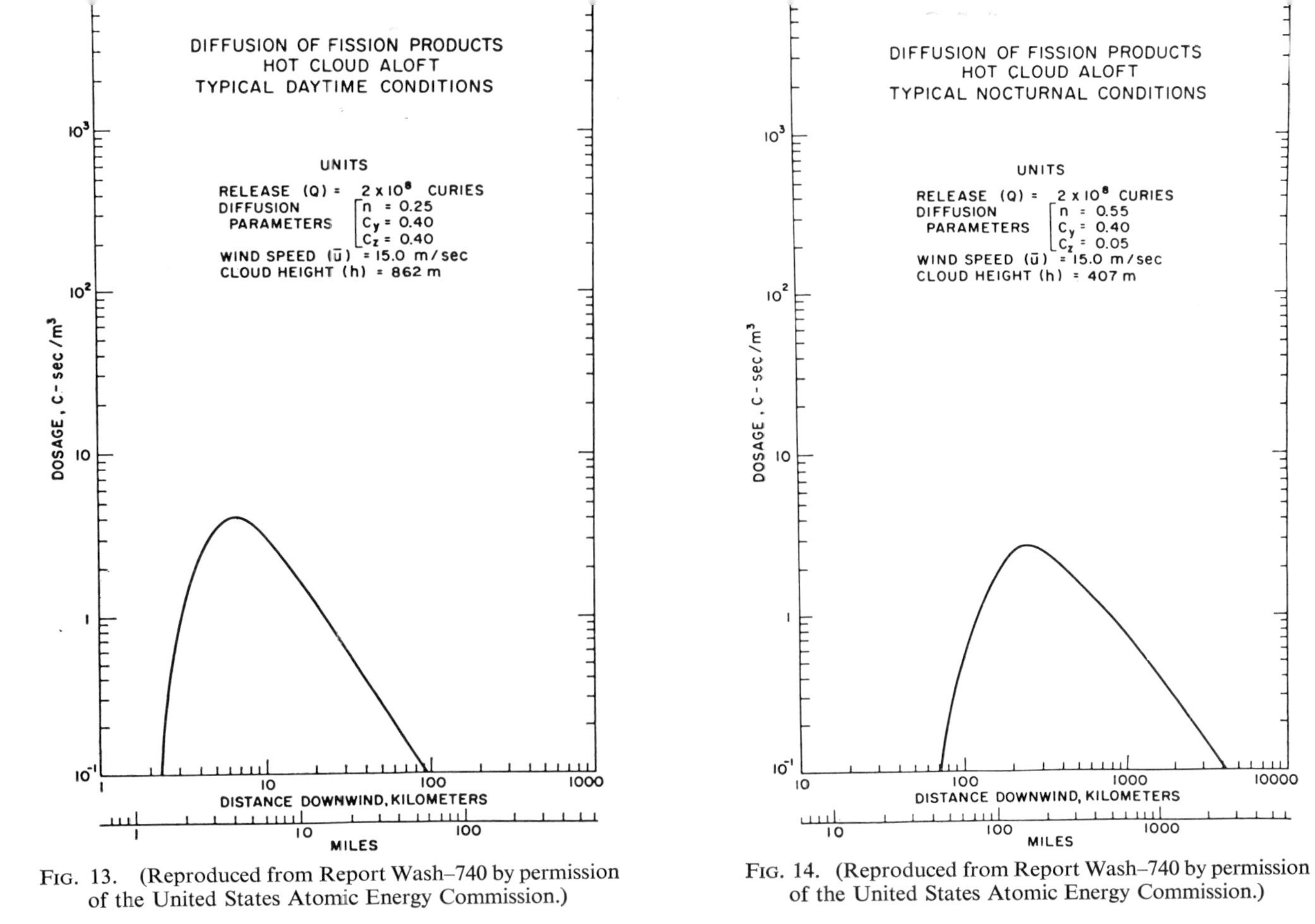

FIG. 13. (Reproduced from Report Wash–740 by permission of the United States Atomic Energy Commission.)

FIG. 14. (Reproduced from Report Wash–740 by permission of the United States Atomic Energy Commission.)

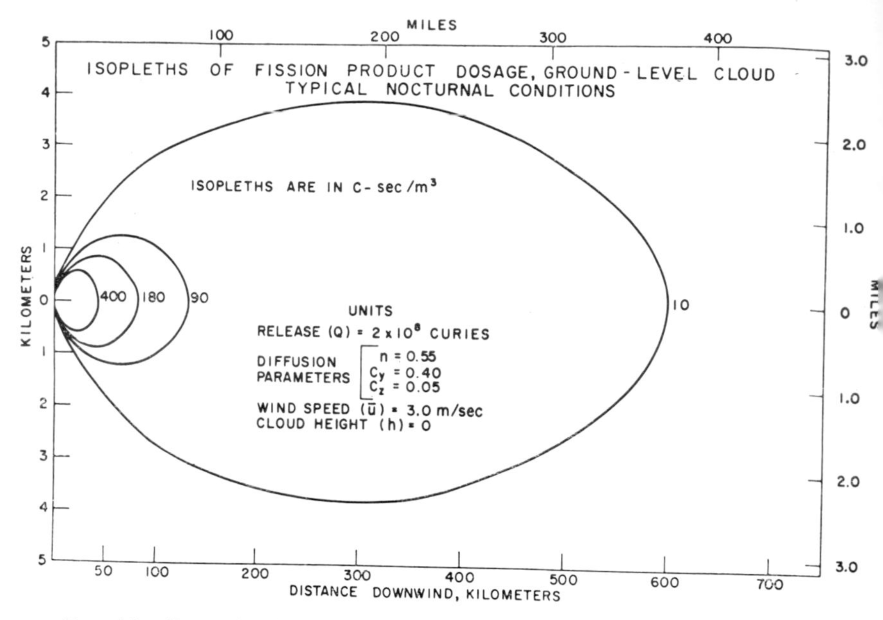

FIG. 15. (Reproduced from Report Wash–740 by permission of the United States Atomic Energy Commission.)

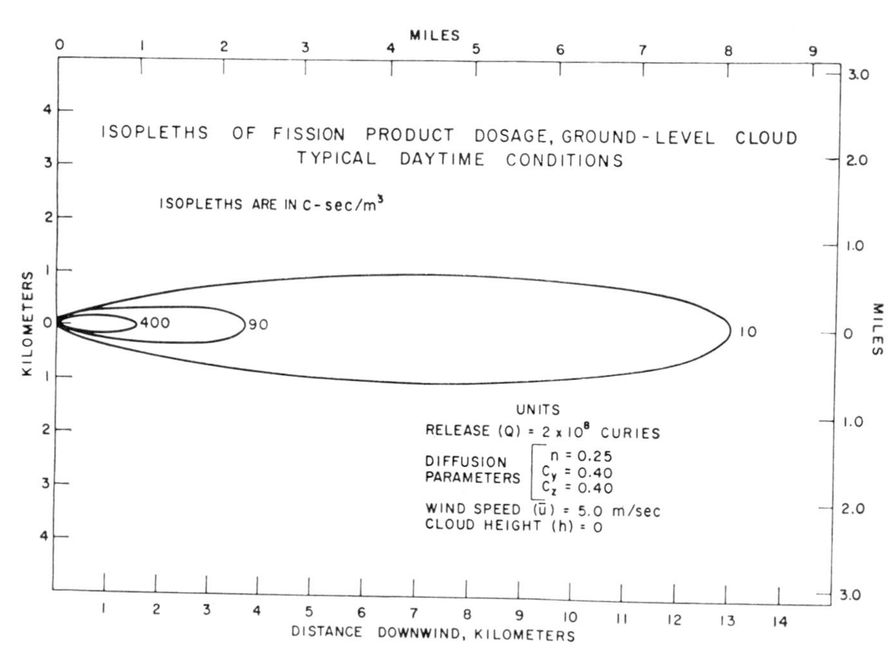

FIG. 16. (Reproduced from Report Wash–740 by permission of the United States Atomic Energy Commission.)

2nd United Nations Int. Conf. on the Peaceful Uses of Atomic Energy, Vol. XI, United Nations, New York. In this study an evaluation of the possible consequences of a hypothetical accident in Enrico Fermi Atomic Power Plant being built at Lagoona Beach, Mich., by the Power Reactor Development Co. was made. This study follows a procedure used for a similar study related

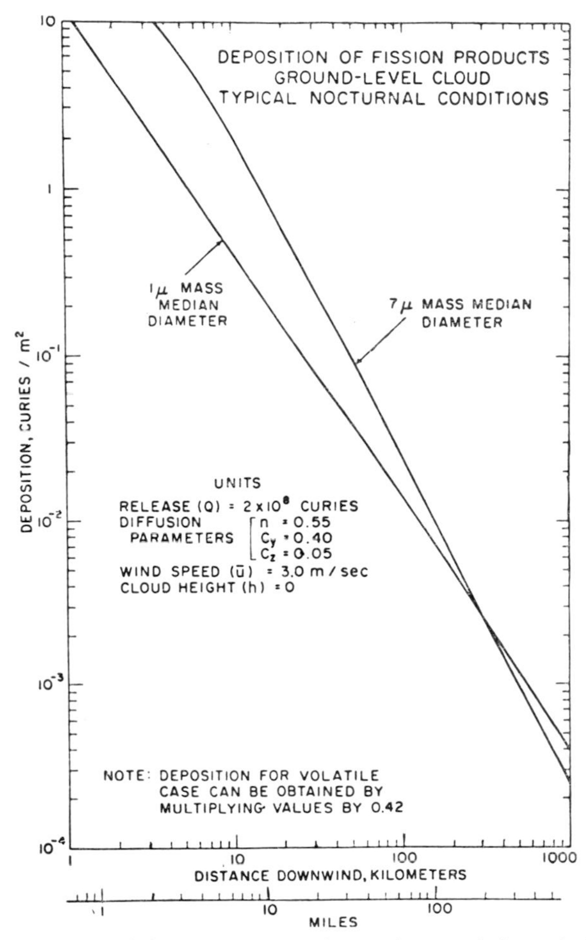

FIG. 17. (Reproduced from Report Wash–740 by permission of the United States Atomic Energy Commission.)

to the research reactor at the University of Michigan. Also in the same volume of those Proceedings is a paper by Leonard, "Hazards associated with fission product release". The extent of possible damages have been estimated in the several Hazards Summary Reports that have been prepared.

From these studies it can be seen that there are variations of orders of magnitude in the estimates of possible exposures and contaminations from a reactor incident depending on the assumptions made in the evaluation. It is of interest to compare these estimates with the consequences (Fig. 2. and 3.) of the release estimated as 20,000 curies of iodine-131, 600 curies of cesium-137, 80 curies of strontium-89 and 9 curies of strontium-90, from the Windscale No. 1 Pile in an incident described below. As a result of this release the use of milk from an area of some 200 square miles extending some 46 miles from the reactor, was temporarily restricted.

REFERENCES

1. Maximum permissible body burdens and maximum permissible concentrations of radionuclides in air and water for occupational exposures, National Bureau of Standards Handbook 69.
2. GLASSTONE, S., Editor, The effects of atomic weapons, U.S. Atomic Energy Commission Report.
3. Selected materials on atomic energy and workmen's compensation, Joint Committee on Atomic Energy, Congress of the United States.
4. Selected material on radiation protection criteria and standards: their basis and use, Joint Committee on Atomic Energy, Congress of the United States.
5. Maximum permissible radiation exposures to man, Addendum to National Bureau of Standards Handbook 59.
6. Theoretical possibilities and consequences of major accidents in large nuclear power stations, Wash–740.
7. MARLEY, W. G., and FRY, T. M., Radiological hazards from an escape of fission products and the implications in power reactor location, *Proc. 1st United Nations Int. Conf. on the Peaceful Uses of Atomic Energy*, Vol. 13, United Nations, New York.
8. PARKER, H. M., and HEALY, J. M., Environmental effects of a major reactor incident, *Proc. 1st. United Nations Int Conf. on the Peaceful Uses of Atomic Energy*, Vol. 13, United Nations, New York.
9. GOMBERG, H. J., A quantitative approach to evaluation of risk in locating a reactor at a given site, *Proc. 2nd United Nations Int. Conf. on the Peaceful Uses of Atomic Energy*, Vol. 11, United Nations, New York.
10. LEONARD, B. P. Jr., Hazards associated with fission product release, *Proc. 2nd United Nations Int. Conf. on the Peaceful Uses of Atomic Energy*, United Nations, New York.

CHAPTER 9

SITE REQUIREMENTS

INTRODUCTION

THE requirements for a reactor site are usually established as the result of compromises made between several conflicting factors including costs, utilization and hazards considerations. The location of the large production reactors for the weapons program at sites remote from major centers of population and on large tracts of land for exclusion of the public was certainly appropriate as was the establishment of the National Reactor Testing Station in a desert region some 40 miles from the nearest city. The decision approving a site without exclusion area on the crowded North Carolina State College

FIG. 1. Reactor building on campus at North Carolina State College. (North Carolina State College photograph.)

campus for a research reactor took into consideration the very low power and inherent stability of the homogeneous reactor as confirmed by extensive operating experience with reactors of this type, the importance of having the installation so located as to be available to students and the faculty, and the urgent need at that time for equipping educational institutions for research

FIG. 2. North Carolina State College Homogeneous Research Reactor. (North Carolina State College photograph.)

and training students in nuclear technology. As mentioned in Chapter 1, the Sodium-cooled Intermediate Reactor was located in the vicinity of Knoll Atomic Power Laboratory, so as to be convenient for the staffs of that and other laboratories in the area. This reactor was designed to operate at high power levels and was of a new and untried type so that complete assurance about safety was not available. Therefore, a large steel pressure vessel was constructed to contain any credible incident thereby permitting the location of this reactor in the vicinity of populated areas without the large exclusion area heretofore associated with reactors of that power level. It was considered necessary to adopt this concept of containment for power reactors for the generation of electricity since the cost of long transmission lines adds appreciably to the cost of power and the centers of electrical loads are usually also centers of population. The large central power reactors operating or being built in the United States are located within containment structures and at distances of the order of 25 miles from large cities. In the United Kingdom large gas-cooled power reactors are built without such steel containment structures on the basis of the inherent safety of the reactor type but the locations are at some distances from large centers of population.

FIG. 3. Naval Reactors Facility at the National Reactor Testing Station. (AEC Idaho Operations Office photograph.)

The policies regarding site location and containment have developed over a period of several years as the atomic energy program progressed from the operation of large reactors for the production of weapons materials to an atomic energy industry with many research, test, and power reactors. However, these policies are not without some inconsistencies. It is generally recognized that the large sites selected for production reactors such as at Hanford only provide time for the evacuation of personnel outside of the exclusion area and do not provide absolute safety since extensive damages considerably beyond the exclusion area could result under some adverse conditions. No site except one removed from populated areas by hundreds of miles would provide nearly absolute protection for the public from the consequences of a major release of fission products from a large reactor. As

Table I

Reactor Site and Containment Data

	Reactor	Power MW (th)	Exclusion* radius (ft)	Design pressure (psig)	Volume (Cu ft) Dimensions (ft)	Leakage†	Shielding‡
1	MITR Mass. Institute of Tech.	1	110	1	190,000 70 × 67	1—	2 ft concrete inside
2	LPTR Livermore, California	1	2400	2	252,000 80 × 77	3	Shell only
3	AFNETR Dayton, Ohio	10	1500	16	590,000 80 × 159	0.00770	Shell only
4	APPR–1 Fort Belvoir, Virginia	10	1000	65	37,000 (net) 32 × 60	0.0006—	2 ft concrete inside 3 ft concrete outside
5	EBWR ANL—Lemont, Illinois	20	2000	15	400,000 80 × 119	0.0013—	1 ft concrete inside
6	WTR Waltz Mill, Pennsylvania	20	1200	2	260,000 70 × 104	1	Shell only
7	GETR Pleasanton, California	33	2800	5	230,000 66 × 105	0.4	Shell only
8	GE–BWR Pleasanton, California	50	2000	45	125,000 (net) 48 × 100	<0.02—	Shell only

9	NACA–TR Sandusky, Ohio	60	3000	2	618,000 100 × 109	0.008	Shell only
10	EBR–11 NRTS—Idaho	62.5	10 miles	24	140,000 (net) 80 × 146	0.04	14 in. concrete inside
11	PWR Shippingport, Pennsylvania	232	2600	52.8	473,000 (net) Multiple units	0.02	Concrete outside Underground
12	ENRICO FERMI Monroe, Michigan	268	2900	32	500,000 72 × 120	0.025	Shell only
13	YANKEE Rowe, Massachusetts	392	2000	31.5	820,000 125 ft sphere	0.0067	Shell only
14	Con. Ed of New York Indian Point, N.Y.	500	1600	23	2,900,000 (net) 194 ft sphere	0.02	Concrete outside
15	Dresden Near Joliet, Illinois	630	2600	29.5	2,570,000 190 ft sphere	0.014—	Shell only
16	SIG West Milton, New York		1 mile	23.8	5,460,000 225 ft sphere	<0.005—	Shell only

* Exclusion radius—the minimum distance to uncontrolled land where settlement or industrial development could occur. At the National Reactor Testing Station (NRTS) the adjacent area is desert unlikely to be settled.

† Leakage is expressed as per cent of volume (measured at STP) per psig. The value given is the design or guaranteed value unless marked — in which case it is the test value.

‡ In some cases the designer has chosen to use a concrete layer inside of the steel shell or as a wall or building surrounding the steel shell to cut down the gamma shine in the place of using a large exclusion radius.

(Reproduced from "U.S. experience with reactor containment and Safeguards," by C. R. McCullough, Vol. 21, *Proc. 2nd United Nations Int. Conf. on the Peaceful Uses of Atomic Energy*, United Nations, New York.)

pointed out by Beck in Report TID-7579[1] if complete safeguards are included in a reactor design against all possible accidents having unacceptable consequences, then it can be argued that any site, however crowded, would be satisfactory, assuming of course that the safeguards would not fail and some dangerous potential accidents were not being overlooked. It can be argued that containment structures do not offer absolute protection since a door in the containment structure could be left open and also it would be prohibitively expensive to build a containment structure to contain any conceivable release of energy. The safety factors used in the design of the containment structure and its shielding as well as the site location and sizes are actually determined as matters of judgment and negotiation after weighing many factors including costs.

The fact that containment is considered necessary for the power reactors constructed near populated areas in the United States implies that there are uncertainties about the safety of these reactors. For example, there is extensive favorable experience with the pressurized water reactor and the basic design of this type is reasonably well established. Also in such reactors there are already multiple barriers preventing the release of fission products including the fuel cladding and the walls of the primary coolant system. The reactor would be shut down in event of failure of either of these barriers and the containment structure around the reactor merely provides a third barrier which is probably more vulnerable to operational error or external damage than either the fuel element cladding or the primary loop.

It can be anticipated that many years of operating experience with nuclear power reactors will be required before this process of evolution in policies regarding sites, exclusion areas and containment will have reached some stability. Operating experience with reactors has been unexpectedly favorable and perhaps this is because such installations are much less prone to accident than was originally predicted. However, a single very unfortunate experience could change the entire outlook regarding public acceptance of reactor hazards and lead to new concepts in reactor location and containment. Perhaps as underground construction techniques improve and the population increases, it will develop that underground locations for nuclear power plants will be preferred since underground locations offer protection against direct radiation and freedom from external damage to a degree that is not feasible to obtain otherwise.

RADIATION EXPOSURE AND SITE CRITERIA

The amount of radioactive material that may be released at a site without exceeding established limits for off-site exposure must form part of the basis for site criteria. BECK in Reference 1 has noted that the design of the plant and its location must be so chosen that the radioactivity released in normal effluents of plant operation (to air, water, earth) will not result in levels beyond the site boundary in excess of maximum permissible levels for continuous exposure.

The U.S. Atomic Energy Commission published in May 1959, a proposed formulation of site criteria. However these have since been revised as proposed guides to identify a number of factors considered in the evaluation of proposed sites recognizing that it is not possible at the present time to define site criteria with sufficient definiteness to eliminate the exercise of agency

FIG. 4. Chalk River reactor building. (Atomic Energy of Canada Limited photograph.)

judgement. The following section is taken from this Notice of Proposed Guides published in the *Federal Register* of February 11, 1961. The basic objectives to be achieved under the criteria set forth in the proposed guides, are:

"(a) Serious injury to individuals off-site should be avoided, if an unlikely, but still credible, accident should occur.

(b) Even if a more serious accident (not normally considered credible) should occur, the number of people killed should not be catastrophic.

(c) The exposure of large numbers of people in terms of total population dose should be low. This problem is being given further study in an effort to develop more specific guides. Meanwhile, in order to give recognition to this concept the distances to very large cities may have to be greater than those suggested by the guides."

Suitable values for the total population dose are controversial. This subjec was recently reviewed by Cottrell in Reference 2 from which the followin has been taken.

"A recent Oak Ridge National Laboratory report, Reference 3, notes that in addition to limiting the maximum permissible individual total-bod exposure, the genetic dose (average dose per person from conception to 3 years of age) to the population must be limited and suggests values of 25 an 0.3 rem, respectively, for these two exposures. The National Committee o

FIG. 5. Penn. State University Research Reactor, State College, Pennsylvania. (Penn. State University photograph.)

Radiation Protection (NCRP) and the International Commission on Radiological Protection (ICRP), in Reference 5, recommend that an emergency total dose of up to 25 rem, occurring only once in a lifetime, not affect the exposure status of the individual.

The total population exposure is derived in part from recommendations of the ICRP,[5] which state that the permissible genetic dose should not exceed 5 rem (exclusive of natural background and medical exposure, which, in the United States, average 4 and 5 r, respectively). No firm recommendations for apportioning the 5 rem are made by the ICRP, but that Commission gives the following apportionment as an illustration:

	Dose, rem
Occupational exposure groups	1.0
Exposure of special groups (principally the public living in the neighborhood of reactors)	0.5
Exposure of the population at large	2.0
Reserve	1.5

In a paper presented at the 1959 Health Physics Society Annual Meeting, C. R. McCullough, AEC Advisory Committee for Reactor Safeguards, proposed 2,000,000 man-rem as the limit on the population exposure from a single reactor maximum credible accident. If it is assumed that there may eventually be one such accident every year and that the population is 2×10^8, the resulting genetic dose per person in 30 years is

$$1 \times \frac{2 \times 10^6 \times 30}{2 \times 10^8} = 0.3 \text{ r}$$

This dose is appreciably less than that allocated by the ICRP for the population at large or the reserve. However, since significant portions of these allocations may be required for routine public exposures, the selection of a value as low as 0.3 r has considerable merit for planning purposes, although it is somewhat conservative at the present time.

This population exposure may be compared with that recently proposed by the British Medical Research Council,[(6)] for application in Great Britain, in which a value of 10 r total-body exposure would be permitted in instances where the number of exposed persons is small compared with one-fiftieth of the total population. "Small" is not defined in this report, but the calculated genetic dose to age 30 for one-fiftieth of the population to receive 10 r would be 6 r if one accident per year in a population of 5×10^7 were assumed. If 10 per cent may be considered a small percentage, the resulting genetic dose would still be twice that proposed by McCullough, even though the emergency exposure specified by the British is a factor of 2.5 less than the recommended NCRP value.

Release of Fission Products

Although standards for site selection are under vigorous study and have been prepared in draft form there remain some areas of controversy about the estimation of rate of release of fission product from a reactor system. If the reactor is operated within a containment structure, the design or test leakage such as 0.1 per cent per day of the containment volume might be a basis for exposure calculations. An estimation of the fraction of the fission products released from the reactor core into the containment structure would first be required.

McCullough in remarks on evaluation of hazards and other site criteria for licensing reactors,[7] mentioned that some exploratory studies had been made using the release of 0.3 per cent of the total fission product inventory and calculating the total amount of radiation which might be received by persons living around the site. The 0.3 per cent was taken as an arbitrary number but one which might be somewhat representative of a release which

FIG. 6. Curtiss-Wright 1000–kW Reactor, Quehanna, Pennsylvania. (Curtiss-Wright photograph.)

might occur in a very serious accident. It was assumed as a first approximation that the total damage can be represented by the radiation dose received per person time the number of people receiving those dosages, since the total damage to the population of two persons getting 100 roentgens each would be the same as 20 persons getting 10 roentgens each for 400 getting half a roentgen each. It was taken as impractical to include dosages lower than $\frac{1}{2}$ roentgen or to use any dosage values greater than 1000. The summation of all the dosages to the population would be expressed in roentgen-units or man-roentgens. This general approach to the reactor site problem is under continuing study.

Some Suggested Site Criteria

One of the more comprehensive treatments of siting criteria is the paper presented at the International Symposium on the Safety of Location of Nuclear Plants,[8] by Farmer and Fletcher of the United Kingdom Atomic Energy Authority which paper is reviewed by Cottrell in Reference 2. Site criteria proposed by Farmer and Fletcher are as follows:

(1) Few persons should live within 500 yards of the site;
(2) In any 10° sector around the site, not more than 500 people should live within 1.5 miles;
(3) No large center of population of 10,000 or more should lie within 5 miles.

The Atomic Energy Commission is understood to have suggested for discussion the following site criteria for small pressurized water reactors:

(1) An exclusion area with a minimum radius of two thousand feet surrounding the reactor.

(2) Edge of site to be five miles from the fringe of population centers. If an intervening area is not likely to be an area of heavy population build up in the near future years (hills unsuitable for residential area, etc.), this distance may be reduced to three miles minimum.

The Atomic Energy Commission published in May 1959, a proposed formulation of site criteria. Although these criteria have not been formally adopted and it has been stated they were intended for purposes of discussion, it does appear that these draft criteria in many ways reflect current practices in some aspects of site evaluation. These criteria as drafted for public comment follow:

"*Factors considered in site evaluation for power and test reactors:*

(a) *General.* The construction of a proposed power or test reactor facility at a proposed site will be approved if analysis of the site in relation to the hazards associated with the facility gives reasonable assurance that the potential radioactive effluents therefrom, as a result of normal operation or the occurrence of any credible accident, will not create undue hazard to the health and safety of the public.

There are wide possible variations in reactor characteristics and protective aspects of such facilities which affect the characteristics that otherwise might be required of the site. However, the following factors are used by the Commission as guides in the evaluation of sites for power and test reactors. The fact that a particular site may be deemed acceptable for a proposed reactor facility when evaluated in the early phases of the project, does not determine that the reactor will eventually be given operating approval, or indicate that limitations on operation may not be imposed. Operating approvals depend on detailed review of design, construction and operating procedures at the final construction stages.

(b) *Exclusion distance* around power and test reactors. Each power and test reactor should be surrounded by an exclusion area under the complete control of the licensee. The size of this exclusion area will depend upon many factors including among other things reactor power level, design features and containment, and site characteristics. The power level of the reactor alone does not determine the size of the exclusion area. For any power or test reactor, a minimum radius on the order of one-quarter mile will usually be found necessary. For large power reactors a minimum exclusion radius on the order of one-half to three-quarter miles may be required. Test reactors may require a larger exclusion area than power reactors of the same power.

(c) *Population density* in surrounding areas. Power and test reactors should be so located that the population density in surrounding areas, outside the exclusion zone, is small. It is usually desirable that the reactor should be several miles distant from the nearest town or city and for large reactors a distance of 10 to 20 miles from large cities. Where there is a prevailing wind direction it is usually desirable to avoid locating a power or test reactor within several miles upwind from centers of population. Nearness of the reactor to air fields, arterial highways and factories is discouraged.

(d) *Meteorological considerations.* The site meteorology is important in evaluating the degree of vulnerability of surrounding areas to the release of air-borne radioactivity to the environment. Capabilities of the atmosphere for diffusion and dispersion of airborne release are considered in assessing the vulnerability to risk of the area surrounding the site. Thus a high probability of good diffusion conditions and a wind direction away from vulnerable areas during periods of slow diffusion would enhance the suitability of the site. If the site is in a region noted for hurricanes or tornadoes, the design of the facility must include safeguards which would prevent significant radioactivity releases should these events occur.

(e) *Seismological considerations.* The earthquake history of the area in which the reactor is to be located is important. The magnitude and frequency of seismic disturbances to be expected determine the specifications which must be met in design and construction of the facility and its protective components. A site should not be located on a fault.

(f) *Hydrology and geology.* The hydrology and geology of a site should be favorable for the management of the liquid and solid effluents, (including possible leaks from the process equipment). Deposits of relatively impermeable soils over ground water courses are desirable because they offer varying degrees of protection to the ground waters depending on the depth of the soils, their permeability, and their capacities for removing and retaining the noxious components of the effluents. The hydrology of the ground waters is important in assessing the effect that travel time may have on the contaminants which might accidentally reach them to the point of their nearest usage. Site drainage and surface water hydrology is important in determining the vulnerability of surface water courses to radioactive contamination. The characteristics and usage of the water courses indicate the degree of risk involved and determine safety precautions that must be observed at the facility in effluent control and management. The hydrology of the surface water course and its physical, chemical and biological characteristics are important factors in evaluating the degree of risk involved.

(g) *Interrelation of factors.* All of the factors described in paragraphs (b) through (f) of the section are interrelated and dictate in varying degrees the engineered protective devices for the particular nuclear facility under consideration, and the dependence which can be placed on such devices. It is necessary to analyse each of the environmental factors to ascertain the character of protection it might afford for operation of the proposed facility and of the kind of restrictions it might impose on the proposed design and operation."

The proposed guides in evaluating proposed site, published in February, 1961 by the U.S. Atomic Energy Commission contain the following definitions:

(a) *Exclusion area* means the area surrounding the reactor, access to which is under the full control of the reactor licensee. This area may be traversed by a highway, railroad, or waterway, provided these are not so close to the facility as to interfere with normal operations, and provided appropriate and effective arrangements are made to control traffic on the highway, railroad, or waterway, in case of emergency, to protect the public health and safety. Residence within the exclusion area shall normally be prohibited. In any event, residents shall be subject to ready removal in case of necessity. Activities unrelated to operation of the reactor may be permitted in an exclusion area under appropriate limitations, provided that no significant hazards to the public health and safety will result.

(b) *Low population zone* means the area immediately surrounding the exclusion area which contains residents the total number and density of which is such that there is a reasonable probability that appropriate protective measures could be taken in the event of a serious accident. These guides do not specify a permissible population density or total population within this zone because the situation may vary from case to case. Whether a specific number of people can, for example, be evacuated from a specific area, or instructed to take shelter, on a timely basis will depend on many factors such as location, number and size of highways, scope and extent of advance planning, and actual distribution of residents within the area.

(c) *Population center distance* means the distance from the reactor to the nearest boundary of a densely populated center containing more than about 25,000 residents.

(d) *Power reactor* means a nuclear reactor of a type designed to produce electrical or heat energy.

Factors to be Considered When Evaluating Sites

In determining the acceptability of a site for a power or testing reactor, these guides specify that the following factors will be taken into consideration:

(a) *Population density and use characteristics* of the site environs, including among other things, the exclusion area, low population zone, and population center distance.

Fig. 7. Sodium Graphite Reactor building, Santa Suzana, California. (North American Aviation, Inc. photograph.)

(b) *Physical characteristics* of the site, including, among other things, seismology, meteorology, geology and hydrology. For example:

(i) The design for the facility should conform to accepted building codes or standards for areas having equivalent earthquake histories. No facility should be located closer than one-quarter to one-half mile from the surface location of a known active earthquake fault.

(ii) Meteorological conditions at the site and in the surrounding area should be considered.

(iii) Geological and hydrological characteristics of the proposed site may have a bearing on the consequences of an escape of radioactive material from the facility. Unless special precautions are taken, reactors should not be located at sites where radioactive liquid effluents might flow readily into nearby streams or rivers or might find ready access to underground water tables.

Where some unfavorable physical characteristics of the site exist, the proposed site may nevertheless be found to be acceptable if the design of the facility includes appropriate and adequate compensating engineering safeguards.

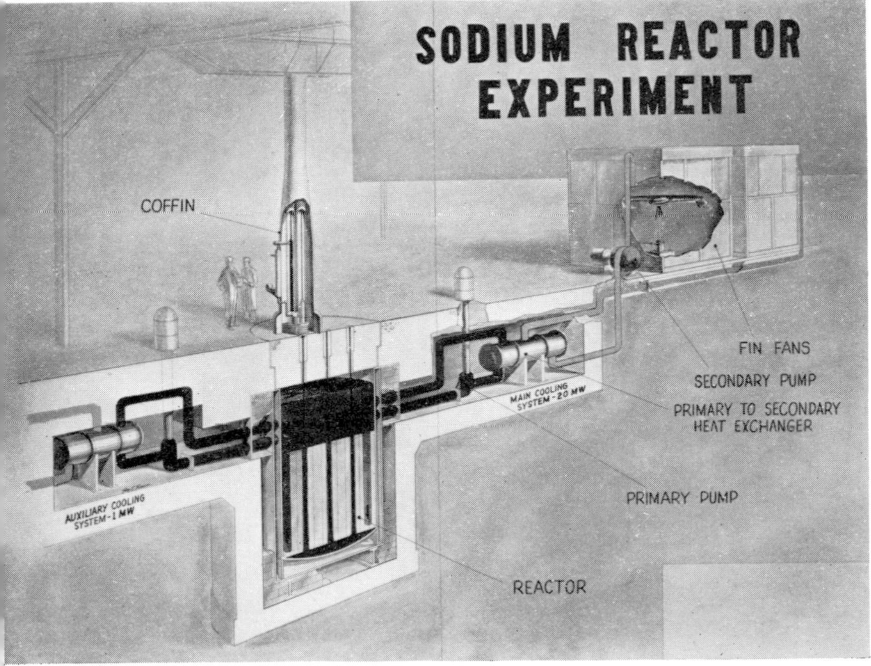

FIG. 8. (Atomics International Division, North American Aviation, Inc. photograph.)

(c) *Characteristics of the proposed reactor*, including proposed maximum power level, use of the facility, the extent to which the design of the facility incorporates well proven engineering standards, and the extent to which the reactor incorporates unique or unusual features having a significant bearing on the probability or consequences of accidental releases of radioactive material.

DETERMINATION OF EXCLUSION AREA, LOW POPULATION ZONE, AND POPULATION CENTER DISTANCE

(a) As an aid in evaluating a proposed site, an applicant should assume a fission product release from the core as illustrated in Appendix "A", the expected demonstrable leak rate from the containment, and meteorological conditions pertinent to his site to derive an exclusion area, a low population zone and a population center distance. For the purpose of this analysis, the applicant should determine the following:

(i) An exclusion area of such size that an individual located at any point on its boundary for two hours immediately following onset of the postulated fission product release would not receive a total radiation dose to the whole body in excess of 25 rem or a total radiation dose in excess of 300 rem to the thyroid from iodine exposure.

(ii) A low population zone of such size that an individual located at any point on its outer boundary who is exposed to the radioactive cloud resulting from the postulated fission product release (during the entire period of its passage) would not receive a total radiation dose to the whole body in excess of 25 rem or a total radiation dose in excess of 300 rem to the thyroid from iodine exposure.

(iii) A population center distance of at least $1\frac{1}{3}$ times the distance from the reactor to the outer boundary of the low population zone. In applying this guide due consideration should be given to the population distribution within the population center. Where very large cities are involved, a greater distance may be necessary because of total integrated population dose considerations.

The whole body dose of 25 rem referred to above corresponds to the once in a lifetime accidental or emergency dose for radiation workers which, according to NCRP recommendations, may be disregarded in the determination of their radiation exposure status. (See Addendum dated April 15, 1958 to NBS Handbook 59.) The NCRP has not published a similar statement with respect to portions of the body, including doses to the thyroid from iodine exposure. For the purpose of establishing areas and distances under the conditions assumed in these guides, the whole body dose of 25 rem and the 300 rem dose to the thyroid from iodine are believed to be conservative values.

(b) Appendix "A" contains an example of a calculation for hypothetical reactors which can be used as an initial estimate of the exclusion area, the low population zone, and the population center distance.

The calculations described in Appendix "A" are a means of obtaining preliminary guidance. They may be used as a point of departure for consideration of particular site requirements which may result from evaluations of the particular characteristics of the reactor, its purpose, method of operation, and site involved. The numerical values stated for the variables listed in Appendix "A" represent approximations that presently appear reasonable, but these numbers may need to be revised as further experience and technical information develops.

THE SHIPPINGPORT SITE

There are many practical problems associated with an actual site which are well illustrated by the considerations that went into selection of a tract for the first power demonstration reactor. This section has been taken largely from "Description of Shippingport atomic power station site and surrounding area with radiation background and meteorological data", USAEC Report WAPD-SC-547, by R. J. McAllister *et al.*, Westinghouse Atomic Power Division.

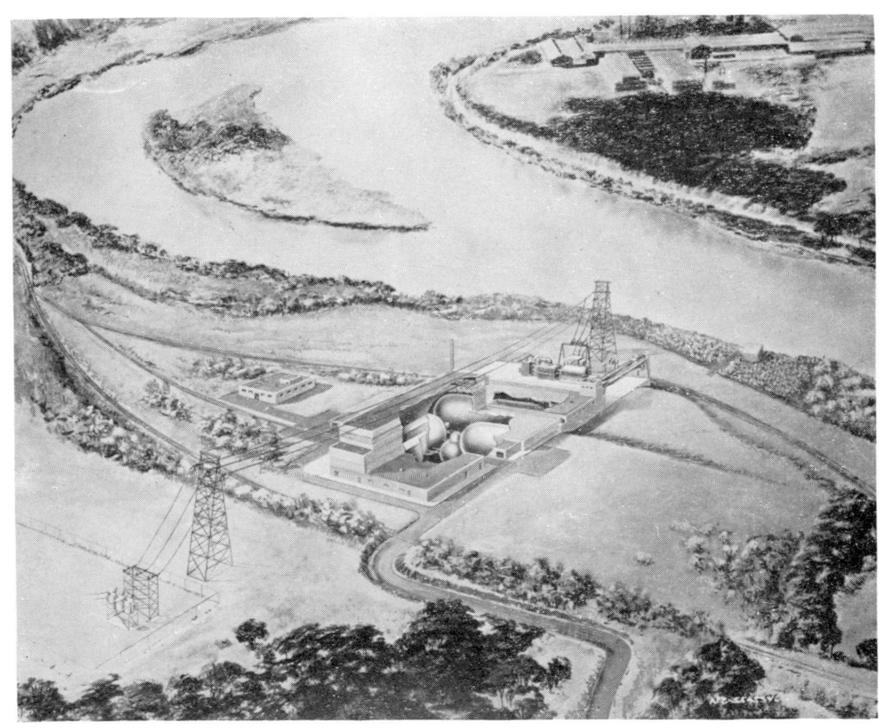

FIG. 9. Artist's sketch of the Shippingport Nuclear Power Plant. (Westinghouse Atomic Power Division Photograph.)

The site of this nuclear power station is on the south bank of the Ohio River in the southwestern portion of Pennsylvania as described in References 9 through 11. The airline distance from the site to Pittsburgh is about 25 miles; to Weirton, West Virginia about 17.5 miles; and to Youngstown, Ohio, about 35 miles.

The tract of land on which the power plant is located consists of over 420 acres and the plot is of irregular shape with a mean width of about 3500 ft and a mean length of about 5500 ft. A strip of land, 100 ft in width and owned by the New Cumberland and Pittsburgh Railway Company, runs

FIG. 10. Artist's concept of Indian Point Nuclear Power Plant. (Babcock and Wilcox Company photograph.)

FIG. 11. The Tower Shielding Facility. (Oak Ridge National Laboratory.)

through the property in a general east–west direction. The site is generally hilly and slopes upward from the river.

Except for the railroad property adjacent to the plant proper, the nearest land not owned by Duquesne Light Company is Phyllis Island. This island is owned by Dravo Corporation and is uninhabitable. The river is approximately 550 ft from the reactor, while the eastern boundary line of the property

FIG. 12. N.S. *Savannah* showing containment vessel. (New York Shipbuilding Corporation photograph.)

is located 1700 ft away. Most of the area within a half-mile from the reactor will be controlled as to its use and occupancy, and this includes all this tract owned by Duquesne. Phyllis Island will also form part of the controlled area. An agreement to that effect was reached between Dravo and Duquesne. Under the terms of this agreement, Dravo agreed not to use or permit the use of the land for any structures, place or area where the public at large can visit or assemble. This agreement is binding on Dravo and any future purchaser or leasee until March, 1994. The property in the area owned by New Cumberland and Pittsburgh Railway and National Transit Companies are not included in the controlled area. Extensive negotiations were conducted with both companies in an effort to reach an agreement with them which would permit the inclusion of their properties within the half-mile circle in the controlled area, however, these negotiations were unsuccessful.

Despite the proximity of the Shippingport Atomic Power Station to the City of Pittsburgh, the population density in the immediate vicinity of the site is relatively low. Within a half-mile radius of the plant there are no residents. There is only one area of concentrated population within a 5 mile radius of the site. This is the Borough of Midland, population 6500, which is located between the 1 and 3 mile radii from the site. The population within a 5 mile radius of the site is approximately 20,000. Between the 5 and 10 mile radii from the site, the population is estimated to be 105,000. Beyond the 10 mile radius and to the east and southeast are the more densely populated Allegheny County and the City of Pittsburgh with its surrounding suburbs. The population of Allegheny County is over 1,500,000.

Within the immediate area of the plant site there are no major industries. The single track freight line of the New Cumberland and Pittsburgh Railway, which traverses the property, is considered of minor importance, since only one or two freight trains travel this line daily. Two oil lines cross the property from south to north near the eastern boundary and several oil storage tanks are located about a half-mile from the plant to the east. The closest major industrial installations are located in Midland about 1 mile northwest of the site. There are four important factories in Midland and these with one other plant upstream and adjacent to the river are the only major industries within a 5 mile radius of the site.

GENERAL SITE REQUIREMENTS

In most cases the decision to construct a nuclear facility by an organization includes a selection of the general areas for the installation. Only in unusual cases such as the site selection studies for the Savannah River plant is a survey of an entire region of the United States required. However, in selecting the final location within a general area for construction, several factors must be evaluated and the procedure of evaluation usually follows the normal practices in industrial plant site selection.

For power reactors large volumes of cooling water are required and this requirement limits most power reactor sites to the immediate vicinity of bodies of water including rivers and lakes. Usually it is preferred to avoid salt water because of corrosion to stainless steel. The quality of the water determines the expense that will be required in water treatment. In considering a supply of cooling water it is necessary to obtain the values for the minimum flow of a stream or the minimum level of a lake used as a source of cooling water. The large power requirements for pumping cooling water also limit the distance from the plant to the water intake and the head through which the water must be pumped.

FIG. 13. Aircraft nuclear propulsion test area at the National Reactor Testing Station. (General Electric Company photograph.)

There are a number of factors affecting construction costs and operating costs at a site including transportation, availability of power during construction, suitability of the ground to support heavy structures, ground waters and the geology of the site. The elevation of the site should be above flood levels.

HYPOTHETICAL SITE DESCRIPTION

The following is a condensation of the "Summary of major characteristics of hypothetical site" developed by the Atomic Energy Commission for design and cost studies of pressurized water, boiling water and organic moderated reactors. This slightly abbreviated description is taken from the Forum manual, "*A uniform procedure for use in evaluation of nuclear power reactors*".[10]

FIG. 14. Camp desert rock located just outside the boundary of the Nevada test site. (USAEC Albuquerque Operations Office photograph.)

(1) TOPOGRAPHY AND GENERAL CHARACTERISTICS

(a) *Location and Total Area*—The site is located on the bank of the North River and 35 miles from Middletown, the nearest city. It is 40 ft above the minimum river level and 20 ft above maximum level. The crib house will be located 1200 ft from the river bank. The site occupies 1200 acres of level terrain and is grass covered.

(b) *Access*—The site is accessible by highway, railroad, air and water as follows:

A 15-mile secondary road to a state highway has been constructed and needs no additional improvements.

A 15-mile spur to be constructed which will intersect the existing railroad.

An airfield 20 miles from the site.

North River is navigable throughout the year. All plant shipments will be made overland, except that heavy equipment such as the reactor vessel and generator stator may be barged to the site.

(c) *Population*—The site is in an area of low population density. Variation in population with distance from the site boundary is:

Miles	Population
0.25	0
0.5	60
1.0	200
5.0	2,700
10.0	8,000
20.0	40,000

The nearest residence is $\frac{3}{8}$ mile from the site.

(d) *Land Use in Surrounding Region*—Five small manufacturing plants, each employing less than 100 persons, lie within 15 miles of the site. Most of the population is centered in eight small towns encompassed by the 20-mile radius. The closest town is 4 miles away. The remaining land is forested or used for cultivated crops.

(e) *Public Water Supplies*—The North River provides adequate water for condenser cooling and auxiliary use. The water temperature is as follows: Maximum 75°F, minimum 40°F, annual average 57°F.
A typical analysis follows:

pH	8
Total dissolved solids	261 ppm
Organic matter	28
Phenolphthalein alkalinity (as $CaCO_3$)	8
Methyl Orange alkalinity (as $CaCO_3$)	162
Hardness (as $CaCO_3$)	152
Total iron (as Fe_2O_3)	0.45

(2) Meteorology and Climatology

(a) *Prevailing Wind Variation*—Prevailing surface winds in the region blow from the *S* through *W* quadrant at speeds varying from 4 to 15 m.p.h. throughout the year. There are no large daily variations in wind speed or direction.

(b) *Average Temperature History*—January is the coldest month and July the hottest with monthly average temperatures as follows:

	January	July
Maximum, °F	34	89
Minimum, °F	15	67
Mean, °F	25	78

90°F dry bulb, 75°F wet bulb temperature exceeded during 5 per cent of total number of summer hours.

Design should be based on a minimum temperature of −10°F and a maximum of 100°F.

(c) *Temperature Inversions*—Surface-based atmospheric inversions occur frequently during summer and early fall nights, but are destroyed quickly by solar heating. Winter and spring inversions are more likely to extend into the day-time. Stagnation periods with steady light winds and a high frequency of inversions are most probably from August to October. A persistent inversion with its base between 1000–4000 ft, wind speeds less than 5 m.p.h. below 5000 ft and clear skies which permit formation of surface-based inversions are characteristic of these periods. The average percentage of time with inversion is lowest in May with 39 per cent and highest in October with 56 per cent. Yearly average is 48 per cent against a national average of 50 per cent.

(d) *Frequency and Severity of Disturbances*—A maximum wind velocity of 100 m.p.h. has been recorded at the site. Thunderstorms occur 46 days a year on the average. Of these 22 per cent occur in June and 91 per cent in the period April–September, inclusive. Hail falls 4.3 days per year, 70 per cent during April, May and June.

(e) *Snow Load*—30 psf shall be used for snow loading.

(3) Hydrology

(a) *Precipitation*—Average annual rainfall is 27.4 in. January is the dryest month with 0.8 in. and June is wettest with 4.3 in. The period April–September accounts for 75 per cent of the total rainfall. Maximum rainfall: 0.5 in. rain in 10 min every two years; 4 in. of rain in 24 hr every 5 years.

(b) *Drainage*—Natural drainage of the site is provided by the land contours. Subterranean water travels toward the river at a velocity of 300 ft per year. The maximum temperature is 75°F with sufficient flow available to prevent exceeding the allowable temperature rise specified by the state.

(c) *Ground Water*—Ground water collects mostly in the weathered shale layer above bedrock. (See Geology and Seismology, below). Adequate ground water for sanitary use and plant make-up is available within 50 ft below grade. Alkalinity of the water (as $CaCO_3$) is 380 ppm. Hardness (as $CaCO_3$) is 481 ppm. A temperature of 52°F is typical.

(4) Geology and Seismology

(a) *Soil Profiles and Load Bearing Characteristics*—Soil profiles show alluvial soil and rock fill to a depth of 8 ft; Brassfile limestone to a depth of 30 ft; blue weathered shale and fossiliferous Richmond limestone to a depth of 50 ft and bedrock over a depth of 50 ft. Allowable soil bearing is 6000 psf and rock bearing characteristics are 18,000 psf and 15,000 psf for the Brassfield and Richmond Strata, respectively. No significant cavities exist in the limestone.

(b) *Seismology*—This is a Zone 1 site as designated by the Uniform Building Code.

(5) WASTE DISPOSAL

(a) *Sewage*—All sewage must receive primary and secondary treatment prior to dumping into the North River.

(b) *Volatile Wastes* (*Radioactive and Toxic Gas*)—Maximum permissible concentrations or dosages shall be as prescribed in:

(i) Standards for protection against radiation, AEC Regulation (10 CFR Part 20) Federal Register Doc. 57–511, filed January 25, 1957.

(ii) National Bureau of Standards, Handbook 69, "*Maximum Permissible Amounts of Radioisotopes in the Human Body and Maximum Permissible Concentrations in Air and Water*".

(iii) In event of conflict between (1) and (2), above (1) shall govern.

(c) *Liquid and Solid Waste*—Maximum permissible activity of water entering North River shall be as prescribed in references under "Volatile Wastes", above. Activity of liquid effluent shall be measured as it leaves the plant. No credit for dilution in the North River will be assumed.

Storage on site for decay will be permissible but no ultimate disposal on site will be made. Assume long term storage in secure tankage and/or commercial rates of sea disposal with shipping site 500 miles from plant.

SPECIAL PROBLEMS

The design of a nuclear-powered merchant ship presents the problem of operating a power reactor without exclusion distance. Therefore adequate shielding of the containment vessel with lead and polyethylene is required for the N.S. *Savannah* to reduce any direct radiation hazards. Also the problem of allowable leakage rates of the containment vessel require careful study. Since the structure will be subject to movement, there will be difficulties with maintaining the leak tightness of closures and points of penetration of the vessel by pipes and cables so that periodic measurements of the containment vessel leakage rate will be required. Control of the ventilation of occupied areas in a nuclear powered ship requires careful study.

Certain special reactor tests where the release of fission products is probable such as the testing of aircraft reactors are done at remote areas of the National Reactor Testing Station and every reasonable means is taken to limit the escape into the environs of any material released from the core. For testing very high temperature reactors such as for nuclear rockets where there will be a substantial release of fission products, an extremely remote site in Nevada is used.

REFERENCES

1. BECK, C. K., Safety factors to be considered in reactor siting, Report TID-7579, June 1959.
2. COTTRELL, W. B., Site selection criteria, *Nuclear Safety*, **1**(2), 2–5 (December 1959).
3. COTTRELL *et al.*, Oak Ridge National Laboratory, Sept. 1959 (Unpublished).
4. Permissible dose from external sources of ionizing radiation, Addendum to National Bureau of Standards Handbook 59, U.S. Department of Commerce, April 1958.
5. PRICE, B. T., HORTON, C. C., and SPINNEY, K. T., *Radiation Protection*, Pergamon Press, London, 1958.
6. Maximum permissible dietary contamination after the accidental release of radioactive material from a nuclear reactor, *Brit. Med. J.*, **1**, 967–969 (1959).
7. Siting of nuclear power reactors, 9–11, Atomic Industrial Forum, Inc., August 1959.
8. FARMER, F. R., and FLETCHER, P. T., Siting in relation to normal reactor operations and accident conditions, presented at International Symposium on the Safety and Location of Nuclear Plants, June 16–19, 1959, Rome, Italy.
9. Description of the Shippingport Atomic Power Station, WAPD-PWR-970 (TID-4500), 13th ed, June 1957.
10. Atomic Industrial Forum Manual, A uniform procedure for use in evaluation of nuclear power reactors.
11. MCALLISTER, R. J., WIRTH, W. B., and HARRIS, T. B., Jr., Description of Shippingport Atomic Power Station site and surrounding area with radiation background and meteorological data, USAEC Report WAPD-SC-547.

CHAPTER 10

OPERATING EXPERIENCE

THE experience to date with the operation of nuclear reactors has been remarkably free from accidents creating any public hazards. This experience extends back to December, 1942 and now includes several hundred reactor-years of operations of all types ranging from criticality facilities to large production and power reactors. It does appear from this record for safe operation that nuclear reactors can be designed and operated so as to be very reliable and stable devices provided that sufficient attention is given to design, construction, training of operators, supervision, testing and inspection. It is quite probable that the extreme care that has been exercised in all details affecting safety has resulted in some excess costs. For example containment for a pressurized water reactor is an item of considerable expense in the construction of this type of nuclear power station. Such a reactor, with which there is now considerable favorable operating experience, could be safely operated at some sites without the usual containment provision. A continuing record of safe reactor operations will probably lead to the acceptance of some power reactors without containment.

The operating experience with nuclear installations has not been quite perfect since human beings and machines which sometimes fail are involved. The history of reactor incidents is useful to study since much of our knowledge of reactor safety has come from the conscientious reporting of the details and circumstances of failures. Some interesting and instructive events were not recorded and have been nearly forgotten. However it is believed that the following review of known reactor incidents is nearly complete.

FUEL ELEMENT FAILURES

There have been several hundred occasions of release of some fission products from leaking fuel elements. In all but a few cases the leakage was detected at an early stage and no significant damages resulted except for the replacement cost of the faulty fuel element. This type of experience has been observed for aluminum-jacketed natural–uranium fuel slugs and aluminum-clad enriched-uranium–aluminum alloy fuel plates of the MTR type. Sometimes fuel elements of a type with which there has been years of satisfactory experience will, for reason of minor unnoticed changes, have a number of failures.

It has been observed that the frequency of fuel element failures increases with power density and low power research reactors are usually free from this difficulty. On a few occasions leaking fuel elements have caused contamination which is very expensive to remove and in a very few cases the swelling of a faulty fuel slug has plugged a fuel tube and caused damages.

Nuclear facilities operating at significant power levels, should be designed to provide for the possible event of a fuel element failure. Design provision may include shielding for the primary cooling system and means for removing radioactive material from the coolant. Instrumentation to detect and locate a leaking fuel element may also be required.

CRITICALITY ACCIDENTS—1

The following description of a fatal accident in a critical mass experiment is taken from Report TID-5360, by D. F. Hayes.

An employee at the Los Alamos Scientific Laboratory was working alone at night making critical mass studies and measurements. Blocks of the tamper material to reflect neutrons were being stacked around a mass of fissionable material. Nearly a critical configuration had been reached when the employee was lifting one last piece of tamper material which was quite heavy. As this piece neared the assembly, the instrument indicating radiation levels showed that fission multiplication would be produced by this additional piece and the employee moved his hand to set the last block at a distance from the pile. In doing so, he accidently dropped the block which landed directly on top of the assembly.

A "blue-glow" was observed and the employee proceeded to knock the assembly apart and the chain reaction was stopped. The individual conducting the experiment received an exposure sufficient to cause death in 24 days and the exposure has been estimated to be equivalent to 590 r. A guard who was seated 12 ft from the experiment received an exposure estimated at 32 r, but suffered no observable permanent injury.

As a result of this incident, a special committee was set up to review the work carried on whenever fissionable material was being assembled under conditions where the assembly might conceivably become critical, and make appropriate safety recommendations. The regulations given below resulted from the work of this committee and are aimed not only at reducing the possibility of a similar accident, but also at moderating the severity should a similar accident occur:

(1) Two special lists shall be prepared naming the persons who are the only ones permitted to do criticality work; one list carries the names of persons who would head up such work.

(2) A minimum of two persons are required to be present in addition to guard personnel. A maximum number of persons permitted is also controlled.

(3) Before starting an experiment, it must be planned, including the mode of operation and behavior for all contingencies.
(4) At least two monitoring instruments must be operating, each giving a fairly audible indication of the neutron intensity at all times during the experiment.
(5) Any change in the assembly involving fissionable material must be made by one person at a time, slowly and readily reversibly.
(6) There shall never be two assemblies, which might become critical, in the same building (room) at the same time.
(7) All operators associated with the critical experiment must be in agreement about the safety procedure all through the operation. If any disagreement arises, operation must stop until an agreement has been reached.

Also as a result of this accident substantial consideration was given to the development of mechanical remote control devices for conducting criticality experiments and in 1946, all such work at Los Alamos was carried out with such remote control systems without exposure to personnel.

CRITICALITY ACCIDENTS—2

The following description of a nuclear incident is taken from Report TID-5360, by D. F. Hayes.

On 4 June, 1945, at the Los Alamos Scientific Laboratory, an assembly designed to measure the critical mass of enriched uranium when surrounded by hydrogenous materials, became super-critical and caused the exposure of three individuals to external radiation. The enriched uranium was in the form of cast blocks of metal and the blocks were stacked in a pseudospherical arrangement in twelve courses in a 6 in. × 6 in. × 6 in. polyethylene box. Voids in the courses were filled with polyethylene blocks of appropriate dimensions. The polyethylene box was supported by a 2 ft. high stool within a 3 ft cubical steel tank. The tank had a 2 in. opening in the bottom through which it could be filled and drained by means of supply and drain hoses attached to a $\frac{3}{4}$ in. tee. The opening in the box was fitted with a shut-off valve, as was the drain hose. A polonium–beryllium source of about 200 mc strength was placed on top of the assembly. A fission chamber and a boron proportional counter were used to follow the experiment.

An immediate superior was absent when the experiment was beginning. According to one of the operators, the water level was raised above the polonium–beryllium source with the supply valve nearly full open. At this point, a slight increase in counting rate was observed, which corresponded with what had been observed previously when the source alone was immersed in water. A few seconds later the counting rate began to increase at an alarming rate. At this point, the supervisor returned and walked to within three feet of the tank and noted a "blue-glow" surrounding the box. Simultaneously, the two operators were hastily closing the supply valve and opening the drain valve. The building was then evacuated.

The three individuals involved received excessive radiation exposure, estimated in the case of two, at about 66.5 rep and the third at about 7.4 rep. The involved individuals were hospitalized for observation but no untoward symptoms appeared. There was no damage to equipment, no loss of active material or local contamination problem.

It seems most probable that the unanticipated increase in activity occurred because water seeped between the blocks of active material, increasing the internal moderation of the assembly. Sufficient heating occurred to melt and deform the plastic bottom of the assembly container, and the water in the bottom of the box was vaporized, thus stopping the reaction. More water would then seep in and the cycle would repeat. It was estimated that three such cycles occurred before the water level in the tank was sufficiently low to prevent further reaction.

The experiment was of poor design in that the rate of addition of moderating material was not restricted in any way beyond the inherent limitation of the supply valve, and the changes in the water level were not readily reversible, since in the first place, the supply and drain valves were at least 15 ft apart; and in the second place, there was a considerable lag between the water level in the tank and that within the assembly. There was no provision made for rapidly dumping the moderator or poisoning the assembly, if a predetermined upper safe limit of reactivity were reached. It is not apparent that any nuclear safety considerations whatever had entered into the planning. It was the opinion of the experimenters that criticality would not be reached during this run.

CRITICALITY ACCIDENTS—3

The following description of a nuclear incident is taken from Report TID-5360, by D. F. Hayes.

On 21 May, 1946, during a demonstration to a group at the Los Alamos Scientific Laboratory, on the techniques of critical assembly, associated studies and measurements, an assembly inadvertently became super-critical and the radiation injuries that resulted caused one fatality and excessive exposure to several individuals. In this demonstration a hollow hemisphere of beryllium was being placed over a mass of fissionable material which was resting in a similar lower hollow hemisphere. The system had been checked with two 1 in. spacers between the upper hemisphere and the lower shell, which contained the fissionable material; the system was sub-critical at that time. Then the spacers were removed so that one edge of the upper hemisphere rested on the lower shell while the other edge of the upper hemisphere was supported by a screw driver. This latter edge was slowly permitted to approach the lower shell. The demonstrator held the screw driver with one hand and with the other hand held the upper shell with his thumb placed in an opening at a polar point. At this time, the screw driver apparently slipped and the upper shell fell in position around the fissionable material. The "blue-glow" was observed, the heat wave felt, and immediately the top shell was slipped off and everybody left the room.

There were eight people present in the room when this demonstration was carried out and of these, two were directly involved in this work leading to the incident. The man who was demonstrating the experiment received a radiation dosage sufficient to result in injuries from which he died nine days later; his exposure was estimated to be more than 1000 r. The man who was assisting in this demonstration received a radiation dosage estimated at 416 r which was sufficient to cause serious injuries and some permanent partial disability. The other six people in the room were exposed beyond the established daily limits with some of them receiving an estimated exposure of more than 100 r. However, there were no observable permanent injuries. Following this accident, all such work was stopped until mechanical remote control devices were designed and fabricated for criticality studies and demonstrations.

CRITICALITY ACCIDENTS—4

With Homogeneous Reactor

The following description is taken largely from Report TID-5360, by D. F. Hayes.

An inadvertent super-criticality of the Los Alamos Homogeneous Reactor occurred in December, 1949, when an employee manually tested the mechanism for two new control rods. The reactor was being remodeled for higher power operation. As part of the required operation, two new control rods had been placed in the system in addition to the three existing control rods. The employee who had built the control rod mechanism wanted to test the comparative fall times of these new rods. He opened the enclosure on top of the reactor and manually lifted the rods, neglecting the possibility that this would affect the reactivity of the reactor because of its higher power arrangement. Before this addition, the other three existing rods were sufficient for safety.

There were no injuries and the employee doing the work received 2.5 r of gamma radiation according to his film badge. There was no damage done to the reactor and no loss of active material.

Normally rods were raised remotely from the control room where the control panel was activated by a key switch. Since the rods were pulled out manually with the panel being off, no equipment was turned on except a direct reading temperature measurement instrument. Therefore, there was no neutron sensitive device to record or warn of a rise in neutron level. It was not observed until after the accident that the reactor temperature had risen about 25°C. The removal of the two rods probably gave an increase in reactivity of about 0.86 per cent producing an initial period of about 0.16 sec. Since the measured temperature coefficient is approximately -0.034 per cent $k/°C$, the observed temperature rise indicates the rods were out sufficiently long so that the reaction was stopped by the negative temperature coefficient.

To avoid any similar incidents in the future, the use of the key to the locked enclosure above the reactor containing the rod mechanism is restricted to two senior individuals in the group. In addition to this administrative arrangement, a special device is added to the rod control mechanism. This device is used whenever the enclosure had to be left open for any period of time. The special device was a pivoted plate provided with a slide which would permit only one control rod to be moved at a time after the plate was locked into position at the top of the control rods.

CRITICALITY ACCIDENTS—5

The following description of a nuclear incident is taken from Report TID-5360, by D. F. Hayes.

A fissionable material assembly, remote from personnel, inadvertently was made super-critical at the Los Alamos Scientific Laboratory on 20 March, 1951. Interactions between two masses of fissionable material in water were being measured at progressive decreasing horizontal separations. Remotely controlled operations had established the desired horizontal separation of the two components and flooded the system. After the final measurements the system was to have been placed in a safe condition with the rapid disassembly mechanism. When this mechanism was actuated the safety monitor indicators went off scale, the neutron counters jammed, and the television viewer indicated steaming. Within a few minutes the indicators and counters returned to operating ranges and indicated a rapid decay of radioactivity. There was no injury, no loss of material and no damage to the facilities.

Unlike normal slow disassembly, the scram rapidly raised one component vertically instead of horizontally separating the two masses. Subsequent tests showed that the center of reactivity of the movable component was below that of the stationary, thus shortening the distance between them as the movable unit rose to the level of the stationary unit. Furthermore, the movable unit when rapidly raised from the water swung toward the stationary unit due to the Bernoulli effect. There probably were several radiation bursts, each stopped by a bubble formation in the water. The total yield was about 10^{17} fissions.

As a result of this experience, "tangential scrams" were outlawed for future design. Also, all assembly machine design was reviewed by several persons rather than being the work of an individual.

CRITICALITY ACCIDENTS—6

Radiation Excursion in a Homogeneous Critical Assembly

The following description of a nuclear incident is taken from Report TID-5360, by D. F. Hayes.

An inadvertent super-criticality of a spherical plutonium-nitrate solution reactor occurred at Hanford on 16 November, 1951, when a safety rod was withdrawn too rapidly during a test.

This accident occurred during a program for measuring the critical mass of solutions and geometries involved, which pertained to chemical separation plant operations from the standpoint of nuclear safety. Criticality experiments were being performed for several container geometries and process reagent concentrations. The fuel consisted of solutions of plutonium-nitrate and the container geometries studied were reflected cylinders, and reflected and bare spheres of different sizes. In the course of the criticality program, it became apparent that it would be desirable to determine experimentally the critical mass of a hemispherical shape. It was decided to make such a measurement by half filling an available sphere and adjusting the concentration to obtain criticality, which was obtained in a spherical segment slightly larger than a hemispherical and nominal 20 in. diameter sphere. Three additional critical points were found as the fuel was diluted and greater fractions of the sphere were filled. The last critical point reached was in a volume 88 per cent of the full sphere. As the critical concentration of the sphere was predictable, it was decided to make the final dilutions for the full sphere as close as possible. This required the total fluid volume be known quite accurately. The method of making this measurement was to add the remaining fuel to the reactor and to determine the total volume by means of the reactor site glass which gave accurate measurement of the fluid volume. The control and safety rods were inserted and were known to be sufficiently strong to easily override the reactivity of the excess fuel addition. The volume measurement was done carefully without incident or significant increase in neutron level. Then, instead of draining the reactor for concentration change, an attempt was made to determine where criticality might occur on the rods. As the total strength of the safety rods was known, it was thought that some additional information as to the required dilution could be determined by this measurement. The control rod was pulled first with very minor reactivity effect. Following this the safety rod was withdrawn intermittently at high speed (2.3 in./sec). The waiting period for the delayed neutron effect of about 15 sec was made just prior to the incident. This was too short a period to determine whether or not the assembly was critical. The operators next heard safety controls actuate, instrument indicators went off scale, scalers jammed, and the portable counter in the control room was off scale. Presumably, a further rod withdrawal had been made and caused the incident.

Although the temperature of the fuel was well below boiling, yet a small amount of fuel was forced out through the gaskets of the reactor assembly and pressures considerably in excess of atmospheric must have existed in the assembly during the incident. These pressures were due to the formation

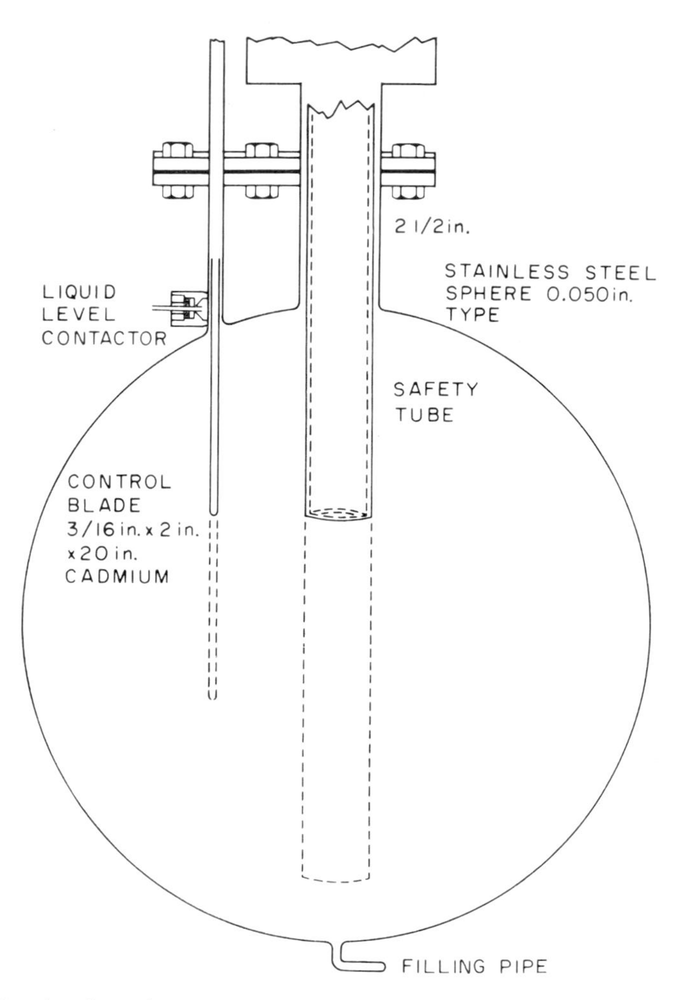

FIG. 1. A schematic diagram of the reactor assembly. (Reproduced from Report TID-5360 by D. F. Hayes.)

of gases in the fuel which caused an expansion of the fuel. This expansion and reduction in density changed the fuel in a way that was of some importance, since an extremely unfavorable change in geometry resulted from fuel expansion in the partly filled sphere. Had the sphere been full in this incidence so that the fuel expansion immediately expelled fuel from this sphere, no incident of this magnitude would have occurred. Extreme caution is, therefore, indicated in experiments involving partial geometries.

This incident was caused by the rapid withdrawal of a strong poison from the reactor. Reactivity was increased at such a rate that the system became prompt-critical before the power level had increased sufficiently to actuate the scram devices. Short periods were then obtained in a time interval which was short compared to that necessary for mechanical safeties to stop the reaction. It is concluded from this experience that safe reactor design requires that no mechanism be employed which will allow rapid withdrawal of a strong poison from the system to eliminate possible errors in the judgment on the part of the operators.

There were no injuries since the assembly was remotely operated. The building was successfully decontaminated except for the test room. Decontamination work was proceeding in the test assembly room on the reactor control assembly. On 4 December, 1951, after the building had been closed down after normal days work, approximately four hours later a fire was discovered by the security patrol and it was immediately apparent that the fire had originated in the contaminated storage room and it had spread to the chemical laboratory exhaust hood filters and the attic. Considerable difficulty was experienced in extinguishing the blaze in the hood filters. It was necessary to break open the filter unit and remove the filter element to put out the flames and prevent further release of plutonium to the atmosphere. Oxygen masks and other protective equipment were required due to the plutonium hazards. The total amount of plutonium release was estimated to be between 1 and 4 g. Most of this came from the burned hood filter and the remainder from the previously contaminated waste. A part of the released plutonium was washed from the building by the water to extinguish the fire and deposited on the ground outside the building. The contaminated ground was covered with sand to prevent the pickup of active material from the ground by winds.

Fire damage obliterated the cause of the fire but it was the opinion of the members of the damage investigating group that the fire originated by spontaneous ignition in the contaminated waste boxed in cardboard cartons in the contaminated waste room. Large quantities of contaminated waste in the form of paper, nitric–acid laden glass wool, rags and miscellaneous material were known to be packaged in this manner prior to accumulation of sufficient number of boxes to require burial.

The building used for the criticality experiment and in which this fire occurred, was built for the purpose of conducting such tests and was thus remotely located and of minimum value. It was, therefore, economically feasible to abandon the structure after the radiation burst and the later fire.

CRITICALITY ACCIDENTS—7

The following description of a nuclear incident is taken from Report TID-5360, by D. F. Hayes.

On 18 April, 1952, at the Los Alamos Scientific Laboratory, an assembly of fissionable material, remote from personnel, inadvertently was made super-critical. Two stacks of fissionable material as discs were being built up step-wise to give a slightly sub-critical assembly. The bottom half of the assembly was moved in position by a hydraulic ram with a 22 in. stroke and the height to which this bottom part was lifted was determined by a precision ground spacer giving a controlled separation distance between the upper and lower parts. The individual stacks were built up by hand in the fixed assembly and then the two stacks were brought together by remote mechanisms.

After two members of the operating crew calculated, from previous steps, that one more disc could be added safely (they thought), the disc was added and the system was assembled remotely by raising the lower hydraulic ram. Radiation indicators went off scale actuating the scram system and a puff of smoke was observed on the television viewer. Within 3 to 5 min the indicators and counters returned to operating range. As a result of the remote operation, there were no injuries. The cause of the accident was an algebraic error when the safety of the final addition of fissionable material was checked. This error would have been detected, if data had been plotted as required by operating rules. Also, it was determined subsequently that the assembly rate of the system exceeded the maximum established rate. The radiation burst was terminated by thermal expansion, with a yield of $1\text{–}2 \times 10^{16}$ fissions.

CRITICALITY ACCIDENTS—8

ZPR-1 Accidental Power Excursion

A critical assembly, designated ZPR–1 and constructed in connection with the development of pressurized water reactors, was being used to calibrate a control rod on 2 June, 1952, when a power excursion caused damage to the core and radiation exposure to some of the operating personnel.

This assembly was in an open tank which was filled with water, when the reactor was to be made critical. The core was an assembly of units each containing 6 zirconium strips and 7 fuel strips. The zirconium strips were each 0.910 in. wide by 0.110 in. thick and were uniformly spaced to occupy 60 per cent of the volume of the elemental core section which was 43 in. long and had a cross section of 1 $in.^2$ The 7 fuel strips each contained 3.2 g of uranium-235 as oxide particles uniformly dispersed in polystyrene. The normal strip size was 0.013 in. by 0.85 in. by 43.0 in. and these strips were fastened to the zirconium plates. A cylindrical core of approximately a circular cross section had been assembled from these elements with 4 blade-type rods acting as safeties near the periphery and the control rods to be tested at the center. The initial loading was 5261 g of uranium-235, however, since it was desired to make measurements with the control rods partly inserted, the fuel loading was increased to 6787 g of uranium-235 and the core then consisted of 324 standard fuel elements.

With the tank filled with water the unit was brought to criticality and measurements were started with the control rods clamped into position. The assembly was then made sub-critical with the safety rods and, with the water still in the reactor tank, the assembly room was entered by the 4 individuals conducting the test. The operator leaned over the assembly and unclamped the central control rod and began to withdraw it to change the assembly for other measurements. By the time the rod had been withdrawn about a foot, a dull thud was heard and a blue light emanation persisting for a fraction of a second was observed. There was lateral displacement of the tops of the fuel elements and water was displaced suddenly from the core about 10 in. above the normal water level as a large bubble with some gas evolution. The operator dropped the control rod and it fell down its guide tube into the core. The people in the room ran out. A water dump valve which had been opened by an instrument trip circuit let the water out of the tank exposing the core partly before the operator left.

Another criticality experiment ZPR-11 which was in operation at the time at a distance of 25 ft and separated from ZPR-1 by a 3 ft concrete wall, was shut down by instrumentation detecting the radiation from this power burst.

The reactor room was contaminated with fission products which entered the control room through cracks around the door separating the control room from the reactor room. Measurements made outside the building showed abnormally high readings which were later found to be due to fall-out from an earlier unannounced weapons test at the Nevada test site. This coincidence of arrival of fall out from a weapons test with the nuclear incident caused some anxiety until the cause of the radioactivity over the general Chicago area was identified.

After a delay of 8 days for decay of activity, the reactor room was decontaminated and the reactor disassembled. Assays indicated that about $\frac{1}{2}$ per cent of the fission product activity was in the water which was retained for 3 months and then dumped.

The reactor was reconstructed with modifications and experiments were made to obtain data for a detailed analysis of the incident. From this and the records of the incident the sequence of events and the exposure level were estimated. A summary of data relating to the excursion is presented in Table I.

The mechanism of shutdown of this reactor was unexpected but very effective. When the power reached a value of several hundred kilowatts, the plastic containing the fuel began to heat in the region of maximum power generation. With all but $\frac{1}{2}$ per cent of the uranium oxide distributed as particles of 10 micron diameter or less, the plastic was heated in a uniform manner so that the plastic temperature increased with the reactor power. Coincidentally, the large particles of uranium oxide of about 40 micron diameter were at a higher surface temperature because the volume to area ratio of these large particles was 8 times that of the small particles. By the time the general temperature of the plastic had reached 80°C, its softening

Table I

Summary of ZPR–1 Excursion

$\Delta k/k$ sub-critical at start	−0.25%
$\Delta k/k$ added before shutdown	1.87%
$\Delta k/k$ maximum	1.62%
Shortest period	0.01 sec
Maximum power level	165 MW
Total number of fissions	1.22×10^{17}
Duration of burst (99%)	50 msec
Total energy release	1040 Whr
Duration of excursion	600 msec
Average temperature reached in plastic at position of peak flux	390°C
Prompt gamma burst	150 r at water surface 8 in. above core
Fast neutron burst	39 r at water surface 8 in. above core
Decay gamma intensity	200 r/sec at top of core one second after shutdown (water down)

point, the temperature of the plastic locally in the region of the large particles had reached temperatures of several hundred degrees and conditions existed for the start of bubble formations with the particles as nuclei. The plastic vapor bubbles caused the plastic strips to grow in volume and force the moderator water out of the core. Since an increase in strip thickness of $\frac{1}{2}$ per cent reduced reactivity by 1 per cent, the reactor became sub-critical within a few seconds after bubble formations began. Even after the nuclear reaction had stopped, the bubble grew and displaced most all the water in the core and forced the fuel elements to move laterally. Since the reactor was sub-critical within a very short time, the dropping of the control rod by the operator was of no importance. Also, since the water was permanently displaced from the core by the foamed plastic, re-occurrence of criticality was prevented. The plastic expansion took place essentially in the last 10 msec of the power burst.

The personnel exposed to radiation were examined by the medical staff and then removed to a hospital. Since at that time the magnitude of the radiation exposure was not known, the patients were placed at absolute bed rest. From studies of film badges and various instruments, it appears that the neutron and gamma exposures to the 4 individuals were 136 rep, 127 rep, 50 rep, and 9 rep. Radiation sickness was not observed in the patients and they were not at any time clinically ill.

CRITICALITY ACCIDENTS—9

Polonium Contamination Problem

A container of a Po–Be neutron source, enclosing nearly 7 curies of polonium, was ruptured in the Oak Ridge Critical Experimental Laboratory resulting in the significant exposure of one person to polonium and the gross contamination of a large area and its contents. This incident occurred in March, 1951, and although not involving a nuclear criticality incident is included because of the problem of safely handling such neutron sources in all criticality experiments and reactor operations. The following description of this contamination problem is taken from Report ORNL-1381 (rev.) by Dixon Callihan and Don Ross dated 12 August, 1952.

A neutron source of approximately $3 \times 10^7 n/\text{sec}$ was required for instrument testing and as a supply of neutrons for multiplication by near-critical assemblies of uranium. The source as made by the Mound Laboratory contained originally nearly 16 curies of polonium although, by the time of the accidental rupture, the polonium content had decayed to about 6.9 curies. The polonium had been vaporized into beryllium powder (0.40) g and was contained in a beryllium tube plugged with a piece of beryllium without threading and sealed by electro-plating with nickle. The users at the Oak Ridge National Laboratory believed the source had a threaded closure and, therefore, in their equipment design lifted the source by the plug. The source was attached to an aluminum rod of a diameter equal to that of the source capsule which was screwed into this rod by a female thread. By this rod and an electric motor the source was moved vertically from the adjoining control room through a guide tube.

No critical experiment was in progress at the time of the accident. The source was being used to test a neutron detection instrument, and in the process, it was being alternately raised and lowered. Apparently, in this operation, the capsule snagged, in some unknown manner, and the threaded plug was pulled out, allowing the capsule to fall about 30 in. into the tank in which criticality experiments were conducted but which contained no water or fissionable material at the time. That an irregularity in the experiment had occurred first became evident when the instrument failed to follow the motion of the source. In order to investigate, a member of the Technical Staff doing the tests entered the room and found the capsule resting within the tank. Believing the capsule had only become disconnected from its support, he climbed into the tank, picked up the capsule with tongs and discovered the open end. It was immediately obvious that the Po–Be mixture had become scattered over the bottom of the tank. Gross personal contamination was confirmed at the checking station located near the entrance to the room with assistance from two of the technical staff and from the Health Physics Inspector regularly assigned to the operations. When the extent of the contamination was realized, within five minutes or less, all personnel were evacuated from the rooms.

During the decontamination operation it was found that there was no one satisfactory way of decontaminating the polonium from metals without removing surface layers. It was found convenient to seal small items, such as pieces of plastic, radiation instruments, and uranium slugs in bags, made of polyvinyl chloride, for disposal or for transfer to a decontamination station. While this work was being conducted, the air-borne activity was the order of a few hundred disintegrations per minute per cubic meter and assault masks with N-11 cannisters were used. It was found that the air-borne contamination was a strong function of the physical activity within the air and, for safe personnel protection, reliance could not be placed on the air samples taken during periods when the area was unoccupied.

The decontamination operation for cleanup that was followed was extremely arduous, requiring the efforts of eight men over a period of two months, during which the experimental program in progress was completely interrupted. Many of the procedures were developed on the spot, mostly by trial and error, and, in final form, were believed to be as good as were available at the time. The medical and clinical findings have indicated no detectable damage was incurred from the radioactivity and the quantity of beryllium associated with the inhaled polonium has been considered too small to be toxic.

As a preventative measure against a re-occurrence of this accident, it is now the policy at the Oak Ridge Critical Experimental Laboratory to enclose all source capsules, as they are delivered, in a suitable metal container which can be sealed.

CRITICALITY ACCIDENTS—10

NRX Reactor Incidents

The following description of NRX reactor contamination incidents has been taken from reports CRR-836 by W. B. Lewis and GPI-14 by D. G. Hurst.

The heavy-water-moderated research NRX reactor at Chalk River was designed to operate at high flux over a large core volume and even though planned in 1944, this reactor is still one of the highest performance research reactors. Much of our knowledge of the behavior of high flux reactors has come from this facility. Considering the advanced design and the limited information available at the time NRX was built its operating history has been remarkably successful; however, there have been some unusual occurances including a major accident, which have provided much information about reactor safety.

FIG. 2. NRX Research Reactor. (Atomic Energy of Canada Limited photograph.)

This reactor is moderated by heavy water contained in a large cylindrical tank (calandria) pierced vertically by a number of aluminum tubes for the fuel and control rods. The natural uranium fuel rods are aluminum clad and cooled with light water flowing in the annular space between the cladding and the cooling water tube. There is an air gap between the water tube and the calandria tube. A central thimble of 5.5. in. diameter and many other provisions for special irradiations have made this reactor particularly useful for in-pile loops and other experiments in reactor engineering and physics. The control and safety systems include shut-off rods, a control rod for fine control, and provisions for adjustment of the level of the heavy water. The shut-off rods were driven out of the reactor by compressed air and held in the up position by electromagnets with a piston to drive the rods back into the reactor. Under normal operating conditions the reactor is stable with a prompt negative temperature coefficient due to Doeppler broadening of the uranium-238 resonances, and a slow negative temperature coefficient associated with an increase in heavy water temperature. However, the coupling between the fuel temperature and the heavy water temperature is quite loose and the only effective heating of the heavy water is by the absorption of radiations, since the air gap insulates the calandria from the fuel. There is also a latent strong positive contribution to reactivity since the cooling water is a poison and boiling of the cooling water will greatly reduce its volume in the core. In this reactor, therefore, there can be a great difference between a transient and equilibrium temperature coefficients.

A major accident occurred on Friday afternoon, 12 December, 1952, during start-up in preparation for reactivity measurements to be made at low power. The reactivity of rods of long irradiation were being compared with that of fresh rods. The preparations included substitution of air cooling for water cooling on a rod of unirradiated fuel and the fact that this rod was fresh was of some significance in regard to the quantity of fission products discharged into the atmosphere. The reactor had not been operated at power for several days and most of the transient fission-product poisons had decayed. Since experiments were to be conducted with the reactor, research physicists were in the reactor control room, but the reactor was under the authority of the reactor branch personnel. The experiment and arrangement in the reactor had been recommended by the physicists and approved in writing by the reactors branch superintendent.

The start-up procedure was to have been normal and the heavy water level was raised to a point for normal operation. An operator in the basement then by error opened three or four bypass valves thereby causing some of the shut-off rods to rise. This movement of the shut-off rods was noticed by the supervisor at the control desk who then went to the basement and corrected the situation. Apparently, however, some of these rods did not drop all the way back into position, although the signal lights at the control desk were cleared. The supervisor then instructed, by telephone, his assistant at the control desk to push specified control buttons; however, his instructions were partly in error, although normally the error would not have resulted in any difficulty. It became evident from the rate of rise of reactor power that the reactor was super-critical at a time when it was believed to be sub-critical. When the control button to insert the shut-off rods was pushed, it was found that all of the rods did not fall back into the reactor because of the earlier error in instructions, and also because of some mechanical difficulties. When it was observed that the reactor power was still increasing, the valve to release the heavy water into its storage tank was opened and after a few seconds the reactor was shut down.

When the first bank of shut-off rods was raised the reactor was more reactive than expected by some 10 mk because some of the shut-off rods were not completely in the reactor. The first bank of rods that was raised by the operator increased the reactivity some 6 mk above criticality and the power of the reactor increased with a doubling time of some 2 sec until the power reached 100 kW, where the reactor trip circuit released the rods, but only one rod fell into the reactor. The slow movement of this rod, which was not accelerated by air pressure due to the earlier error in instructions, allowed the power to increase to some 20 MW, at which point it appears that boiling of the cooling water occurred, thereby increasing reactivity by some 2 mk at least. Then the reactor power continued to rise to a power of between 60 and 90 MW before the reactor was shut down by the dumping of heavy water and perhaps other mechanisms.

Evidence for damage to the reactor was not immediately obvious. An operator reported hearing a rumble and seeing a spurt of water up through the top of the reactor. Another operator saw water flowing down from the reactor into the lower header room. Instrumentation soon indicated the release of activity to the air and this was later identified as coming from the vaporization and discharge through the stack of fission products from the air-cooled rod of fresh uranium. The site contamination might have been more serious had this rod been highly irradiated uranium.

Since the reactor contained fuel with a considerable fission-product inventory, it was considered unsafe to turn off all the cooling water which was now flowing from the reactor into the basement of the building, although the flow was cut back to the minimum point considered necessary to prevent fuel melting. This cooling water was highly contaminated and was temporarily stored in storage tanks until emergency measures were taken by the construction of a pipe line to an isolated area where the water was disposed of into the soil which retained effectively the fission products.

From disassembly of the reactor, it was found that when the air-cooled rod melted, the aluminum cladding ran down the rod and congealed forming a barrier on top of which the molten uranium formed an ingot within the calandria tube without damage to this tube. In the areas where damage to the calandria was severe, it appeared that the cooling water played a part in determining the type of damage. Steam under pressure was the cause of the initial rupturing of the outer sheaths of the rods. Flying chips from this rupture may have penetrated the calandria tube, but perhaps the reactions between uranium and steam or water was the direct cause of damages. Some of the residual metal showed evidence of considerable chemical reaction. The conditions required for the initiation of the energetic aluminum–water reaction may have existed in local areas, although evidence for such a reaction is inconclusive. By some mechanism a considerable volume of gas was generated suddenly at one stage of the accident.

Although the damage to the reactor was extensive and the problem of cleaning up the contaminated areas were unprecedented, the decision was made to restore the reactor and a great deal of valuable experience was obtained in this restoration process. It required a few weeks to do some preliminary clean up of the site and remove the undamaged fuel elements. Then the damaged fuel channels were cut out and removed. Finally the entire calandria was removed and buried. The building was decontaminated even though this required the chipping away of contaminated concrete surfaces. Finally the reactor was restored and placed back in operation in February, 1954.

Another accident that occurred in July, 1955, in the NRX reactor is of interest to reactor safeguards. A fuel rod made of plutonium–uranium alloy melted through the calandria tube and some decontamination of the calandria was required. However, the damages were repaired by replacing the damaged tube. This incident is of particular interest because the dispersal of the plutonium–alloy fuel into the heavy water from the rod geometry reduced the self shielding and suddenly increased reactivity so that the reactor power increased by some 50 per cent within a period of one or two seconds. This sudden reactivity increase, possibly by fuel dispersal, is a problem of interest to the safety of several types of reactors.

CRITICALITY ACCIDENTS—11

The following description of a nuclear incident is taken from Report TID-5360, by D. F. Hayes.

On 2 February, 1954, at the Los Alamos Scientific Laboratory, an assembly of fissionable material, remote from personnel, was inadvertently made excessively super-critical. The incident occurred in the course of an extensive study of the properties of the super-critical radiation burst produced by an assembly of fissionable material. This experiment was known as "Godiva", since it was a bare assembly without reflectors. This study involved operations of the type that the normal regulations were designed to protect so the study was covered by specific procedures. A reference check of critical conditions preceded each super-critical burst. To obtain rapidly sufficient power for a delayed critical check, it was customary to set the control rod at the position of minimum reactivity and to insert a reactivity booster in the form of a fissionable material slug. This time, when the booster was inserted, the radiation indicators and the assembly temperature recorder went off scale and scrams were actuated. The resulting shock separated parts of the assembly and damaged steel supporting members. There was no injury and the property loss called for repair to the assembly only at an expenditure of $600.00.

Apparently, the control rods had been run to their wrong extreme; i.e. to the position of maximum reactivity before the booster was inserted. An interlock to prevent this condition had been omitted in the course of remodeling the assembly and, contrary to regulations, only one crew member was in the control room during the operation that led to the accident. There was a single burst, terminated by thermal expansion. The yield was 6×10^{16} fissions, about three times that of the largest planned burst.

CRITICALITY ACCIDENTS—12

Radiation Excursion in an Experimental Homogeneous Critical Assembly

The following description of a nuclear incident is taken from Report TID-5360, by D. F. Hayes.

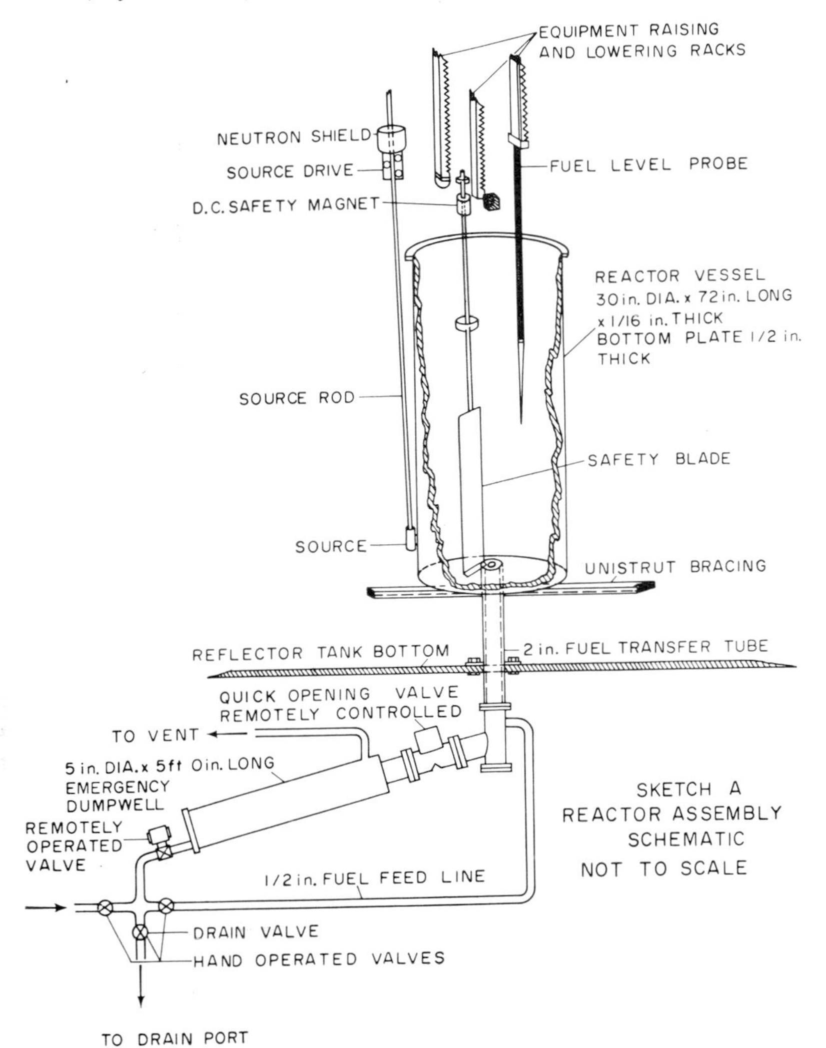

Fig. 3. Apparatus for Homogeneous Reactor Criticality Experiment. (Reproduced from Report TID-5360 (Suppl.) by D. F. Hayes.)

An inadvertent excess reactivity occurred due to mechanical failure in a homogeneous assembly at the Oak Ridge National Laboratory on 26 May, 1954. The incident occurred during a series of experiments designed to study the criticality conditions of uranium–water solutions in annular cylindrical containers. The outer container of the annulus was a 10 in. inside diameter by 6 ft high aluminum cylinder having a $\frac{1}{16}$ in. thick wall. The bottom of the container was $\frac{1}{4}$ in. thick aluminum with an off-center 2 in. inside diameter aluminum pipe for fuel solution flow. The 10 in. cylinder sat on a plexiglass table and was secured by three bolts welded to the outside of the cylinder. The inner container of the annulus was a 2 in. outside diameter by 6 ft long aluminum pipe having a $\frac{1}{16}$ in. thick wall lined on the inside by a $\frac{1}{8}$ in. thick cadmium cylindrical shell and filled with water. The inner container or tube was positioned at the low end of it by a pin which fitted into a recess in the bottom of the 10 in. cylinder. The upper end was held by downward pressure from a spider, the legs of which were bolted to a flange on the top of the 10 in. cylinder. The legs of the spider were 1 in. aluminum angles welded about the center. The assembly was contained in a 9 ft × 9 ft cylindrical tank which could be filled with water to provide neutron reflector, if desired.

The cause of the accident was the displacement of the center tube, effectively a poison rod, to a region of less importance. This displacement resulted from a dislocation of the positioning spider by a pin, used to connect sections of the liquid level indicator rack, protruding beyond the side of the rack and engaging a leg of the spider as the indicator was raised. Removing the compressional force from the top of the central tube allowed it to fall against the inside of the 10 in. cylinder. Although the displacement was small it was sufficient to cause a large increase in the effective neutron multiplication.

The safety system apparently operated normally and the reaction was stopped automatically. All personnel in the building during the incident were protected by a minimum of 5 ft of concrete shielding so no serious exposures were incurred. While the total number of fissions during the transient have been estimated to be 10^{17}, there was no evidence of violent boiling in the system or of mechanical breakdown due to energy release.

ACCIDENT—13

Reactor Fuel Leak from a Homogeneous Reactor

The following description of this incident is taken from Report TID-5360, by D. F. Hayes.

The stainless steel core of a water boiler type reactor at North Carolina State College developed a leak in May, 1955, allowing the fuel solution to leak into the surrounding chamber and the cooling water. The first indication of trouble was a rise in pressure of the core. Further investigations revealed the presence of fission products in the surrounding chamber and the cooling water. The gases were carefully bled off the chamber and core and out the stack. The cooling water was drained and the fuel solution was transferred to storage containers.

There were no injuries or exposures to significant radiation as a result of this event; however, the reactor decontamination required extensive work and the facility was largely reconstructed. The cause of the accident was attributed to corrosion of the stainless steel cooling coil by the solutions and the corrosion was believed to have been augmented by chlorides left in the core as a result of previous operations.

ACCIDENT—14

Experimental Breeder Reactor I Core Melt Down

The transient test being conducted with EBR-I on 29 November, 1955, resulted in the melting of some of the fuel elements and release of fission products into the cooling system with minor leakage of some gaseous fission products into the reactor room. As a precaution the room was evacuated until a quantitative measurement of the activity was made and the contamination was found to be minor.

The reactor had been operated for four years through two core loadings and the plant had been found to be quite stable and largely self-regulating. It was known that, if the coolant flow rate through the core was changed, a prompt positive metal temperature coefficient of reactivity was observed. A decrease in flow rate from 45 to 17 gal/min gave a fuel temperature rise of 10°C and a rise in power. It is believed that the increase in temperature caused inward bowing of the fuel rods and an increase in reactivity. A second phenomena observed in EBR-I was an oscillation in power, if the coolant flow was reduced when the reactor was operating at power. If the coolant flow was reduced at full power to a value around $\frac{2}{3}$ of the design value, the oscillatory behavior became rather violent. The mechanism which was thought to cause the oscillatory behavior was a negative power coefficient of reactivity that was delayed some time of the order of 10 sec, if a single delayed negative coefficient was postulated.

The reactor was scheduled to be placed on stand-by early in 1956, since most all the significant experiments that were practical to perform with this core had been completed. Measurements of the transient temperature coefficients were to be experiments to be performed and were known to be difficult with a significant chance of core damage. The reactor was to be placed on a short period without coolant flow to measure the temperature coefficient during a fuel temperature rise of 500 to 600°C. The resulting temperature of the uranium was close to that at which uranium metal and stainless steel form a eutectic at 725°C. Because of this and the rapid rate of temperature rise, the reactor had to be shut down within one second. In other experiments of this series at longer periods it has been possible to interrupt the power excursions by using the motor-driven control rods which subtracted reactivity slowly. However, in this final test the operator repeated the use of the slower motor-driven control rods until the scientist conducting the experiment recognized the situation and pressed the rapid shut-off button and, simultaneously, the automatic power level trips activated the shut-off rods. The delay in time, of up to two seconds, permitted the reactor power to overshoot and heat the fuel elements so that alloying of uranium and steel and uranium melting occurred, and there was extensive damage to the core. After an extended period of time until radioactivity had decayed appreciably, the core was removed and after examination, sent to chemical processing. Besides the data on temperature coefficients, these experiments also provided valuable information on reactor behavior during melt down and on the behavior of fuel elements when melting in liquid sodium. No unforeseen or catastrophic processes occurred.

ACCIDENT—15

The following description of a nuclear incident is taken from Report TID-5360 (supplement), by D. F. Hayes.

A homogeneous UO_2F_2 water-moderated critical assembly at the Oak Ridge National Laboratory was put on a prompt critical period by an over–addition of fuel to the assembly on 1 February, 1956. Before reaching the critical point, the hand-operated valve for adding fuel was turned off. However, fuel continued to be added to the reactor due to air pressure in the line. Although the automatic safety system operated, assuring termination of the burst, considerable fuel was displaced from the reactor. The number of fissions in the burst was estimated to be 1.6×10^{17}.

No serious exposures resulted from this incident since all personnel were shielded by a minimum of five feet of concrete. There was no significant property damage and all the uranium was recovered.

Although it was determined that this incident was initiated by over-addition of solution to the reactor, later measurements showed that fuel would continue to be added to the reactor for several seconds after the control switch was placed in "drain" position if insufficient time were allowed for the operating pressure to be vented from the line.

ACCIDENT—16

Volatile Radioactivity in a KEWB Reactor Building

The following description of this contamination incident is taken from Report TID-5360 (supplement), by D. F. Hayes.

The KEWB reactor test facility was constructed at the Santa Susana site of Atomics International for kinetic experiments on homogeneous reactors. On 4 January, 1957, contamination of the reactor test building occurred during a series of routine operating reactor tests which involved the use of auxiliary apparatus. A number of the staff entered the test building on schedule, 15 minutes after shutdown, to measure the core pressure. The radiation survey at the door indicated no unusual radioactivity in the valve gallery. At the time this was being done, another staff member checked the remote continuous gas sampler and found it above breathing tolerance. The test building was vacated immediately after about 10 sec of occupancy. The employees internal dose was less than 50 mr. No measurable external body radiation was received during the incident.

The apparent cause of the contamination was a malfunction of equipment which had operated satisfactorily for a period of seven months. The leak in the system appeared to be caused by the failure of the vacuum pump. The pump was replaced and the necessary decontamination was completed without difficulty.

ACCIDENT—17

The following description of a nuclear incident is taken from Report TID-5360, by D. F. Hayes.

A fissionable material assembly, known as Godiva, operated remotely, produced on 12 February, 1957, at the Los Alamos Scientific Laboratory, a neutron burst greater than intended which resulted in the rupture of the assembly. This neutron-producing assembly was to be used to irradiate uranium-carbide graphite samples. The samples were to be heated in a shielded furnace, exposed to a "prompt" burst of neutrons and then transferred to a counter for evaluation. The experiments were conducted at an isolated site in a building separated from the control room (and all personnel) by a quarter of a mile. Mechanisms were operated remotely. Although excessive radiation, to the degree experienced in this accident, does not occur routinely, it is not entirely unexpected in this type of work. In these experiments a prompt burst was produced by bringing together sections of uranium-235 to form a sphere. The exact degree of criticality was determined by a pre-set control of reactivity rods, also enriched with uranium-235. The assembly was expected to achieve the condition slightly above the prompt-critical condition which had been found to produce a burst of 10^{16} fissions or a total energy of about 100 watt-hours. The heat of the reaction normally expanded the metal and reduced the reactivity below prompt-critical thus terminating the burst before the parts were mechanically separated.

On the occasion of the accident, preliminary bursts had been produced and in the process of lowering the top safety block, the unexpected burst occurred which was estimated to have produced 1.2×10^{17} fissions, or about 12 times the normal energy of a prompt burst, and twice the energy of the previous incident which occurred in 1954. The energy was great enough to tear the uranium parts from the assembly and distort the steel rods in the frame. The uranium was deformed and there was much surface oxides. There were no personal injuries or over-exposures since the facility was operated remotely. Radiation levels in the building were high initially and, therefore, clean-up procedures were delayed 2½ days until scrubbed down completely without unnecessary exposure to clean-up personnel. The total damages were estimated at $2400.00 consisting largely to damage of the frame and electronic components. The building was placed back in operation within eight days. The burst was believed to have been caused by an accidental movement towards the assembly of an incidental neutron reflector (the furnace for heating the sample), causing additional reactivity and a neutron burst above that expected. The neutron reflector may have been jarred by the action of pneumatic cylinders which operate the assembly. It was decided to replace Godiva by another assembly which had ridged mountings of uranium and a more inflexible geometry.

ACCIDENT—18

The Windscale Incident

The following description of the events leading up to and the consequences of overheating and a fire in an air-cooled graphite-moderated production reactor is taken from *Accident at Windscale No. 1 Pile on 10th October, 1957*, London, Her Majesty's Stationery Office, and "District Surveys following the Windscale Incident, October, 1957," by H. J. Dunstee, H. Howells and W. L. Thompston, Vol. 18, *Proc. 2nd United Nations Int. Con. on the Peaceful Uses of Atomic Energy*, United Nations, New York.

On 8 October, 1957, during certain special operations of the air-cooled graphite-moderated plutonium production reactor, excessive fuel element temperatures resulted in the rupture of the fuel element cladding and fire. The resulting damage to the reactor, together with other factors, led to the retirement of the facility. Also the quantities of some of the volatile fission products that were released from the fuel elements and that passed through the exhaust air filters were of some significance, and considerable information was obtained about environmental hazards.

The accident occurred during a semi-routine heating of the graphite in the reactor to a higher temperature than during normal operation in order to release part of the energy stored in graphite by neutron irradiation. At the time the reactors were built, it was known that energy accumulates in graphite and it changes in dimensions and thermal properties in a reactor. However, little information was available on the conditions necessary to release the energy and restore the graphite partly to its original condition; however, spontaneous release of stored energy occurred in this reactor in September, 1952, while the pile was shut down and the cooling fans turned off. The release of stored energy resulted in a significant rise in the temperature of the graphite; however, only part of the stored energy was released and then only in certain sections of the reactor. The results were beneficial and conditions for the controlled release of stored energy were developed. This was a self-sustaining annealing operation initiated by starting up the reactor with the cooling air shut off so that the graphite temperature increased to the point that the stored energy was being released. Usually two periods of nuclear heating were needed to obtain the desired extent of stored energy release and this process had been repeated some eight times in the period between 1952 and the accident in 1957.

The events leading up to this accident included the preparation for a stored energy release and the turning off of the main air blowers. The first period of nuclear heating was followed as usual by a partial stored energy release and the operators thought that the process was stopping and decided to raise again the temperature by nuclear heating. The temperature of the uranium fuel elements was increased for a few minutes at a rate, during this period of nuclear heating, that was several times the rate permitted during normal operation; however, the instruments did not record during the 15 min period of nuclear heating a temperature of the uranium in excess of the maximum permitted during normal operation. When the rapid temperature rise of the uranium was noted, the control rods were inserted, however, since the thermocouples with which the maximum temperature was measured were not positioned in the region of maximum uranium temperature during this transient heating, excessive temperatures probably were reached and it appeared from subsequent events that the cladding on some of the fuel elements was ruptured at this time. A contributing factor to the rapid temperature rise was a low reading of the power meter in the region of small powers, although the instrument was correctly calibrated for normal power.

Following the second nuclear heating, the operator was not aware that damage had been done, and the graphite temperature gradually increased through the next day from the release of stored energy. However, the uranium in the fuel elements with broken cladding was oxidized and led to the heating of other cartridges which caught fire and some of the graphite was ignited in a region which included about 150 fuel channels. When excessive graphite temperatures were observed, the dampers to the chimney were opened to permit air to circulate and cool the graphite, and the temperature rise was halted. Also, an increase in radioactivity at the top of the stack was observed as the air started to circulate as would be expected under normal conditions. However, when the graphite temperatures started to rise again and the dampers were again opened, excessive radiation was observed at the top of the stack and the operator became aware of the existence of ruptured fuel elements in the reactor. At this point, it was found that the instrumentation for measuring activity in individual fuel channels for the location of leaking fuel elements was inoperative. In order to locate the source of activity, a charge plug in the charge wall in front of the reactor was removed and it was seen that some of the fuel was at a red heat. It was found that the cartridges could not be discharged from these hot channels because of distortion and then an attempt was made to put out the fire with carbon dioxide, but this was found to be ineffective. Finally, the decision was made to flood the reactor with water, and after declaring a state of emergency and warning personnel in the area to stay indoors and to wear face masks, the water was turned on. After an hour, the fire subsided; after 24 hours of flooding the reactor was cold.

The release of a considerable amount of radioactive material into the atmosphere was indicated by the radiation instruments in the area and procedures to determine the extent of this release and the radiation hazards to the people in the surrounding countryside were initiated. The filters at the top of the stack were not designed for high efficiency of particle removal of all sizes and, therefore, some particulate material passed into the atmosphere although the volatile fission products made up the largest part of the material that escaped as indicated by later estimates of the magnitude of the release as shown in Table I.

Table I

Estimated Release of Radioactive Material from the Windscale Reactor

Iodine-131	20,000 curies
Cesium-137	600 curies
Strontium-89	80 curies
Strontium-90	9 curies

Small quantities of other radioactive materials were released including ruthenium-103 and 106, zirconium-95, cerium-144, and polonium-210.

The wind at the start of the release was light and variable and there was a later change to a definite wind flow from the north and west. The distribution of material was, therefore, in a complex pattern with some of the material falling into the Irish Sea. The distribution of fission products was indicated by the ground measurement of gamma radiation and studies of iodine-131 in milk.

The hazards from external radiation, inhalation and from ingestion of activity from contaminated food and water were determined. A measurement directly under the plume from the stack at a distance of 1 mile down-wind indicated a direct gamma exposure level of only 4 mr per hour. Measurements of gamma radiation from ground deposition showed that the exposure would only be 10–20 mr during the week following the incident which level is not significant considering the rapid decay of this activity. The major hazards were from the contamination of food (milk) from the contaminated area. Measurements showed that the activity from iodine-131 increased from a trace to between 0.4 and 0.8 microcuries per liter as a result of this incident and the decision was made to restrict the distribution of milk from an area of over 200 square miles. A few areas of high contamination remained under restrictions until 23 November, 1957. The limiting value for contamination of milk by iodine-131 during this period was 0.1 microcuries per liter.

In addition to milk, the grass and vegetable products including potatoes, cabbage, kale, turnips and lettuce were sampled for strontium isotopes with none being found sufficiently high to represent any danger, although some grass samples contained sufficient radioactive strontium to require monitoring of the milk produced in such areas for some time. Water samples from drinking water supplies were also analysed for activity and in no case was the level of contamination sufficient to constitute a hazard.

No worker received serious radiation exposure or had to be confined for medical attention as a result of this accident. The highest exposure recorded was 4.6 r. Some workers had high thyroid iodine activity of 0.5 μc as compared with the then recommended maximum level of 0.1 μc but this value is for continuous exposure and the activity in the thyroid of the workers decayed rapidly.

From the extensive studies that were conducted it was concluded that it is in the highest degree unlikely that any harm was done to the health of anybody. The damage to the reactor was, however, extensive and would have been difficult to repair. For this and other reasons, it was decided not to put the Windscale No. 1 reactor back into operation.

ACCIDENT—19

The Contamination of the NRU Reactor

The following description of this incident is taken from the Report AECL No. 850, by J. W. Greenwood.

The NRU Reactor at Chalk River, Ontario, is moderated and cooled with heavy water and operates at a very substantial power level of 2000 MW with a remarkably high neutron flux in the neighborhood of 3×10^{14} neutrons/cm^2/sec. The reactor uses natural uranium fuel in the form of flat metal bars clad with aluminum and installed in fuel-rod assemblies. Each bar is 10 ft long and the tube containing the family of 5 bars is $2\frac{1}{2}$ in. outside diameter. About 200 fuel-rod assemblies are installed in the core of the reactor and each assembly weighs about 120 lb. An unusual design feature of this reactor includes provisions for changing fuel rods during full-power operation. In order to accomplish this while still maintaining continuous cooling of the fuel during all stages of the removal, a complex mechanism is provided in the permanent tube that leads from the reactor core through the upper shielding to the deck plate above each rod. A heavily shielded removal flask is provided for moving the fuel element to the storage area and lowering the irradiated rod down an elevator shaft which leads to a water-filled storage bay. This shielded flask had a built-in heavy water circulating system which cools the fuel until just before it is discharged into the storage pool. The time required to drain the heavy water out of the flask chamber and lower the fuel into the water of the storage bay is so short that no overheating of the fuel element reoccurs.

The reactor is controlled by 16 absorber rods which combine the functions of control and emergency shut-off. These rods are held at the top by electromagnets which can release the absorbers and allow them to fall freely into the reactor for emergency shut down. The magnets are also raised and lowered by motor-driven screws, thereby, raising and lowering the absorbers to control reactivity. The reactor can be started up, brought to power and held at power under complete electronic control without human intervention. The supervisor sets the controls to determine the demanded final power, maximum permissible linear rate of change of power, and the maximum logarithmic rate of change. The logarithmic rate can be set at any value between 1 per cent/sec and 4 per cent/sec. The linear rate can be set between 0.2 and 2.0 MW/sec. The leveling off begins when the power is about 10 per cent below the demanded power. Signals proportional to apparent reactor power are provided by ion chambers, but provisions are made to apply temperature corrections that keep the average temperature of the exit cooling water constant at a pre-set value. Most of the components, such as ion chambers, power supplies, and amplifiers and relays in the control system and shut off systems, are in quadruplicate being arranged in four parallel channels to convey signals. An important feature of this arrangement is that when a signal in one channel disagrees with a signal in the other channels, the channel carrying the aberrant signal is disconnected automatically and an alarm is given. If there are two or more aberrant channels, the reactor will shut down. In this way testing and maintenance of individual channels is facilitated, shut downs due to instrument fault are reduced and the safety of the reactor is greatly enhanced.

The NRU reactor first went critical on 3 November, 1957, and for the ensuing 6 months various start-up experiments were carried out as the power was slowly raised up to even a few megawatts beyond the nominal design power of 200 MW. For several weeks before this incident, the reactor had been operated most of the time between 100–200 MW and for a few hours at a somewhat higher power. The difficulties usually experienced in initial reactor operation had been largely overcome during this 6 months period. Included among these difficulties had been a few cladding failures that had allowed some contamination of the heavy water. Also the chart of reactor power displayed an almost continuous record of very small and erratic variations of power. This is found to arise from random lateral movement of the control rods, caused by circulation of the coolant moderator. The effect was reduced by installing shrouds around some control rods, but spikes corresponding to power changes as fast as 2 MW/sec were nevertheless still observed frequently.

On 23 May, 1958, the reactor had been operating steadily without shut downs of any kind for exactly one week, during which the power of the reactor had been brought up to around 200 MW without incident. In the early morning of Friday, 23 May, the shift supervisor noticed that the fuel temperatures were close to the normal alarm point of 170°F and a few alarms occurred early in the morning. It was, therefore, decided to hold the reactor power at about constant without further increases. Sometime later, the reactor shut down automatically as a result of excessive linear rate-of-rise of power on all of the 4 identical channels, indicating a genuine rate-of-rise greater than 5 MW/sec. The shift supervisor could find no explanation for the scram and there was no indication of high activity in any of the coolant streams nor in the area of the building. Because the shift supervisor was unable to discover any reason for a rate-of-rise scram and had no reason to anticipate a danger in re-starting the reactor, he prepared to return it to power under automatic control. His first four start-up attempts were terminated by automatic trips arising as secondary effects from one of the reactor safety features and it was not until the fifth approach to power that the start-up procedure proceeded normally during the first five minutes. When the reactor again tripped as a result of excessive linear rate-of-rise of power, unlike the early trip, this one was accompanied by many alarms, indicating high activity in several parts of the building as well as in the heavy-water system and fuel moderators. The trip was preceded by an unexpected flashing of light, indicating that 3 control rods were momentarily in their uppermost position even though it was unreasonable to expect them to be up in these particular circumstances. It was clear to the shift staff that a fuel fire had occurred and this condition was verified by a check of the radiation level at the top of the reactor, which had been found to be over 100 r/hr. After this activity had decreased, it was found that 3 fuel rods displayed high levels of radioactivity and one of these rods was successfully removed. Attempts to remove the second rod by normal methods failed because of the damage which the rod apparently had suffered and modifications were made to the fuel removal flask to accommodate the damaged rod. The non-standard procedure that was demanded by the circumstances also necessitated the moving, not only of the fuel but also the entire plug from the top of the fuel hole. One unforeseen result of this operation was that the heavy-water coolant drained out of the removal flask into the reactor. If more time had been available, there would probably not have been any serious consequences, but unfortunately, the damaged rod jammed after being raised partway into the flask and was without cooling for about 10 min. At the end of this time, repeated attempts to free the rod were eventually successful, probably because the rod melted or was torn into two or three pieces. As the flask moved toward the fuel storage area, a 3 ft piece of the fuel fell out and burned in the open air before being extinguished by wet sand. This task was made particularly arduous by the necessity of wearing respirators and was hazardous since the radiation field directly over the pit exceeded 1000 r/hr. Nevertheless, the highest radiation exposure suffered by any individual up to this point was only 5.3 r. It was later found that about 4 ft of the rod had fallen to the bottom of the reactor vessel, while the remainder was still in the removal flask. The top of the reactor was badly contaminated with uranium oxide, and although the burning uranium was extinguished within 10 or 15 min the entire reactor area was seriously contaminated.

Over the next two months some 600 men took part in the clean-up, including men of the Armed Forces and outside contractors. Average radiation dosages dropped gradually from 2000 mr for those employed in the early weeks to negligible amounts. By the end of July, both surface and airborne contaminations throughout the reactor building had been reduced to negligible levels as a result of intensive vacuum cleaning, mopping and wiping. Fragments of fuel had been removed from the bottom of the reactor vessel by means of several ingenious tools developed for the purpose. By the end of August the reactor was operating again. When the very small amounts of contamination were released outside the reactor building, and because of weather conditions favoring close-in fall-out, detectable activity was confined to an area about the size of 100 acres. Most of this area was uninhabited forest just outside the P project boundaries.

The most plausible reason found for the failure of the fuel element is that a defect in the aluminum cladding near the bottom of the element permitted water to enter and force its way up between the cladding and the uranium. During the high-power operation, water would be prevented from entering by the generation of fission gases, but on shut down an appreciable volume of water might be sucked into the sheath of an unbonded uranium bar. The return to high power would then generate steam in this enclosed space sufficient to cause a failure of the fuel cladding. It has also been found that the structure of the reactor is such that a pressure transient within the reactor vessel, such as might accompany the violent bursting of the fuel element, could easily produce the various signals that were observed by the reactor operator including the spurious "up" signal for the control rods. The reason for the initial tripping of the control system probably was associated with a weak spring on a "low power" selector switch which was used to initiate an automatic start up. If this switch was not fully turned from "low power" to "normal" so that some electrical contact was not made, the reactor power would behave in exactly the anomalous manner that was observed.

ACCIDENT—20

Accidental Nuclear Reaction in Y-12 Enriched Uranium Recovery Area

A quantity of highly enriched uranium solution leaked from a small diameter tank through a pipeline into another small diameter tank in the Oak Ridge uranium-235 recovery area. Both of these tanks by their size were

Fig. 4. Drum in which critical incident occurred. (Reproduced from "Accidental Radiation Excursion in the Y-12 Plant," Report Y-1234.)

of the "always safe" type and a nuclear reaction could not occur even though the tanks were filled with enriched uranium solution. During a routine leak testing of the system following a monthly clean-out for inventory, two of the three tanks in the recovery area were partly filled with water. An employee unaware that enriched uranium had leaked into the tank and piping, then drained what he thought was water into a 55 gal drum. The enriched uranium solution in the tank nearest the valve preceded the water into the drum, which was large enough to permit a "non-safe" configuration for the concentrated solution of uranyl nitrate. The operator stayed in the vicinity of the drum while the solution was being drained and after about fifteen minutes sufficient solution accumulated in the drum to reach criticality and there was an initial nuclear burst, which, however, did not discharge the contents of the drum and the nuclear system oscillated for approximately twenty minutes until the additional water flowing into the drum finally stopped the reaction. The radiation monitor alarm siren sounded within a few seconds of the initial burst and the building was evacuated. This accidental nuclear excursion occurred at approximately 2.05 p.m. on 16 June, 1958, and by 5.00 that afternoon a radiation survey team established that the incident had taken place in a drum and at approximately 9.30 that night a cadmium scroll was inserted in the drum to poison any further nuclear reactions, and clean-up procedures were started.

Eight employees were in the vicinity of the drum at the time of the incident and five of these were exposed to what has been described as a medium dose of radiation as follows:

Employee "A"	365 rad—461 rem
Employee "B"	270 rad—341 rem
Employee "C"	339 rad—428 rem
Employee "D"	327 rad—413 rem
Employee "E"	236 rad—298 rem

Following the incident these men and three others who received smaller exposures were hospitalized and special medical attention was provided and all were released from the hospital by 30 July, 1958.

ACCIDENT—21

KIDRICH INSTITUTE ACCIDENT

The natural-uranium-fueled, heavy-water moderated, unreflected critical assembly at the Boris Kidrich Institute of Nuclear Science, Yugoslavia, became critical inadvertently on 15 October, 1958, and caused significant radiation exposures to eight members of the technical staff. One fatality resulted. The following description of this incident is taken from the review by A. D. Callihan in *Nuclear Safety*, Vol. I, No. 1, pp. 38–40.

The core of the assembly is a lattice of natural uranium rods in heavy water. Control is primarily effected by adjustment of the level of the heavy water that is transferred by a pump from a storage tank located below floor level. Water can be pumped into the tank at two rates corresponding at near critical, to 3.75 and 1.2 per cent Δk/min. The neutron detection instrumentation is conventional, consisting of 3 BF_3 proportional counters with associated scalers and recorders. There is at least one logarithmic meter.

The safety system consists of a "control key" through which power is apparently supplied to magnetically-supported, gravity-actuated cadmium safety rods, each reportedly being sufficient to hold the system sub-critical separately even with the tank completely filled with heavy water. Dose rate meters for both thermal and fast neutrons and for gamma radiation give audible signals of radiation fields above preset rate levels. Suitable circuitry is provided to shut down the reactor by dropping the rods from radiation signals. It was apparently possible to start the reactor in the normal manner with the instrumentation de-energized. The assembly is unshielded.

At the time of the accident personnel was reported to have engaged in a "sub-critical experiment" at a distance of 3 to 8 m from the tank. No monitoring, safety, or control instrumentation was operating. By a means which is not mentioned, heavy water was transferred (possibly pumped) from the storage tank into the assembly tank through a vertical distance of at least four meters, without cognizance by any of the persons present. The resulting super-critical assembly was first detected by an odor of ozone. The immediate actions, which no doubt stopped the reaction, are not described. The duration of the energy release is indicated to be the order of ten minutes from a record of an air monitor some 540 m distant, and the magnitude was evaluated at 2.4×10^{18} fissions. The power versus time pattern is not reported and, presumably not determinable.

The estimated average whole-body dose was 683 rem, which was made up of 49 rem from thermal neutrons, 223 and 116 from epithermal and fast neutrons respectively, and 295 rem from gamma rays. Body areas nearest the point of incidence of the neutrons experienced of course exposures estimated to be two or three times the whole body average.

The patient who received the greatest exposure died a month after the accident from complications in the respiratory and gastro-intestinal tracts. Five other patients who were treated at the Curie Foundation in Paris, are recovering.

ACCIDENT—22

HTRE-3 Excursion

The following description of a power excursion in the aircraft reactor experiment, HTRE-3, is taken from a review by E. P. Epler in *Nuclear Safety*, Vol. I, No. 2, p. 57–59.

The reactor attained criticality at the National Reactor Testing Station on 24 October, 1958, after which extensive tests were taken, including evaluation of the controls, to provide information for safe operation of the system. This series of tests was terminated abruptly about four weeks later, following a power excursion that melted several fuel elements and resulted in the release of a large quantity of fission products.

Just before the incident the reactor was operating with greatly reduced coolant flow so that measurements of heating rates could be made. Several of the ionization chambers, including one of the three safety ionization chambers, had been replaced with heat-rate sensors. Signals from either of the two remaining ionization chambers would, however, produce a scram. During the afternoon of 18 November, 1958, the reactor was operated at 60 kW, and the reactor behavior was as anticipated. At the conclusion of this run, preparations were made for repeating this test run at 120 kW, the highest power up to that time. Since all ionization chambers were inserted to the full-in position, it was necessary to alter the circuit constants in the two linear-flux channels in order for the servo to operate over the range of 10 to 100 per cent of full power. The scram level was correspondingly set at 180 kW.

The final run was started, and then proceeded in a normal manner up to the power range. Control was then switched over to the servo to increase the power level on a 20 sec period. The servo seems to have been switched on too soon because the flux increased stepwise from 4 to 10 per cent of full power. The period recorder trace shows that the recorder pin was being driven at full speed until a period of 8.5 sec was reached, suggesting that the period scram set at a 5 sec period, was narrowly averted. The flux was then held constant at 10 per cent of full power for approximately 20 sec, and thereafter it proceeded to rise on a 10 sec period rather than on the 20 sec period that the servo was programmed to produce. The only nuclear instruments that remained on scale when the reactor reached 30 per cent of full power, were the period recorder and the linear-flux recorder. Either of the two uncompensated ionization chambers could have actuated the safety circuit, although only one of these supplied a signal to the linear-flux recorder and the servo channel. During the next time interval of approximately 20 sec the flux continued to rise, with the period recorder indicating very nearly 10 sec until the desired power (80 per cent of full power on the linear-flux recorder), was substantially reached. At this point, instead of holding at constant power, the linear-flux recorder showed the flux level to be dropping rapidly, and the period recorder started to drive down scale. The servo, seeing a negative error signal, continued to withdraw shim rods. This state of affairs continued for approximately 20 sec, at which time the reactor shut down by means of a reactivity loss of more than 2 per cent due to the redistribution of fuel resulting from melting and/or a temperature scram due to the melting of the thermocouple lead wires. As the reactor scrammed, the recorded flux level first increased and then decreased. The operator also scrammed the reactor manually but his action was preceded by perhaps three seconds by the automatic scram.

The primary cause of this incident is attributable to the insufficient voltage applied to the chamber terminals. The uncompensated ionization chambers were designed to be operated at 1500 V, but the voltage of the power supply had erroneously been set at 746 V. In addition, electrical noise filters which had been introduced during the testing of two previous reactors at the same site, had been retained in the chamber power supply. The filter resistor alone would have limited the maximum chamber current to 0.746 mA, whereas 0.84 mA were required to produce a scram. Also the linear-flux circuitry was unable to indicate the true reactor period.

This malfunction is attributed to coincident human factors and not basic design of the reactor or the instrumentation. The failure to remove the filters from the chamber power source and the failure to set the specified voltage at the chamber power source constitute the human elements. The elimination of either of the above would probably have prevented the incident. A possible contributory cause was the operator's decision to go to previously unobtained powers on automatic servo control. It is believed that the operator might have recognized the malfunction had the reactor power been increased in small steps under manual control. Also it is a well-recognized principle that safety equipment should be used for no purpose other than safety. Since the same chambers are used for both safety and control, it is, therefore, no coincidence that the safety system failed at the same time as the servo which precipitated the incident.

ACCIDENT—23

The following description of a nuclear incident at Los Alamos is taken from a review by G. E. Guthrie in *Nuclear Safety*, Vol. I No. 1, pp. 37 and 38.

This criticality incident occurred in a chemical processing plant during a physical inventory. It is believed that the operator had washed plutonium-rich solids into a vessel containing diluted aqueous and organic solutions and after removal of most of the aqueous solution the materials were being washed from the bottom of the tank with nitric acid into the solvent treating tank in which the accident occurred. This 225 gal stainless steel tank already contained a caustic stabilized emulsion. Shortly after the starting of the motor to initiate the expected mild non-nuclear reaction between the chemicals, the operator observed a "blue flash", which was also observed by a second employee in an adjoining room. The operator then ran out of the building and told a second employee, "I am burning up".

Perhaps because of an incorrect belief that the "burning" feeling had resulted from acid exposure, the employee was led to a shower. The employee went later (within 15 min) into deep shock and regained consciousness 6 hr later and remained rational and comfortable until nearly the time of his death, about 35 hr after exposure which was estimated at 12,000 plus or minus 50 per cent rem. Two employees other than the operator received radiation exposures ranging up to 118 rem. These employees were not seriously injured. The total property damaged was reported as negligible.

The investigating review committee attributed the accident to the handling of several batches of material instead of one at a time as in normal procedures.

ACCIDENT—24

Sodium Reactor Experiment Incident

The following description of an incident involving melting of fuel elements is taken from a review by W. B. McDonald and J. H. DeVan in *Nuclear Safety*, Vol. I, No. 3.

The SRE reactor at Santa Susana, Calif. was built to develop this reactor concept for civilian power application. The reactor is moderated by graphite, cooled with sodium, and contains 43 channels for fuel elements in the center of the moderator cans. The design outlet temperature of the sodium is 960°F. The graphite is canned with zirconium to prevent sodium from entering the void spaces of the graphite. The fuel elements in the SRE are fabricated in clusters of seven rods, each rod consisting of a 6 ft high column of uranium slugs in a thin-walled stainless-steel tube. A thermal bond between the uranium and the stainless-steel jacket is obtained by using NaK in the annulus, and helium is contained in a space above the NaK to allow expansion of the NaK and to serve as a reservoir for fission gases escaping from the uranium.

On 24 July, 1959, the Sodium Reactor Experiment (SRE) was shut down to investigate abnormalities which prevailed in the operations during power run 14. A subsequent preliminary examination revealed that extensive damage had been sustained by several fuel-element clusters during this power run.

During run 8 a large spread existed in the fuel-channel exit temperatures; this spread was attributed to the high oxygen content in the sodium which caused oxide plugging in the process tubes. Action was taken to reduce the oxygen content, and the fuel (which had been running excessively hot) was removed, washed, and then returned to the reactor. During run 8 a sample of the core cover gas indicated that Tetralin, an organic compound used as an auxiliary coolant, had entered the primary system.

During runs 9 to 11 the fuel-channel exit temperatures continued to show large spreads but tended to improve as the various runs proceeded. At the end of run 11, fission-product contamination was detected in the sodium, but it was not excessive. Run 12 exhibited no abnormal behavior. During this run the sodium outlet temperature was raised to 1065°F for a short period, and steam was produced at 1000°F.

In run 13, which was a high-temperature run with a 1000°F sodium outlet temperature, after an initial scram as a result of an abnormal sodium flow rate, the reactor was returned to normal operating conditions. Several unusual situations then arose: the reactor inlet temperature started a slow rise; the log mean temperature difference across the intermediate heat exchanger started to increase, indicating changes in the heat-transfer characteristics; a thermocouple in a fuel slug in channel 67 showed an increase from 860 to 945°F; some of the fuel-channel exit temperatures showed slight increases; and the temperature difference across the moderator abruptly jumped 30°F. Later examination indicated that a reactivity increase of about 0.3 per cent occurred over a period of about 6 hr and then increased about 0.1 per cent over the next three days of operation. By 2 June, 1959, it was obvious that the heat-transfer characteristics of the system had been impaired. Since the sodium oxygen content had been low at the start of the run, it was decided that a Tetralin leak was causing the trouble instead of excessive oxygen in the sodium. Subsequently, Tetralin was detected in the pump casing of the main primary pump; and, after the run was terminated, a leak was found in the thermocouple well in the freeze seal of the pump. Examination of the fuel by television camera indicated the fuel to be "slightly dirty but in good condition".

On 4 June, 1959, while the fuel element from core channel 56 was being washed, a pressure excursion occurred which severed the fuel hanger rod and lifted the shield plug out of the wash cell. It is postulated that the hydrocarbons (from the Tetralin leakage) could cause sodium to be trapped in the hold-down tube on the hanger rod by blocking the sodium drain holes, and the sodium could cause the reaction. As a result of the "wash-cell incident," no further washing of elements was done. Prior to run 14 the Tetralin-cooled freeze-seal in the sodium pump was replaced by a NaK-cooled unit, and Tetralin and naphthelene, a product of Tetralin decomposition, were removed from the primary system.

During the initial phases of run 14, some of the abnormal conditions noted in run 13 were again observed (i.e. large fluctuations in the temperature difference across the moderator and divergence of the fuel-channel exit temperature); however, reactor operation continued at low power (1 MW) until a scram was caused by loss of auxiliary primary sodium flow. Shortly after re-establishing operation, a sharp increase in air activity was indicated in the reactor room, followed later by a sharp increase in the stack activity. Since the radiation level over channel 7 had reached 25 r/hr, the reactor was gradually shut down. The activity escape was traced to the sodium-level coil thimble which was replaced by a shield plug. After returning the reactor to operation, no significant activity was noted in the high-bay area.

On 13 July, 1959, a series of negative and positive reactivity excursions was observed; one of these excursions resulted in a 7.5 sec period. The reactor was scrammed manually. It is estimated that the reactor reached a peak power of 24 MW(t). The cause of the reactivity changes is not known, but investigations are being made in an attempt to explain them.

From the preliminary findings of the Atomics International Committee, it may be concluded that:

(1) The fuel-element failures resulted indirectly from leakage of Tetralin into the primary sodium circuit. The mechanism of failure is thought to

have been either the blockage of coolant passages or the fouling of fuel elements by the products of Tetralin decomposition, which caused subsequent overheating of some fuel elements.

(2) The fuel-element temperatures rose sufficiently to induce eutectic melting between the uranium and the iron in the type 304 stainless-steel fuel cladding.

(3) Complete melting of the cladding around 10 of the 43 fuel assemblies in the reactor is now known to have occurred. The resultant loss of cladding support led to a complete separation of the top and bottom halves of these 10 assemblies. In every case the zone of fracture was between one-third and two-thirds of the length measured from the top of the elements.

(4) The iodine released at the time of the fuel-cladding failure appears to have been retained very effectively by the sodium coolant.

(5) The reactor excursion during run 14 is attributed to the rapid addition of reactivity and to the failure of the setback circuit to initiate proper corrective action. This circuit is actuated mechanically by means of a cam in the reactor period recorder; a notch in the cam trips a switch that initiates reduction of reactor power by proper rod action. Although the circuit was set to operate when the period reached 10 sec, examination of the mechanism indicated that it operated satisfactorily only if the period decreased at a fairly slow rate and that, when the period decreased rapidly, the switch would fail to open. Modifications have been made to assure satisfactory operation at all rates of change of the reactor period.

(6) The difficulties encountered at the SRE are not attributed to the use of sodium as a coolant but rather to the impurities that were introduced into the coolant.

ACCIDENT—25

Radiation at the MTR

The following description of this radiation exposure incident is taken from Report TID-5360 (supplement), by D. F. Hayes.

Six employees were working on the top of the materials testing reactor adjacent to the reactor tank opening and two men were present as observers and advisors. When a highly radioactive reactor component was placed in a position where it was not adequately shielded because of the lowered water under the reactor tank, these eight men received radiation exposures. The moving of the component and the coincident lowering of the water level were both done to facilitate the insertion and removal of experiments in the reactor. Eight employees received gamma radiation exposures ranging from 2.5 r to 21.5 r, which were not sufficient to cause illness to any of the men, although they were placed under routine medical observation. As a result of this experience, full-time radiation monitoring service is provided at the top of the reactor when workmen are present.

ACCIDENT—26

IDAHO CHEMICAL PROCESSING PLANT CRITICALITY INCIDENT

The following description of a nuclear incident on 16 October, 1959, is taken from a review by J. W. Ullman and J. P. Nichols, in *Nuclear Safety*, Vol. I, No. 3.

The Idaho Chemical Processing Plant (ICPP), is a shielded facility remotely located at the National Reactor Testing Station for the recovery of highly enriched fissionable material from irradiated fuel elements.

At the time of the incident, the plant was processing stainless-steel fuels. Since the capacity in the initial recovery steps (dissolution and first extraction cycle) for this type of fuel is lower than in the balance of the process (second and third extraction cycles), uranium is accumulated and stored before the final operations are performed. The solution was stored in two banks at a concentration of approximately 170 g of uranium per liter of solution. Each bank consists of eight interconnected 5 in. diameter 10 ft high pipes. The two banks have a common drain line which is provided with a jet suction line to a waste header for transferring decontamination solutions. Prior to the incident each of the banks contained approximately 42 kg of uranium.

Approximately 200 l. of solution containing about 170 g of the 91 per cent assay uranium per liter flowed from the storage banks through a diversion spout to a waste tank that contained about 600 l. of aqueous solution of negligible uranium content. The diversion spout had been set to this almost empty receiver prior to the incident because its companion vessel was almost full. Both 5000 gal vessels are 50 ft below grade and are covered with a 4 ft thick concrete deck.

The nuclear excursion occurred in the almost empty receiver. The system went critical and returned to sub-critical under conditions unknown to, and unsuspected by, operating personnel. The actual mechanism or duration of the excursion is not determinable, but it is known that the reaction was shut down by the expulsion of 600 l. of solution (of the 800 l. then in the vessel) presumably into the almost full waste receiver. The vessels did not rupture, but the diversion spout was forced back into the position that caused it to drain into the almost full receiver.

The first indication that there was an abnormality in the process building came as a result of the spread of gaseous and air-borne beta and gamma contamination from the waste tank area, possibly through vent lines and drain connections, to areas where radiation monitors were located. Alarms actuated by these monitors resulted in the evacuation of the building. Contamination was also spread outside the building by air-borne particulate and/or gaseous beta and gamma activity which entered the unfiltered vessel off-gas system and was discharged through the local plant stack. At the time of the building evacuation, the radiation field outside the building and for

130 yd to one side of it was 5 r/hr or greater. Forty-five minutes after the evacuation, the air-borne activity had dissipated sufficiently for personnel to re-enter the building and shut down the equipment.

Seven of 21 persons directly involved in the incident received significant external radiation exposure, with a maximum individual exposure of 50 rem.

ACCIDENT—27

SL-1 REACTOR ACCIDENT

On 3 January, 1961, an explosion and release of fission products occurred in the SL-1 reactor at the National Reactor Testing Station, Idaho, resulting

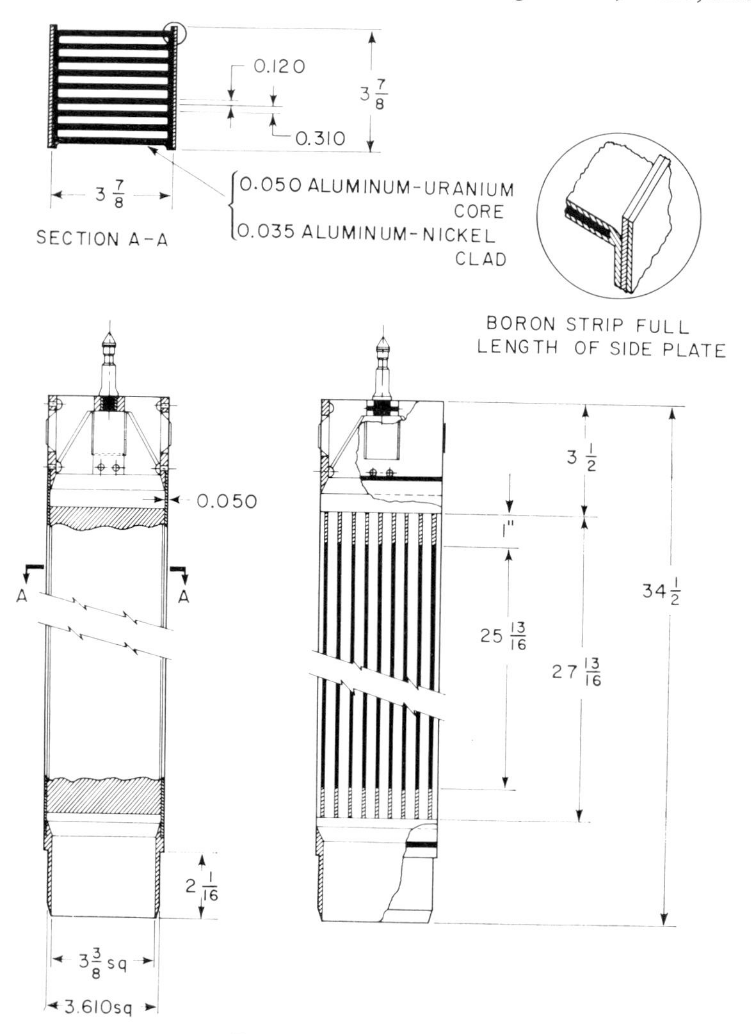

FIG. 5. SL-1 Fuel Element.

in the death of three persons, damage and contamination of the reactor installation and some release of radioactive material outside the reactor building. The following description of this event is taken largely from the *Interim Report on SL-1 Incident, 3 January, 1961* by the Atomic Energy Commission General Managers' Board of Investigation and from Dr. Frank Pittman's remarks at the public meeting held on 24 January, 1961.

The SL-1 is a direct-cycle, natural-recirculation boiling-water reactor formerly designated as the Argonne Low Power Reactor (ALPR). It was designed for 3 MW heat power and produced 200 kW of net electricity and 400 kW thermal energy for space heat. The reactor was constructed, starting in 1957, as part of the program for the development of nuclear power plants for remote military installations. Initial criticality was achieved on 11 August, 1958 and was followed by a series of critical experiments with different core loadings up to a full 59 element core. On 24 October, 1958, the SL-1 achieved full-power operation. Since February, 1959, the reactor has been in more or less routine operation to develop plant performance characteristics and to train military personnel in plant maintenance and operation.

Fuel elements

The core loading consists of plates of highly enriched uranium–aluminum alloy clad with 35 mils of X-8001 aluminum (0.4 per cent nickel). Criticality was achieved with 10 fuel element assemblies containing a total of 3.5 kg of enriched uranium. The reactor was to operate for three years on a fuel loading and be capable of 3 MW heat power output at the end of this period. To provide for this requirement, the fuel loading was increased to the 40 fuel assemblies with 40 full length and 16 half-length boron-containing strips fastened to the fuel assemblies so that the excess reactivity could be controlled by five control rods. On the 16 fuel assemblies in the center of the core, a full length burnable boron poison strip is spot-welded to one side plate and a half-length strip is spot-welded to the other side plate as shown in Fig. 6.

The other fuel elements have a full length boron strip only on one side plate. These poison strips are aluminum nickel containing boron-10. The half-length strips are 21 mils thick and the full length strips are 26 mils. The core contained 23 grams of boron-10 when initially loaded.

Excess reactivity

The core was designed to provide 15 per cent excess reactivity, with the burnable poison strips holding 11.2 per cent of the reactivity. The reactivity was to increase by 10 per cent as the boron-10 was burned out over the core lifetime. The fission products provided an additional negative reactivity of up to 2 per cent. The combined excess reactivity or the reactivity held by the control rods was calculated to be 3 per cent at the beginning of core life and to increase to $3\frac{1}{2}$ per cent in just under one year of normal power operation before decreasing to the end of core life, which was calculated to be over three years at normal power. The reactivity changes calculated for the

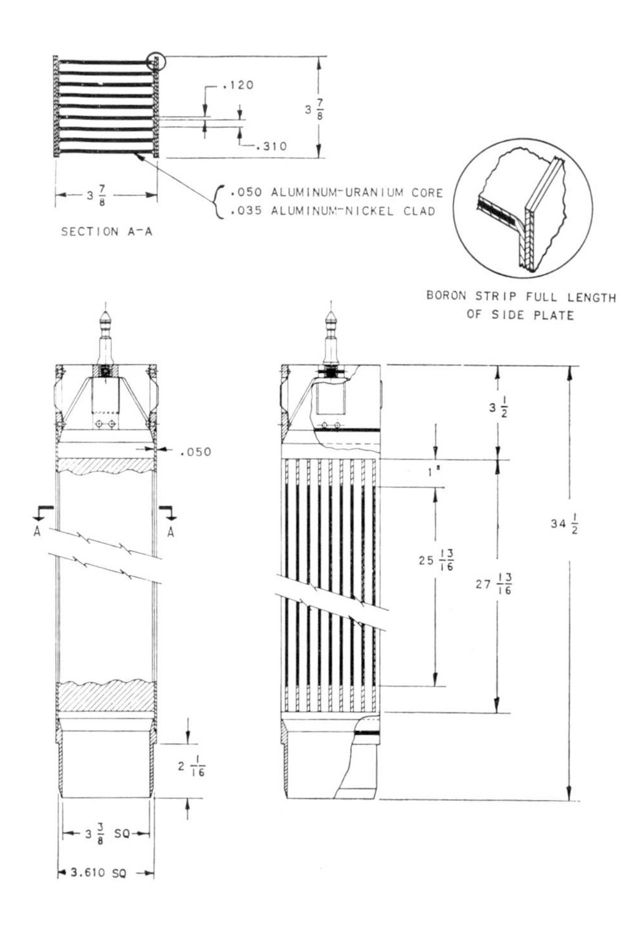

FIG. 6.

core life are shown in Fig. 7. At the time of the accident, the reactor had been in operation for over two years and had produced 931.5 MWd of thermal energy equivalent to about 40 per cent of the core life. The core loading at the time of the accident is shown in Fig. 8 where the channels loaded with fuel assemblies are designated by R. Each fuel assembly contained nine fuel plates.

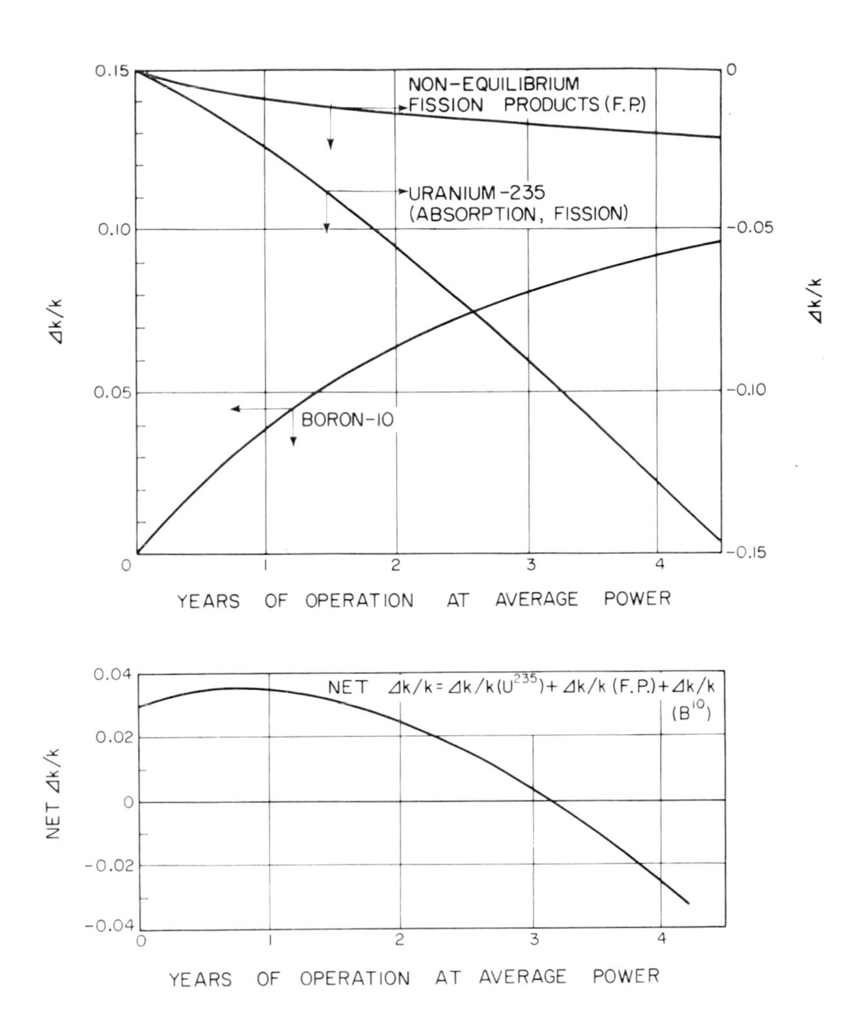

FIG. 7. Reactivity variation during core lifetime.

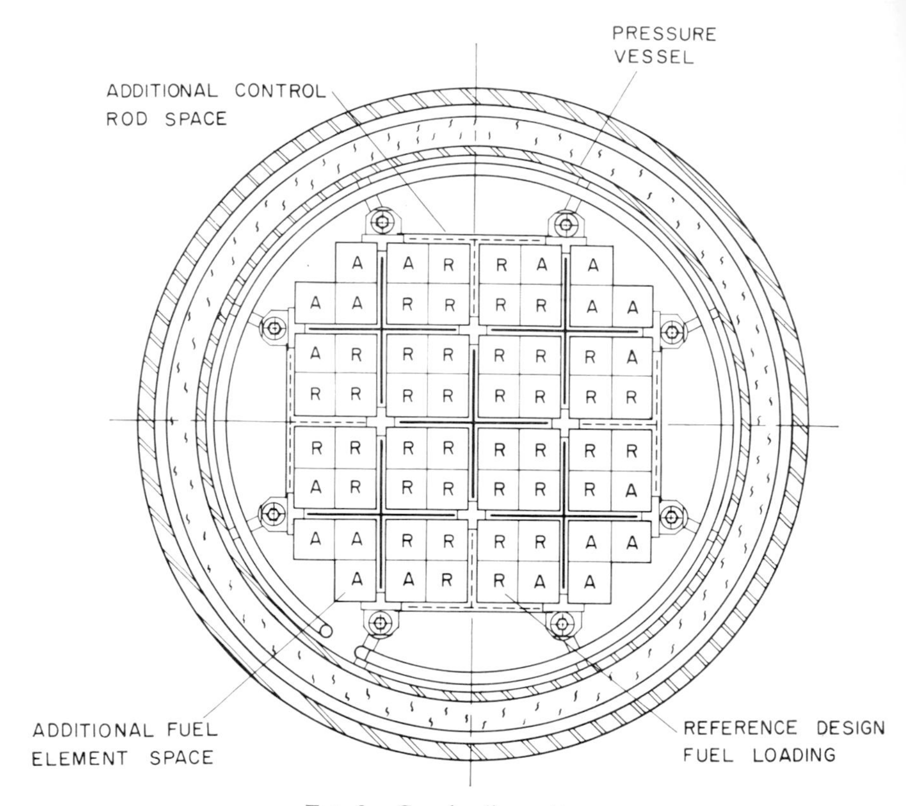

FIG. 8. Core loading pattern.

Control rods

The reactor was controlled by five rods made of 60 mil cadmium sheet clad with 80 mils of aluminum nickel. These five control rods are cruciforms with each blade of the cruciform being 7 in. wide and 34 in. long. The details of control rod construction are shown in Fig. 9. There are spaces for four T-shaped control rods which were to have been used with larger core loadings in the future. The edges of the unbonded aluminum cladding were welded; however, the top center of the envelope was left unwelded to permit the escape of steam since it could not be assured that the welding would not leak.

There are cruciform shroud tubes open at top and bottom within which the control rods moved. The blades of the rod are 0.22 in. thick and the corresponding channels in the shroud tube were initially one-half in. wide. As will be discussed later, there is evidence that the shroud channels may have been compressed by bowing of the boron strips on the fuel assemblies as one of the causes of control rod sticking.

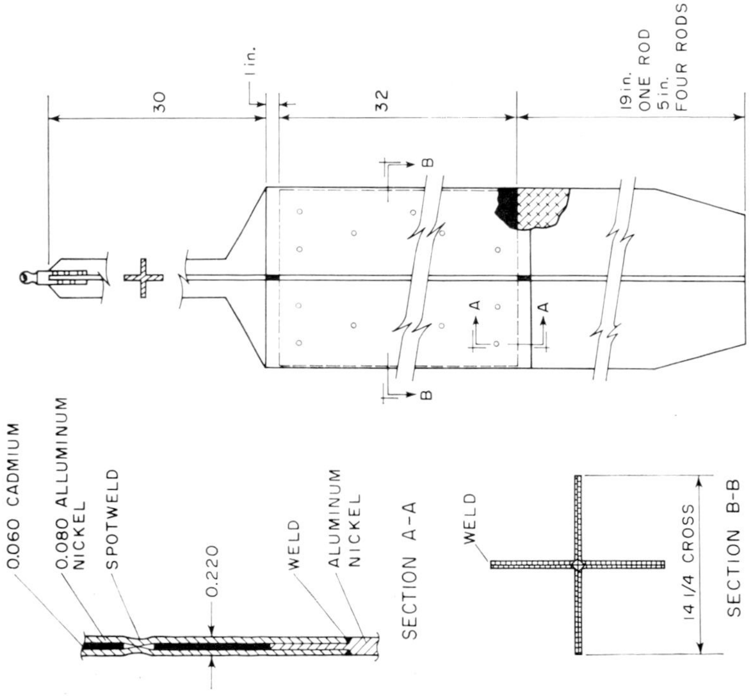

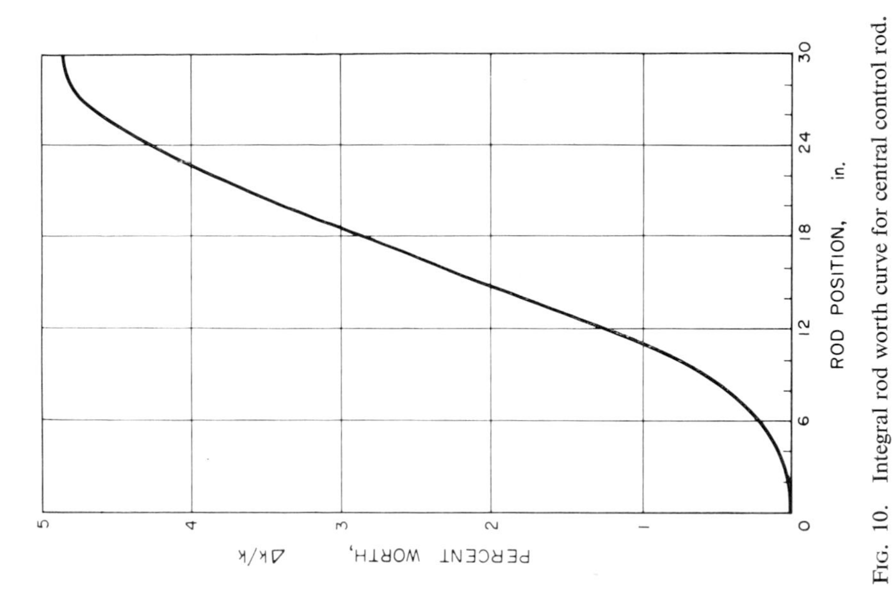

FIG. 10. Integral rod worth curve for central control rod.

0.060 CADMIUM
0.080 ALLUMINUM NICKEL
SPOTWELD
0.220
WELD
ALUMINUM NICKEL
SECTION A–A
WELD
14 1/4 CROSS
SECTION B-B
30
1 in.
32
19 in. ONE ROD
5 in. FOUR RODS

FIG. 9. Cruciform type control rod.

Control rod drive mechanisms

An extension of each control rod projected through the top head of the reactor pressure vessel into the sealed housing containing the rack and pinion gear drive and the coiled-spring shock absorber as shown in Fig. 11. This extension projected through the shock absorber so that its top was normally

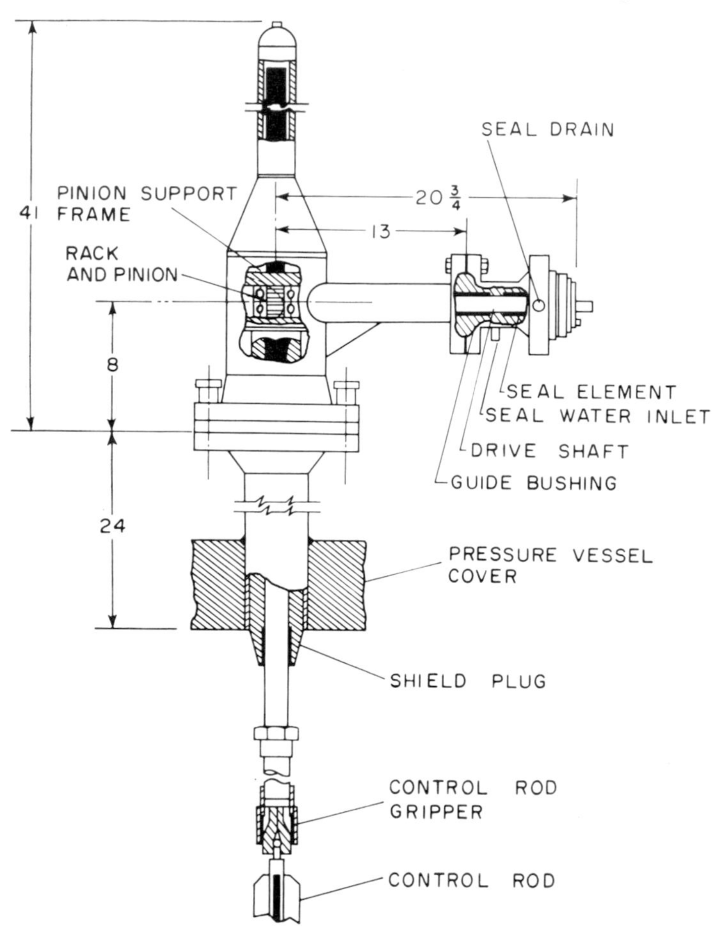

FIG. 11. Control rod drive.

above the top of the spring. A washer, larger in outer diameter than the spring diameter, was attached by a nut and cotter key to the top of the extension. When the control rod was scrammed, the energy acquired during free fall was absorbed by compression of the spring by the washer.

When the drive mechanism was disassembled, the housing was removed, the pinion gear was removed, the nut and washer were removed from the top of the extension, and by use of a special tool, the rod was allowed to be lowered 3 in. below the normal, fully down position. At this point, the rod became suspended by dogs on the rod extension that came to rest on the top of the shroud tube. In this position, the cadmium of the center rod extended about 7 in. below the core, so that this rod could be lifted by 7 in. from this point without causing an appreciable change in reactivity. During

FIG. 12. Control rod drive rack.

reassembly, the control mechanism had to be lifted temporarily to about 2 in. above its normal fully-down position, but even in this temporary elevated condition, the cadmium extended slightly below the core and the lifting to this point did not change the reactivity significantly.

According to the design, complete removal of the three outer control rods would not make the reactor critical. Removal of the central control rod a distance of 19 in. would make the reactor just critical and further rapid removal of this rod above 19 in. was known to be capable of causing a serious nuclear excursion. The motor drive for this rod was geared during operation to permit the maximum speed of 1.8 in./min to preclude rapid withdrawal.

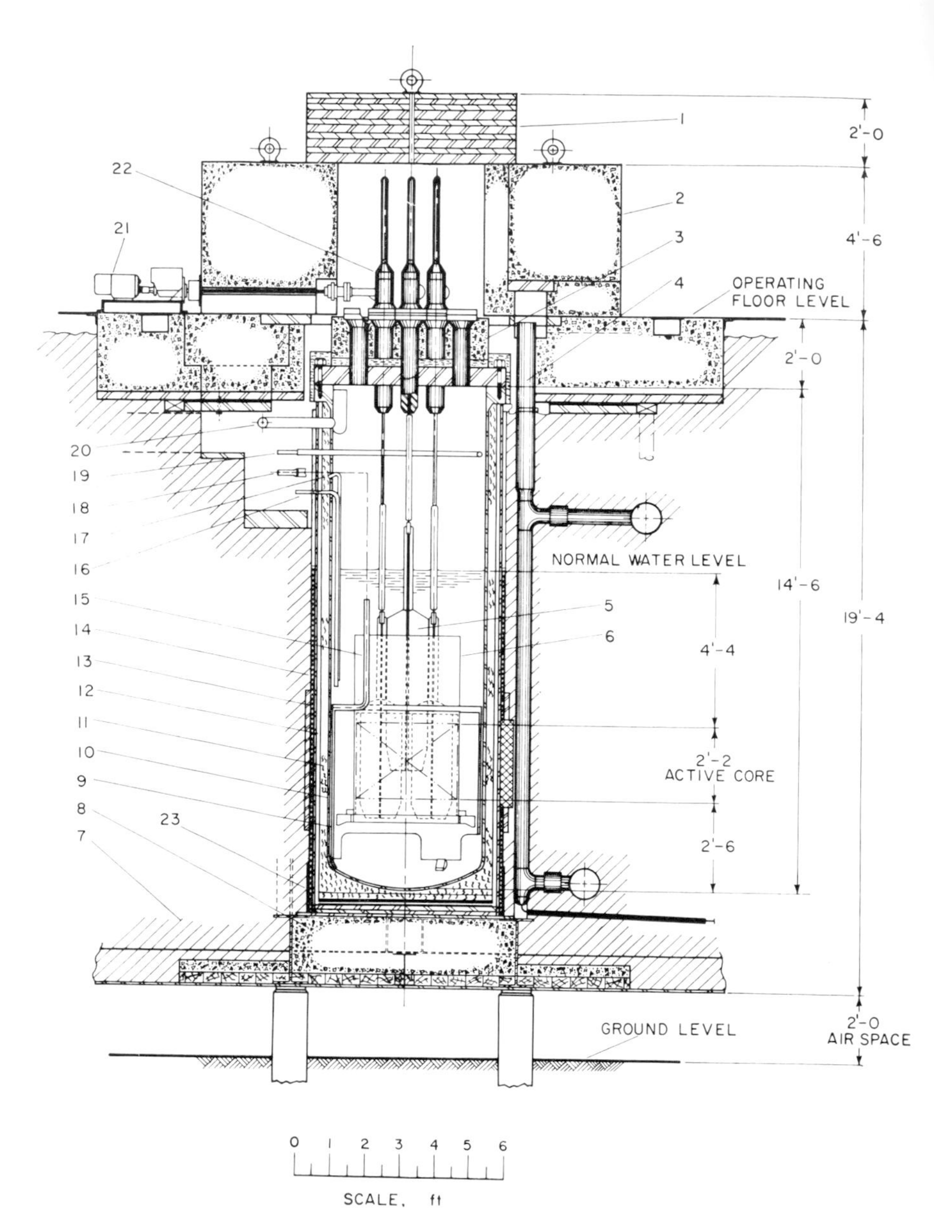

FIG. 13. Vertical section of reactor installation.

The control rods are connected to extension rods and racks and are driven by the pinions in the control drive mechanisms located on the head. The mechanisms are driven by shafts which extend through shielding blocks to motors located on the outside of the shielding blocks. These drives and their controls are described in some detail in Chapter 4, "Control and Safety Systems." Over the head of the reactor vessel is a metal enclosure filled with metal punchings and gravel to provide shielding. A top shield cap rests on the shielding blocks as shown in Fig. 13.

Reactor building

The reactor and most of the plant equipment are located in a cylindrical building made from steel plate most of which is $\frac{1}{4}$ in. thick. The building had normal doors and the structure was not considered to provide the degree

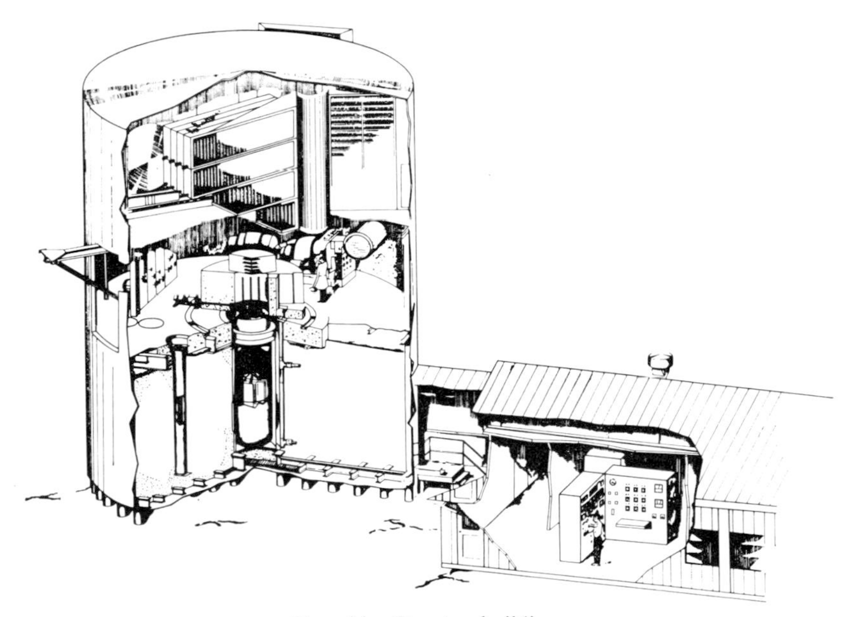

FIG. 14. Reactor building.

of containment that would be appropriate for a reactor near a populated area. However, since there was subzero weather at the time of the accident, it can be assumed that the building was tightly closed. The reactor vessel, the fuel storage well and demineralizer are located in the lower third of the building and are shielded with gravel. A recirculating air-cooled condenser is located in the upper third of the building. The middle third of the building contains the turbine generator, feed water equipment, switch gear and shielding block located around the pressure vessel head. The operating floor is shown in Fig. 16.

FIG. 15. Reactor site.

FIG. 16. Operating floor.

Location

The service and administrative buildings are shown in Fig. 15. The reactor is located in the south end of the National Reactor Testing Station about one mile north of Route 20 in an area used for the testing of other Army reactors. The reactor location is designated by ALPR, see Chapter 1, Fig. 12. The site is some 40 miles across the desert from the nearest major center of population, Idaho Falls, and just over five miles from the small community to the south of Atomic City.

Conditions prior to the accident

The use of boron strips was intended to prolong the life of the core. Early in the operation of the reactor, it was noted that these strips were beginning to bow outward, away from the fuel plates, in the sections between the attachment-weld points. This outward bowing of the strips, located in the narrow spaces between individual fuel elements and between fuel elements and the shroud, increased steadily with time to a point where removal of elements in the central region of the core could be accomplished only with great difficulty. Concurrently, the reactivity began to increase more rapidly than expected until September of 1960, when a net gain of 2 per cent over expected value had been observed. At this time, three fuel elements were removed from their core positions for inspection as had been done previously. Exertion of great force was required to remove the elements, and the boron strips remaining on the elements were damaged and shattered in the process. It was estimated at that time that 18 per cent of the boron had been lost from the core. Typical debris pumped from the bottom of the reactor is shown in Fig. 17. It was observed during the inspection that when the core was disturbed, clouds of white flocculent material streamed into the water from the boron strips. When the reactor core had been reassembled, it was estimated that the shutdown margin was then only about 2 per cent (the exact margin being uncertain).

A decision was made to increase the shutdown margin by the addition of six cadmium strips, each welded between two thin aluminum plates, into the unused T-shaped control rod shroud tubes on the sides of the core. The cadmium strips were inserted into position in November of 1960. As shown in Fig. 19, the shutdown margin was increased but it was about 1 per cent less than expected. The total shutdown margin was then estimated to be 3 per cent. Positions of the control rods at the final shutdown on 23 December indicated that just prior to shutdown the margin had not changed. It is noted that these six cadmium strips were encased in 2S aluminum. All other aluminum in the reactor is alloyed with nickel for corrosion resistance at high temperatures. The aluminum plates were suspended by aluminum strips anchored on the top of shroud tubes.

From early operations onward, intermittent and increasing difficulty was encountered in the free movement of the control rods. At least over the

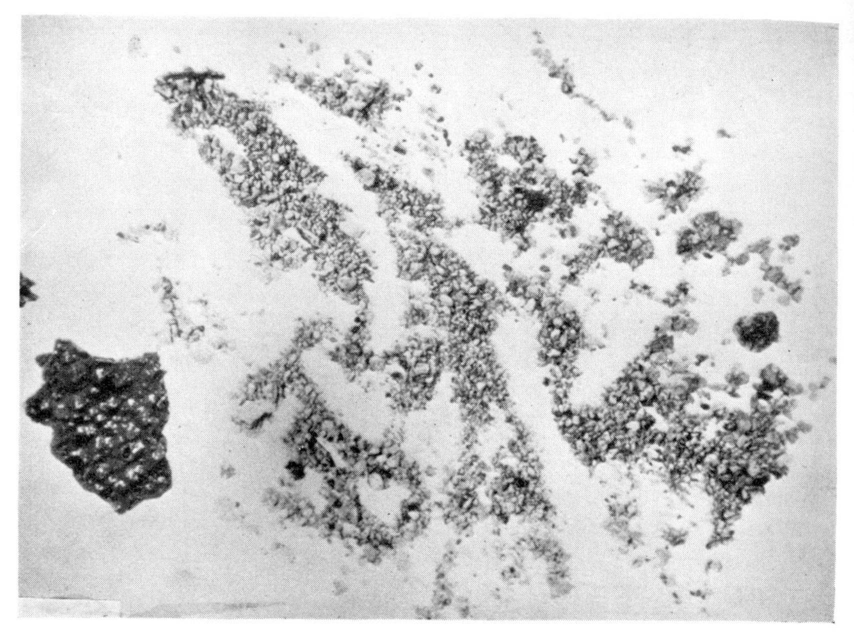

FIG. 17. Debris recovered from bottom of reactor vessel.

first year of operation and possibly in large measure thereafter, the difficulty arose from the friction in the seals through which the drive rod shafts penetrated the rack and pinion gear housing on top of the reactor. It was frequently found that one or another of the rods would not fall freely when

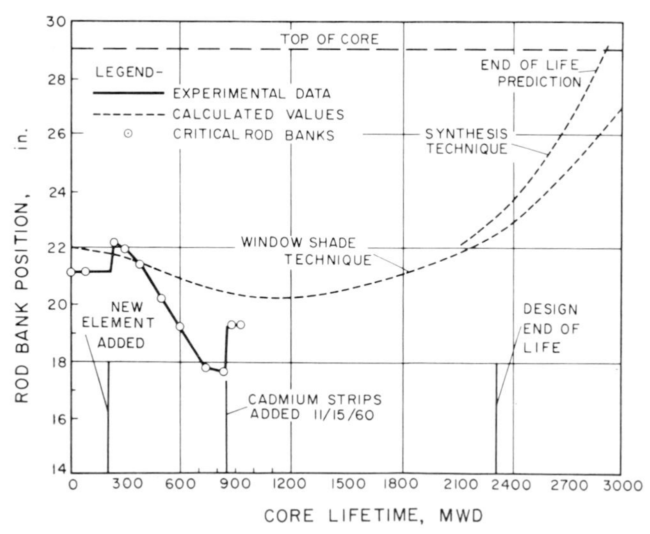

FIG. 18. Rod bank position with equilibrium xenon concentration at 2.56 MW.

the magnetic clutch drive was disconnected in a scram or single rod drop test. Later, the rod sticking problem, both on insertion and on withdrawal (when manual assistance to the motor drive was sometimes used), became accentuated. Over the last two months of operation, approximately 40 occurrences of rod sticking were recorded in the Operations Log. During the last period of reactor operation, the rods were assisted manually in their withdrawal, and on scram at shutdown, three rods had to be driven in by their motors.

High power operation

The reactor was operated at a power level of 3 MW thermal up to November, 1960. At that time, in order to test out a new condenser loop of higher capacity (a component which had been developed for another proposed reactor and for which performance data were desired), approval was given to operate the reactor at 4.7 MW thermal. Over the period to 23 December, 1960 (the final shutdown), the reactor was operated for a dozen short runs at 4.7 MW thermal. The effects of this increased power level were an increase of about 5° in operating water temperature, some increase in radiation levels, and some increase in turbulence in the core. It was found that the reactor could not be operated at this level with the usual control rod pattern (all rods equally out) without "chugging" instability. Hence, operation at 4.7 MW thermal was accomplished with the outer four rods almost fully withdrawn and the center one in as far as necessary. This pattern achieved some flux-flattening, and was made necessary, at least in part, by the loss of boron from the central region. Permission had been requested to operate the reactor of 8.5 MW thermal and this was under consideration, but had not been approved, at the time of the accident.

Sequence of events

After having been in intermittent operation for slightly more than two years, the SL-1 was shut down on 23 December, 1960. It was planned that maintenance on certain components of the whole reactor system would be performed during the succeeding twelve days and the reactor would again be brought to power on 4 January, 1961. While maintenance work on several auxiliary systems of the plant was completed during the period, the only work planned for the reactor core was the insertion of 40 cobalt wire flux measuring assemblies into fuel element channels of one quadrant of the core. Access to the core through nozzles in the head of the reactor vessel required removal of the concrete shielding blocks and of the control rod drive assemblies. This portion of the work was begun during the early morning hours of 3 January, 1961. When the day crew arrived on 3 January, 1961, disassembly had been completed. Installation of the flux measuring assemblies was accomplished during the day shift.

The crew of the next shift (4.00 p.m. to midnight, 3 January) consisted of three military personnel: a chief operator, an operator, and a trainee. This crew and the following one were assigned the task of reassembling the control rod drives and preparing the reactor for start-up.

The control room log book contained a single entry by the 4.00 to 12.00 shift as follows:

> "Pumped reactor water to contaminated water tank until water level recorder came on scale. Indicate plus 5 ft. Replacing plugs, thimbles, etc., to all rods."

The water level had been raised to the top of the reactor vessel for shielding during the installation of the flux mapping wires. A necessary step in the preparations for start-up was the lowering of this water level. The crew was connecting the control rods to the drive mechanisms when the accident occurred. This operation involved manually lifting each control rod which weighed some 100 lb for making these connections.

At approximately 9.00 p.m. on 3 January, alarms indicating a possible fire at the SL-1 sounded at the Central Facilities Fire Station, the AEC Security Station, and other fire stations within the National Reactor Testing Station. Fire and security personnel arrived at the SL-1 at approximately 9.10 p.m. and observed no indication of a fire or other damage. They were unable to arouse the SL-1 crew. Upon entering the adjacent reactor support building, they found high level radiation of 25 roentgens per hour (r/hr). There were higher radiation fields of the order of 1000 r/hr on the reactor operating floor. Two of the crew were found on the operating floor. One of the crewmen was still living at that time. He was removed by approximately 11.00 p.m. and shortly thereafter, was pronounced dead by a physician who responded to the emergency call. On subsequent entry into the reactor room, the body of the third crewman was located in the ceiling structure directly over the reactor. Because of the high radiation levels, it was not possible to remove the second crewman until 5 January and the third crewman until 9 January.

Nature and extent of the accident

(a) *Reactor building.* There was only minor physical damage to the reactor building. The reactor room ceiling directly above the reactor was buckled. Two of the shield plugs were driven upward out of the nozzles in the head of the reactor vessel and penetrated the reactor room ceiling. A peeling back of a portion of this ceiling indicated that some additional parts of the reactor system had been projected into the fan room area above.

(b) *Reactor.* Preliminary estimates have been made from observed damage to the top of the reactor vessel and from calculation of energy needed to propel certain components to observed locations of an internal pressure of

500 psi. The normal working pressure in the reactor was 300 psi. The sheet metal covering on top of the reactor head was bent upward allowing dispersal of some of the gravel, steel punching and pelletized boron shielding material. Shield plugs, and other parts of the control rod drive mechanisms, which had been placed in the reactor nozzles, were blown upward by steam or water.

Motion pictures provided the first interior views of the reactor vessel. Severe damage to the core was disclosed including bending, wrenching, and expansion of the core. The pictures showed that four of the five control rod blades appear to be partially or fully in the core, and that one appears to be entirely out of the core. The motion pictures were obtained with a motor actuated camera suspended from a crane boom. The crane operator, at ground level, was directed by telescope-equipped observers peering into the second-story reactor room from a wooden tower 200 ft distance. The distance and remote manipulation were necessary to shield the workers from the high levels of radioactivity that persisted at and near the top of the reactor.

(c) *Contamination.* As was mentioned above, the radiation level within the operating room was initially in the range of 500 to 1000 r/hr and the rate of decay of the radiation was very slow indicating thc presence of accumulated fission products from prior operations. The radiation level at the road into the site was initially the order of 0.25 r/hr. The levels in the SL-1 area soon after the accident are shown in Fig. 19.

Early in thc morning after the incident, aerial checks for airborne contamination were made with no indication being found of significant quantities of activity being dispersed from the site being found. Soil samples from within the fenced area around the reactor showed gross fission product contamination. Iodine activity surveys were made at a control point which is three-quarters of a mile from the SL-1 reactor and at Atomic City which is 5.3 miles south of the SL-1 site. On 3 and 4 January, measurements showed iodine-131 concentrations to be less than 10^{-9} microcuries per milliliter at Atomic City and on 14 and 16 January, this reading had dropped to a value of 3.5×10^{-11} at Atomic City and on 19 and 20 January to 2×10^{-11}. Analysis of vegetation on the site indicates a total air-borne release of iodine-131 to be one curie from the time of the accident up to noon of 4 January. An additional three curies escaped during the next six days. Throughout this period, temperature inversions averaged 17 hr per day. The airborne activity level three-quarters of a mile from the plant on 14 to 16 January showed an iodine level of half of the maximum permissible concentration for a 168 hr week.

Approximations of the radiation level outside the reactor core indicate that less than one per cent of the fission product inventory of the core may have escaped from the reactor, and most of this radioactive material was in the reactor room.

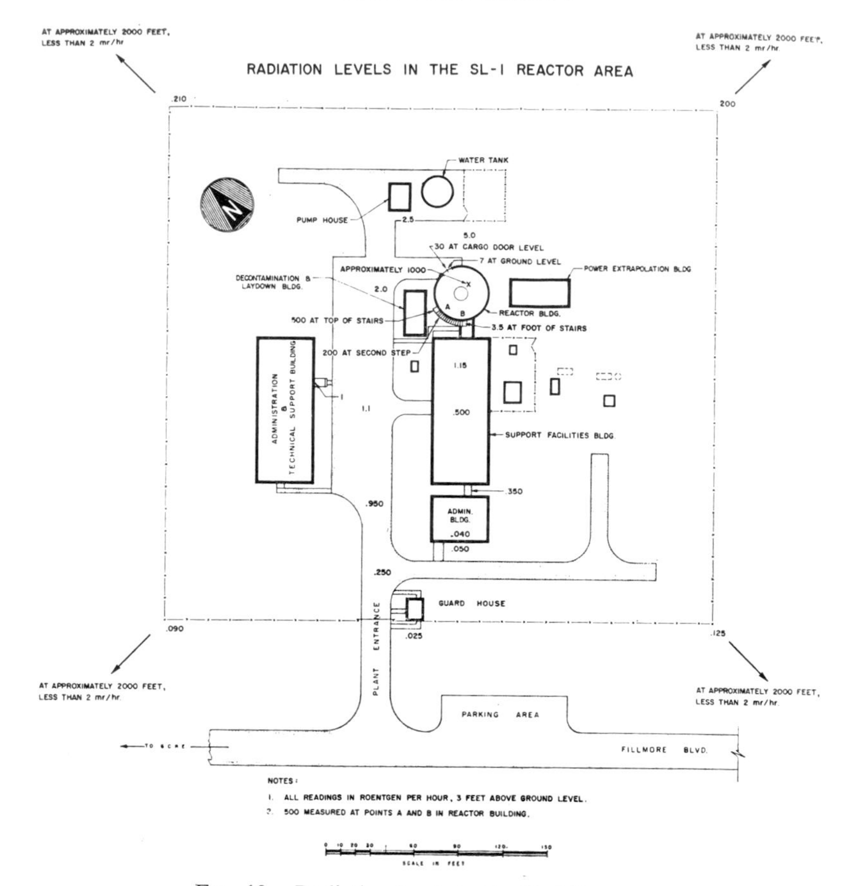

FIG. 19. Radiation levels in the SL-1 Reactor area.

Cause of the accident. There is evidence that a nuclear energy release of the order of 50 MW-sec (1.8×10^{18} total fissions) took place. This evidence includes the measurement of yttrium-91 activity and neutron activation of material in the area. The amount of blast damage is not considered inconsistent with an energy release of this magnitude in a reactor.

A possible cause of this accident is considered to be associated with the raising of a control rod manually in the procedure for connecting these rods with the drive mechanisms. The operators were instructed not to raise the control rods beyond the point required; however, the condition of reactivity of the core may have been different than expected due to changes in the properties of the reactor between 23 December, 1960 and 3 January, although there is no other evidence that such changes had taken place. However, if there had been a loss of boron or cadmium from the core, substantially less withdrawal of the central control rod would have produced criticality.

It is noted that when the control rods were disconnected from their drives as was the condition just prior to the accident, a pressure surge in the reactor core could have caused the rods to be driven rapidly further out and thereby lead to a large nuclear excursion of the magnitude observed.

FIG. 20. Top of reactor following accident.

Other conceivable initiating events, though at the present their likelihood appears to be low include:

(a) A hydrogen explosion or other chemical reaction in the reactor vessel which drove the central control rod out of the core by a pressure increase.

(b) Addition of water to the core which had become partly dry.

REFERENCES

1. HAYES, D. F., A summary of accidents and incidents in atomic energy activities, June 1945 (through December 1955), TID-5360.
2. BRITTAN, R. O., HASTERLICK, R. J., MARINELLI, L. D., and THALGOT, F. W., Technical review of ZPR-1 Accidental transient, the power excursion, exposures, and clinical data, ANL-4971.
3. CALLIHAN, D., and ROSS, D., A review of a polonium contamination problem, ORNL-1381 (Rev.).

4. LEWIS, W. B., An accident to the NRX Reactor on 12 December, 1952, AECL-232.
5. HURST, D. G., The accident to the NRX Reactor, Part 11, AECL-233.
6. GREENWOOD, J. W., Contamination of the NRU Reactor in May, 1958, AECL-850.
7. HAYES, D. F., Summary of incidents with radioactive material in Atomic energy activities, January–December 1956, TID-5360 (Supplement).
8. *Accident at Windscale No. 1 Pile on 10th October, 1957*, London, Her Majesty's Stationery Office.
9. DUNSTEE, H. J., HOWELLS, H., and THOMPSTON, W. L., District surveys following the Windscale incident, October 1957, *Proc. 2nd United Nations Int. Conf. on the Peaceful Uses of Atomic Energy*, Vol. 18, United Nations, New York.
10. Accidental radiation excursion in the Y-12 plant, ORNL Report Y-1234.
11. CALLIHAN, A. D., Kidrich Institute accident, *Nuclear Safety*, Vol. 1, No. 1, pp. 38–40.
12. EPLER, E. P., HTRE-3 excursion, *Nuclear Safety*, Vol. 1, No. 2, pp. 57–59.
13. GUTHRIE, G. E., Los Alamos criticality accident, *Nuclear Safety*, Vol. 1, No. 1, pp. 37–38.
14. MCDONALD, W. B., and DEVAN, J. H., Sodium reactor experiment, *Nuclear Safety*, Vol. 1, No. 3, pp. 73–75.
15. ULLMAN, J. W., and NICHOLS, J. P., Idaho Chemical Processing Plant criticality incident, *Nuclear Safety*, Vol. 1, No. 3, pp. 75–77.

APPENDIX A

Example of a Calculation of Reactor Siting Distance*

Example of a calculation of reactor siting distances:

(1) The calculations of this Appendix are based upon the following assumptions:

(a) The fission product release to the atmosphere of the reactor building is 100 per cent of the noble gases, 50 per cent of the halogens and 1 per cent of the solids in the fission product inventory. This release is equal to 15.8 per cent of the total radioactivity of the fission product inventory. Of the 50 per cent of the halogens released, one-half is assumed to adsorb onto internal surfaces of the reactor building or adhere to internal components.

(b) The release of radioactivity from the reactor building to the environment occurs at a leak rate of 0.1 per cent per day of the atmosphere within the building and the leakage rate persists throughout the effective course of the accident which, for practical purposes, is until the iodine activity has decayed away.

(c) In calculating the doses which determine the distances, fission product decay in the usual pattern has been assumed to occur during the time fission products are contained within the reactor building. No decay was assumed during the transit time after release from the reactor building.

(d) No ground deposition of the radioactive materials that leak from the reactor building was assumed.

(e) The atmospheric dispersion of material leaking from the reactor building was assumed to occur according to the following relationship:

$$X = \frac{Q}{\pi u \sigma_y \sigma_z}$$

where Q is rate of release of radioactivity from the containment vessel, the ("source term,"):

X is the atmospheric concentration of radioactivity at distance d from the reactor

u is the wind velocity

σ_y and σ_z are horizontal and vertical diffusion parameters resp.

(f) Meteorological conditions of atmospheric dispersion were assumed to be those which are characteristic of the average "worst" (least favorable)

* Reproduced from 26 Federal Register, 1224, 11 February, 1961, "Notice of Proposed Guides".

weather conditions for average meteorological regimes over the country. For the purposes of these calculations, the parameters used in the equation in section (e) above were assigned the following values:

$$u = 1 \text{ m/sec};$$
$$\sigma_y = [\tfrac{1}{2}C_y^2 d^{2-n}]^{\frac{1}{2}};$$
$$\sigma_z = [\tfrac{1}{2}C_z^2 d^{2-n}]^{\frac{1}{2}};$$
$$C_y = 0.40;$$
$$C_z = 0.07;$$
$$n = 0.5$$

(g) The isotopes of iodine were assumed to be controlling for the low population zone distance and population center distance. The low population zone distance results from integrating the effects of iodine 131 through 135. The population center distance equals the low population zone distance increased by a factor of one-third.

(h) The source strength of each iodine isotope was calculated to be as follows:

Isotope	Exclusion Q (curies/megawatt)	Low population Q (curies/megawatt)
I^{131}	0.55	76.4
I^{132}	.68	1.40
I^{133}	1.19	18.5
I^{134}	.72	.91
I^{135}	1.04	5.4

These source terms combine the effects of fission yield under equilibrium conditions, radioactive decay in the reactor building, and the release rate from the reactor building, all integrated throughout the exposure time considered.

(i) For the exclusion distance, doses from both direct gamma radiation and from iodine in the cloud escaping from the reactor building were calculated, and the distance established on the basis of the effect requiring the greater isolation.

(j) In calculating the thyroid doses which result from exposure of an individual to an atmosphere containing concentrations of radioactive iodine, the following conversion factors were used to determine the dose received from breathing a concentration of one curie per cubic meter for one second:

Isotope	Dose (rem)
I^{131}	329
I^{132}	12.4
I^{133}	92.3
I^{134}	5.66
I^{135}	25.3

(k) The whole body doses at the exclusion and low population zone distances due to direct gamma radiation from the fission products released into the reactor building were derived from the following relationships:

$$D = 483 \frac{B\,e^{-\mu r}}{4\pi r^2} \int t^{-0.21}\, dt$$

where D is the exposure dose in roentgens per megawatt of reactor power

r is the distance in meters
B, the scattering factor, is equal to $[1 + \mu r + \mu^2 r^2/3]$
μ is the air attenuation factor (0.01 for this calculation)
t is the exposure time in seconds.

In this formulation it was assumed that the shielding and building structures provided an attenuation factor of 10.

(2) On the basis of calculation methods and values of parameters described above, initial estimates of distances for reactors of various power levels have been developed and are listed below.

Power level (thermal megawatt)	Exclusion distance (miles)	Low population zone distance (miles)	Population center distance (miles)
1500	0.70	13.3	17.7
1200	.60	11.5	15.3
1000	.53	10	13.3
900	.50	9.4	12.5
800	.46	8.6	11.5
700	.42	8	10.7
600	.38	7.2	9.6
500	.33	6.3	8.4
400	.29	5.4	7.2
300	.24	4.5	6
200	.21	3.4	4.5
100	.18	2.2	2.9
50	.15	1.4	1.9
10	.08	.5	.7

APPENDIX B

Estimates by the Reactor Safeguard Committee of Radiation from a Cloud of Fission Products (Reproduced from "Summary Report of Reactor Safeguard Committee," Wash-3 (Rev.) by permission of the U.S. Atomic Energy Commission)

An example of the methods of calculation of the radiation hazard presented by the vaporization of all a reactor's fission products is presented here. These methods are given in detail and the principles may be applied with suitable changes and corrections, to the discussion of reactors of several designs. In each case, the worst one of probable events is assumed. No safety factors are included.

Radiation hazard is calculated from the basis of a reactor in normal operation. It is assumed that for one of the several reasons which have been discussed in the main body of this report, the chain reaction gets out of control, the "*k*" of the pile is increased about 2 per cent above the critical value and the power level increases by a factor of 2.718 in each period of about 0.1 sec. The temperatures of the reactor parts will rise rapidly until factors such as negative temperature coefficient or mechanical disruption of the pile stops the reaction. In any case, it is assumed that a considerable part of the fission activity of the reactor will have been carried into the air in the form of a cloud before the reaction stops.

We estimate that radiation dosage due to this cloud by first calculating the total power output of the fission products in this cloud and then evaluating the total exposure received by an individual as the cloud passes by or over him. The level of fission product activity after the disruption of a pile which had been operating at a steady level before the accident, is given approximately by the empirical formula:

$$\text{fission product activity in kilowatts} = 0.1\,\frac{\text{previous steady power in kilowatts}}{(\text{time in seconds after accident})^{0.2}}$$

We assume that of this activity 50 per cent is present in the radioactive cloud. Thus we get:

$$0.05\,\frac{\text{previous steady power in kilowatts}}{(\text{time in seconds after accident})^{0.2}}$$

Negligible for the present purpose is the activity of the additional fission products produced during the period of run-away pile operation.

We have to multiply the formula for fission product activity by the following factors in order to calculate the level of radiation hazard attained:

(1) $$\text{Concentration factor} = \frac{1}{\text{volume of cloud}} = \frac{1}{\text{cloud thickness} \times \text{length} \times \text{width}}$$

According to meteorological evidence, the spread of a smoke cloud is roughly one seventh the distance of downstream travel, under a wide variety of wind conditions. Thus we have:

$$\text{Concentration factor} = \frac{7}{\text{distance from pile} \times \text{cloud thickness} \times \text{length}}$$

(2) The time of exposure, which is given by the formula:

$$\text{time} = \frac{\text{cloud length}}{\text{wind velocity}}$$

The product of expressions (1) and (2) is evidently independent of the cloud length. In other words, the integrated radiation dosage at a given position will be the same—granted constant wind direction—whether the uranium burns up and goes into the air in 1 min or 1 hr.

(3) The conversion factor between integrated radioactive power in kilowatt seconds per cm^3 (the product of (1) and (2)), and accumulated radiation in roentgen units. This factor is

$$\frac{10^3 \text{ volt ampere seconds} \times 3 \times 10^9 \text{ franklins/coulomb}}{30 \text{ volts per ion pair}} = 10^{11}$$

The factor just given will apply to a cloud so large that as much energy is delivered up per cm^3 and per sec in the form of ionization as is produced by radioactive decay per cm^3 and per sec. This condition will be satisfied if the thickness of the cloud is great compared to the range of the radioactive radiations and if, furthermore, the receptor is in the cloud.

We take as a rough average range for the beta particles, 5 ft, and 1,000 ft for the gamma rays, and note that roughly equal fractions of the energy discussed above go off as beta and as gamma radiation. If the receptor is on the ground, if the cloud extends to the ground and is greater in thickness than the range of the beta and gamma rays, and if finally none of the radioactive substances are deposited on the ground, then the dosage obtained by the receptor is roughly 0.6 times the dosage within the cloud. The factor 0.6 is obtained by considering that the receptor is exposed to gamma rays only from one-half of all directions (i.e., from above) and to beta rays of a 5 ft range from somewhat more than one-half of all directions.

We find for the accumulated radiation, by taking account of factors just discussed, the result:

$$\text{Accumulated radiation in roentgens} = 2 \times 10^{10} \frac{\text{previous steady power in kilowatts}}{\underset{\text{(in cm)}}{\text{(cloud thickness)}} \times \underset{\text{(in cm/sec)}}{\text{(wind velocity)}^{0.8}} \times \underset{\text{(in cm)}}{\text{(distance from pile)}^{1.2}}}$$

The result just obtained can be applied in the form in which it stands, but may be put into a more reasonable form by a little further analysis. From the present formula it follows that a receptor at a given distance, say 10 miles, from the pile, subject to radiation from a cloud of fission products of a given thickness, will receive an integrated radiation almost inversely proportioned to the speed with which the cloud passes by him. He will get about twice as great a dosage from a cloud which moves at $\frac{1}{2}$ mile an hour as from one traveling at 1 mile an hour. By admitting the possibility of a cloud of arbitrarily low velocity we should, therefore, admit an irradiation of arbitrarily great intensity. This result is unreasonable, as: (a) from a meteorological standpoint it is extremely unlikely that at the time of the accident a wind velocity will occur as low as $\frac{1}{2}$ mile per hour, and (b) if, contrary to all reasonable expectation, the cloud should actually move with such a low velocity, it would reach the 10-mile distant receptor only after the lapse of 20 hr. A time so long would seem adequate to notify individuals to evacuate. On both accounts we exclude irradiation at a 10-mile distance from a cloud moving so slowly.

On the other hand, a 3-mile an hour wind is not improbable in view of the meteorological conditions observed at Hanford. The 3.3 hr required in this case for the cloud to travel 10 miles will only barely allow notification of hazard.

In the light of this discussion we adopt the following point of view. We accept 3 hr as a critical time. We consider a receptor at a given distance from the pile catastrophe:

(a) If the wind is so slow that more than 3 hr are required for the cloud to reach the receptor then indeed the possible accumulated exposure will be greater than we are about to calculate. It is, however, possible to notify and move the people out of the way of the cloud. It is recognized that this evacuation itself will be difficult and hazardous.

(b) If the wind is so fast that much less than 3 hr are required for the cloud to reach the receptor, then the exposure time is short and the dosage is reduced. Consequently the worst case will be that in which a time just in the neighborhood of 3 hr is required to reach the receptor:

$$\text{Wind velocity in miles per hour} = \frac{\text{miles from pile to receptor}}{3\text{ hr}}$$

On this basis, our formula for exposure takes the form:

$$\text{Accumulated irradiation in roentgens} = 43\,\frac{\text{operating level in kilowatts}}{\text{cloud thickness in feet} \times (\text{distance from pile in miles})^2}$$

At an exposure to 300 r units, it is to be expected that acute sickness will follow in each individual. Approximately 15 per cent are expected to die within the next 2 or 3 months. The rest are expected to recover. It is unknown to what extent the life-span of the "recovered" individuals will be shortened. We have taken as the basic figure of our calculations 300 r. This evidently corresponds to most serious danger. Setting the left-hand side of the last expression equal to 300, we get

$$\text{exclusion distance in miles from pile} = \left(0.144\,\frac{\text{operating level in kilowatts}}{\text{cloud thickness in feet}}\right)^{1/2}$$

For a more specific estimate of the exclusion distance some value is required for the possible minimum thickness of the cloud.

The foregoing argument shows that in our present state of knowledge we cannot possibly recommend settlement of population closer to a pile than this distance.

APPENDIX C

Estimates by the Reactor Safeguard Committee of Radioactive Contamination by Precipitation "Summary Report of Reactor Safeguard Committee," (Reproduced from Wash-3 (Rev.) by permission of the U.S. Atomic Energy Commission)

In the following an estimate is given of the probable upper limit of ground contamination by β and γ active substances following dispersal of the active content of a reactor in the atmosphere.

Assumptions

It will be assumed that the whole active content of a reactor is carried up in the atmosphere and that all of this active content will be deposited on the ground at a distance of M miles. In traveling the distance of M miles it is assumed that the radioactive cloud will have been widened out by the turbulence of the air currents to cover a circle of radius of $M/14$ miles. This gives an angular spread of the radioactive cloud of 1/7 of a radian, which seems to be a reasonable lower limit of observed angular spreads.

The assumption that the radioactive material in the cloud is deposited in a circle as described will definitely give an upper limit of the possible contamination. This, however, is not the case in a thunderstorm, where masses may be sucked in from the side and deposited in the active region of the thunderstorm. Thus during such a storm activity may be concentrated on a smaller area than the one over which the radioactive cloud was originally dispersed.

We assume that prior to the explosion the reactor has been steadily operating at a power of k kilowatts. Under these conditions there will occur in the reactor $3 \times 10^{13}\ k$ fissions/sec. We consider the radioactivity of the fission products at a time t after the explosion. This radioactivity has been expressed in *The Rate of Decay of Fission Products* ((MDDC-1194) by Way and Wigner), on page 22, by various formulae for various times. We select the time interval between 20 min and 3 days, because this time interval is most relevant for the acute effects of the radioactivity. For this time interval one finds that each fission releases $15t^{-0.28}$ MeV γ radiation per second if the reactor had been running at one fission per second It also releases $9t^{-0.26}$ MeV of β radiation per second.

The total activity of the reactor will be deposited over an area of $\pi(M)^2/(14)$ miles^2, or $4.15 \times 10^8\ M^2$ cm^2. Thus one finds that per cm^2 there is

$$\frac{15t^{-0.28} \times 3 \times 10^{13}k}{4.15 \times 10^8 M^2} = \frac{1.1 \times 10^6 t^{-0.28}k}{M^2}$$

MeV γ radiation released per second at a time t after the explosion. Similarly one finds that

$$\frac{9t^{-0.26} \times 3 \times 10^{13}k}{4.15 \times 10^8 M^2} = \frac{6 \times 10^5 \times t^{-0.26}k}{M^2}$$

MeV of β radiation released per cm^2-sec.

It is of interest to compare the assumption of total deposition of radiation to events following explosion of atomic bombs. One of the bomb tests on Eniwetok was followed by a rainstorm on Kwajalein. The actual path that the air took in going from Eniwetok to Kwajalein was 840 miles. The radioactive products, however, were quite dispersed by the time they arrived at Kwajalein. In fact, the course of events was probably that the top of the cloud above an altitude of 20 miles was carried westward, then some of the radioactivity fell to lower levels and was carried by northeasterly winds to Kwajalein. A great dilution under such conditions may well be expected. One actually finds that in the heavy rainstorm at Kwajalein only 1/500 as much activity was deposited as could have been expected if all activity had been deposited at the distance of 840 miles, in accordance with the formulas* given above.

Irradiation Produced

One MeV of β or γ radiation produces approximately 30,000 ion pairs. If the radioactive material is deposited on the ground, only one-half of this activity will go in the upward direction and cause ionization of the air. Taking this factor into account, as well as the obliquity of the radiation, and assuming that the average range of γ rays in air is approximately 200 m, one finds that near the ground approximately 10^{-4} of the radiation will be deposited in the air for each one cm height. For the β rays one may assume that approximately 10^{-2} of the radiation will be dispersed per one cm height. Thus, if one has on the ground one γ disintegration/cm^2-sec of one MeV energy, one finds that in the air 3 ion pairs/cm^3-sec are formed. Furthermore, under these conditions one β ray/cm^2-sec with one MeV energy gives rise to 300 ion pairs/cm^3-sec. Since an r unit corresponds to 2×10^9 ion pairs/cm^3, one finds that one MeV γ radiation/cm^2-sec (on the ground) will give rise to 1.3×10^{-4} r units per day, and a similar amount of β radiation to 1.3×10^{-2} r units per day.

Converting the time unit into days and taking into account all the above factors, one finds the following situations for the explosion of a reactor which

* Converted to the case of instantaneous irradiation.

has been operating at k kilowatts: At M miles distance and after the elapse of d days we find that the γ radiation on the ground gives rise to $5.9d^{-0.28}$ k/M^2 r units per day, whereas the β radiation gives rise to $420d^{-0.26}$ k/M^2 r units per day.

Since most of the β radiation does not penetrate clothing and since one should not reasonably assume that more than 1/10 of the body will be exposed to the β radiation, one might reduce the effects of the β radiation by a factor of 10. Thus, one finds the effective radiation to be $42d^{-0.26}$ k/M^2 r units per day.

Example

In order to illustrate the above formulae we assume for

d	1 day
k	1000 kW
M	10 miles

This gives 59 units per day γ radiation. This γ radiation will constitute an immediate hazard, which can effectively be reduced by prompt decontamination. The effective β radiation on the other hand will amount to 420 r units per day. Thus, a full day's radiation would amount to a most dangerous dose. In evaluating the effects of the β rays one must bear in mind that covering of free body surfaces by gloves, masks, and goggles will effectively eliminate this danger. One, nevertheless, must count on damage to health and probably on some casualties as well if the activity described above is deposited in a densely populated area.

APPENDIX D

Estimates by the Reactor Safeguards Committee of the Energy Developed in an Exploding Reactor (Reproduced from "Summary Report of the Reactor Safeguard Committee," Wash-3 (Rev.) by permission of the U.S. Atomic Energy Commission)

Numerous discussions on the effects of a possible reactor explosion have resulted in estimates of the energy yields. These estimates vary from case to case according to the nature and structure of the reactor. In the following discussion a few points of view and a few simple formulae are summarized which may prove useful as a guide for crude calculations in which the energy developed in reactor explosions is estimated. No attempts at a refined analysis can be made here. It must, furthermore, be remembered that the nuclear explosion may be followed by a chemical reaction which may release considerably more energy than the nuclear explosion itself.

We assume that a reactor has by some means become supercritical, and that after this its energy content increases as $e^{t/\tau}$, where τ is a multiplication period. It is useful to distinguish two cases: (1) $R/\tau \ll c$; and (2) $R/\tau \gg c$. Here R is the radius of the active core of the reactor (in all of the subsequent discussion we assume for simplicity that the core is spherical), and c is the average sound velocity within the material of the reactor.

(1) $\underline{R/\tau \ll c}$. If this condition holds, one may consider the reactor in a first approximation as a system in equilibrium in which the pressure developed by the reactor is contained by the stresses set up in the reflector that surrounds the reactor. We introduce $\delta\Delta R/R$, where ΔR is defined by the assumption that the multiplication stops when the core radius expands to $R + \Delta R$. In all of the discussions we shall assume that δ is very small compared to unity. If our condition $R/\tau \gg c$ holds, δ will be given by the approximate expression $\delta = kp$; where k is the coefficient of compressibility of the reflector, defined as

$$-\rho \frac{d(1/\rho)}{dp}$$

and p is the pressure developed in the reactor. This expression is approximate, mostly because it assumes elastic behavior of the reactor. From p one can calculate the energy from a knowledge of the equation of state of the core materials. This assumption of elasticity will seem reasonable at least in most cases in which the core materials are not vaporized. Under this condition the energy content per unit volume is quite small (of the order of 0.1 of the pressure), and the energy can promptly be estimated from the pressure.

We shall not discuss the case of an inelastic container, but we shall briefly consider a case where the reflector containing the reactor breaks. In this case (still assuming $R/\tau \ll c$) one will expect that the expansion of the reactor will occur sufficiently rapidly to counter-balance further increase in pressure due to the progress of the general reaction. Both in this case and in the case where the reflector does not completely confine the reactor, one may assume that only as much energy will be developed as is necessary to disassemble at least part of the reactor. In many reactors this will mean the boiling off of a few per cent of the reactor material. In the slowest reactors disassembly may occur through melting.

(2) $R/\tau \gg c$. In this case the energy liberation is regulated by an entirely different mechanism. Under these conditions a shock wave will move from the reactor into the surrounding reflector. The velocity of the shock wave will increase exponentially with time and most of the expansion required for disassembly will take place during a single period. From this condition one finds that the pressure p built up in the reactor is equal to

$$P = \frac{R^5 \delta}{V^2_\tau} (\rho_c \rho_r)^{1/2}$$

where V is the volume of the core and c and r are the average densities of the core and of the reflector. From this formula one obtains the total energy E developed

$$E = \frac{R^5 \delta}{\tau^2} (\rho_c \rho_r)^{1/2} \frac{\varepsilon}{p}$$

where the dimensionless quantity ε/p is the energy per unit volume divided by the pressure within the core. It has been mentioned above that for the condensed phase ε/p is considerably less than unity. If the core materials have been vaporized, ε/p lies roughly in the region between 2 and 5. As long as vaporization has not taken place, the total energy evolved is relatively small, not only because p is small but also because ε/p is small. Vaporization is connected with a great increase in energy. In order to estimate the energy developed in a very rapid manner it is well to remember that this energy is roughly equal to that developed by a TNT explosion of a mass equal to the core mass, provided the energy developed in the general reactor is just sufficient to vaporize the core material.

The approximate formula given above needs modification if the core is not homogeneous and if the distance between fuel elements in the core, which we call r, is large enough so that $r/\tau < c$. In this case, the initial expansion of the fuel elements will not decrease the reactivity and effective expansion takes place only after the shocks emitted from the various fuel elements collide. Thus the effective expansion of the core is slower and a greater energy liberation is to be expected than is indicated by the above equations. The condition $r/\tau > c$ is not satisfied in any reactor that has been built or designed. Furthermore, in the case of rodlike fuel elements, expansion along the axis of the rod tends immediately to reduce the activity of the reactor even before the shocks moving perpendicular to the axis of the fuel rods collide.

In some cases, the initial reaction will actually enhance the reactivity. Such reactors are called autocatalytic. Autocatalytic reactors are inherently more dangerous than such reactors which do not possess this feature, and the possibility of an autocatalytic process must be particularly carefully investigated. Autocatalytic action may be due to motion of materials (e.g. ejection of absorbers), or to a change in the behavior of materials with temperature (e.g. broadening of resonance lines), or to increase of thermal-neutron energies (e.g. heating of neutrons to a temperature slightly above an absorption resonance).

Examples

In the following we shall give two examples to illustrate the energy production in cases in which the simple considerations of elasticity or rupture of the reflector are not the governing factors.

The following characteristics of the reactor are assumed:

$$\text{Radius } R = 8 \text{ cm}$$

$$(\rho_c \rho_r)^{1/2} = 15$$

$$\delta = 0.005$$

$$\tau = 10^{-6} \text{ sec}$$

These figures refer to a fast reactor in which the condition $R/\tau > c$ is satisfied, and which is approximately 2 per cent prompt supercritical. Substitution into the above formulae gives

$$\frac{R^5 \delta}{\tau^2} (\rho_c \rho_r)^{1/2} = 2 \times 10^{15} \text{ ergs}$$

and we obtain $E = (\varepsilon/p) \times 2 \times 10^{15}$ ergs.

The energy developed is sufficient to vaporize the substance. The values of ε/p for dense vapors are not well known. Setting $\varepsilon/p = 5$ is likely to be an overestimate. This leads to an energy value of 10^{16} ergs. For purposes of comparison we note that 4×10^{16} ergs is one ton of TNT equivalent. Thus, the energy which we obtain for the reactor is equivalent to $\frac{1}{4}$ ton of TNT.

As the second example we shall consider the behavior of a pile under runaway conditions. This reactor is autocatalytic because evaporation of the water will increase the multiplication factor by approximately 2 per cent. Even under these conditions it remains true, however, that $R/\tau > c$. The reactor is likely to disassemble by melting and partial vaporization of the slugs. This may result in dispersion of some U in the graphite and also in the movement of some slug material from the center toward the periphery of the reactor. The greatest energy release is likely to be connected with the partial vaporization of the slugs. Even so, an immediate energy release of several tons of TNT is quite likely. Burning of slug materials and possibly of the graphite may increase this energy release to a very considerable extent. This burning, however, will not take place in an explosionlike fashion.

APPENDIX E

Some Reports to the Atomic Energy Commission by the Advisory Committee on Reactor Safeguards

Honorable JOHN A. MCCONE,
Chairman,
U.S. Atomic Energy Commission,
Washington 25, D.C.

SUBJECT: *Humboldt Bay Power Plant—Pacific Gas and Electric Company*

Dear Mr. McCone:

At its twenty-fourth meeting, 10–12 March, 1960, the Advisory Committee on Reactor Safeguards reviewed the proposed 200 MW (thermal) boiling water reactor and vapor suppression containment for the Pacific Gas and Electric Company at Humboldt Bay, California.

This reactor and its containment concept had previously been reviewed by the ACRS at its September and November, 1959, meetings and by the ACRS Subcommittee meetings of 29 October, 1959, and 25 February, 1960. The Committee reviewed the Preliminary Hazards Summary Report and subsequent Amendments Nos. 1 to 6, referenced below. The Committee had the benefit of advice from the AEC Staff and others.

Presupposing continued generally favorable experience with boiling water reactors of this type, it is the opinion of the Committee that the conceptual design of this boiling water reactor is adequate for this site with conventional pressure vessel type of containment.

Because of the high population density relatively close to this site and other unfavorable site factors, it is essential that the reactor be well contained.

It is not clear how much of the total reactor system will be housed within the vapor suppression chamber. The information so far provided does not demonstrate the suitability of the steam condensing system. Further tests are necessary.

The Advisory Committee of Reactor Safeguards believes that while the concept has merit it has not yet been demonstrated that the vapor suppression system proposed can be relied upon to protect the health and safety of the public at this site.

Sincerely yours,

(SGD.) LESLIE SILVERMAN

Chairman, Advisory Committee on Reactor Safeguards

REFERENCES:

1. Preliminary Hazards Summary Report—Humboldt Bay Power Plant Unit No. 3, 15 April, 1959.
2. Amendment No. 1 to Application of Pacific Gas and Electric Company, 20 July, 1959.

3. Amendment No. 2 to Application of Pacific Gas and Electric Company, July, 1959.
4. Addenda A and B—Amendment No. 3 to Application of Pacific Gas and Electric Company, September 1959.
5. Amendment No. 4 to Application of Pacific Gas and Electric Company, November 1959.
6. Amendment No. 5 to Application of Pacific Gas and Electric Company, 30 November, 1959.
7. Amendment No. 6 to Application of Pacific Gas and Electric Company, 29 January, 1960.

Honorable JOHN A. MCCONE,
Chairman,
U.S. Atomic Energy Commission,
Washington 25, D.C.

SUBJECT: *Experimental Low Temperature Process Heat Reactor Project (ELPHR)—Point Loma, California*

Dear Mr. McCone:

The Advisory Committee on Reactor Safeguards has studied the proposed site for the Experimental Low Temperature Process Heat Reactor Project which is to be adjacent to the demonstration Saline Water Plant on Point Loma. The data furnished the ACRS in the report referenced below is preliminary in nature and dealt in general terms with the conceptual design of the reactor. This report deals primarily with the site and its characteristics.

The Committee considers Point Loma to be a poor site because of unfavorable meteorology and high population density, aggravated by recreational and fisheries aspects, and lack of ocean dilution. The close proximity of the San Cabrillo Monument area with its numerous visitors and its proposed enlargement with the probability of an increased number of visitors add to the unfavorable features. The experimental nature of the proposed installation contributes to our lack of assurance.

The Committee believes that it would be unwise at the present time from the safety point of view to locate this reactor at this site.

Sincerely yours,

(SGD.) LESLIE SILVERMAN

Chairman, Advisory Committee on Reactor Safeguards

REFERENCES:

SL-1760—Site Evaluation Report for the Experimental Low Temperature Process Heat Reactor Project, 29 January, 1960.

Other pertinent information is contained in:

Report on the Collection, Treatment and Disposal of the Sewage of San Diego County California, September, 1952.

TID-7580—Proceedings of the 1959 Symposium on Low-Temperature Nuclear Process Heat held at Atomic Energy Commission Headquarters Building, Germantown, Maryland, 1 October, 1959.

15 January, 1959

Honorable JOHN A. MCCONE,
Chairman,
U.S. Atomic Energy Commission,
Washington, D.C.

SUBJECT: *Carolinas Virginia Nuclear Powerplant, Parr, S.C.*

Dear Mr. McCone:

At its 12th meeting in Washington, 11–13 December, 1958, the Advisory Committee on Reactor Safeguards reviewed the characteristics of the site proposed for this reactor. The Committee considered this reactor site again at its 13th meeting on 8–10 January, 1959. Data concerning the site were presented in the reports referenced below and discussed orally by the Hazards Evaluation Branch.

The project is for the investigation and development of a pressure tube type reactor of 60.5 megawatts (heat) capacity. Heavy water will be used as an unpressurized moderator while the light water coolant is pressurized. Containment is contemplated. The reactor will be connected with an existing steamplant.

The proposed site is located adjacent to a company-owned village of Parr, S.C., with a population of 58.

It is concluded that a reactor of this general type and power level as presently proposed including provision of suitable containment could be operated at this site without undue hazard to the health and safety of the public. The advisability of abandoning the village of Parr should be decided prior to the beginning of operation.

Sincerely yours,

C. ROGERS MCCULLOUGH,

Chairman, Advisory Committee on Reactor Safeguards

REFERENCES:

Proposal for a power demonstration reactor by Carolinas Virginia Nuclear Power Associates, Inc., August 1958.

Hazards Evaluation Report to the Advisory Committee on Reactor Safeguards, 2 December, 1958.

U.S. Weather Bureau Comments on Preliminary Site Evaluation, Nuclear Powerplant Parr, S.C., 8 December, 1958.

12 January, 1959

Honorable JOHN A. MCCONE,
Chairman,
U.S. Atomic Energy Commission,
Washington, D.C.

SUBJECT: *Consolidated Edison Reactor*

Dear Mr. McCone:

At its 12th meeting, 11–13 December, 1958, the Advisory Committee on Reactor Safeguards reviewed the containment proposed for the nuclear power station being built for the Consolidated Edison Co., at Indian Point, N.Y. Members of the Hazards Evaluation Branch and representatives of the Consolidated Edison Co., Babcock & Wilcox Co., and the Vitro Corp. of America participated in the discussion. Documents pertinent to this proposal are referenced below. The Committee considered this proposal further at its 13th meeting, 8–10 January, 1959.

The proposed steel sphere and the heavy-walled concrete building enclosing it will provide the most nearly complete containment presented to the Committee in any reactor project to date. The Committee is satisfied that this containment will provide protection to the public. This favorable opinion relates to the proposed containment structures only. Information regarding the safety of the reactor itself and its operation will be considered later.

Sincerely yours,

C. ROGERS MCCULLOUGH,

Chairman, Advisory Committee on Reactor Safeguards

REFERENCES:

(1) Evaluation of Potential Radiation Hazard Resulting from Assumed Release of Radioactive Wastes to Atmosphere from the Proposed Buchanan Nuclear Powerplant, April 1957.

(2) Core design and Characteristics for the Consolidated Edison Reactor, 18 August, 1958.

(3) Report on Hazards Analysis and Design for Containment Vessel for the Consolidated Edison Reactor 29 August, 1958.

(4) Division of Licensing and Regulation Report to the Advisory Committee on Reactor Safeguards on Consolidated Edison Co. of New York Indian Point Reactor, 3 December, 1958.

12 November, 1958

Honorable JOHN A. MCCONE,
Chairman,
U.S. Atomic Energy Commission,
Washington, D.C.

SUBJECT: *Westinghouse Testing Reactor* (*WTR*)

Dear Mr. McCone:

During its 11th meeting, 7 November, 1958, the Advisory Committee on Reactor Safeguards reviewed the Westinghouse testing reactor. The WTR is a water-moderated and cooled heterogeneous reactor located at Waltz Mills, Pa., and nearing completion under a construction permit issued by the Commission. The Westinghouse Co. is now requesting a license to operate this reactor at a power of 20 megawatts. For its review, the ACRS was furnished Westinghouse report, WCAP-369 (revised), and discussed the reactor with the Division of Licensing and Regulation and with Westinghouse personnel.

In many respects the WTR is similar to the materials testing reactor for which 8 years of operating experience is available. Thus both the characteristics of this type of reactor and the operating problems associated with its testing function are well known. Like the MTR, the WTR will also be operated in conjunction with a critical facility with which the reactivity of new experiments can be determined with fair precision. In addition, the WTR is housed in a large steel vessel designed to contain, with nominal leakage, the fission products which might be released in a severe reactor accident.

The Advisory Committee on Reactor Safeguards concludes that the Westinghouse testing reactor can be operated without undue hazards to the health and safety of the public.

Sincerely yours,

C. ROGERS MCCULLOUGH,

Chairman, Advisory Committee on Reactor Safeguards

REFERENCES:

WCAP-369 (revised).
Amendment No. 8 to license application, 29 September, 1958.
Amendment No. 9 to license application, 30 October, 1958.
HEB Staff Analysis, 7 October, 1958.

4 November, 1957

Honorable LEWIS L. STRAUSS,
Chairman,
U.S. Atomic Energy Commission,
Washington, D.C.

SUBJECT: *Dresden Nuclear Power Station—Enclosure*

Dear Mr. Strauss:

This letter constitutes an interim report of the Advisory Committee on Reactor Safeguards on the reactor for which construction permit No. CPPR-2 has been issued to the Commonwealth Edison Co.

The Committee reviewed the design characteristics of the containment vessel being constructed for this reactor.

It is the opinion of the Committee that the proposed containment vessel for the Dresden Nuclear Power Station will contain the maximum credible accident postulated by the applicant. On the basis of the information reviewed to date by the Committee, it seems reasonable that this postulated accident is the most serious which might occur.

This opinion is based upon the technical information regarding the containment vessel made available to the Committee covering (a) design specifications, (b) construction methods, (c) methods of testing, and (d) general procedures for the proper arrangements for reinspection and interim testing the details of which will be supplied before operation.

Sincerely yours,

C. ROGERS MCCULLOUGH

Chairman, Advisory Committee on Reactor Safeguards

21 October, 1958

Honorable JOHN A. MCCONE,
Chairman,
U.S. Atomic Energy Commission,
Washington, D.C.

SUBJECT: *Commonwealth Edison Co.—Dresden Nuclear Power Station*

Dear Mr. McCone:

At its 10th meeting on 16 October, 1958, the Advisory Committee on Reactor Safeguards reviewed the application of the Commonwealth Edison Co. for an operating license for its Dresden nuclear power station. The general design and containment of this reactor were discussed with representatives of the company and members of the Hazards Evaluation Branch. The review of the plan of operation and any proposed emergency procedures was deferred.

The Committee does not see any problems of concern relating to the possible hazards of this nuclear powerplant, but the Committee has not had time to complete its consideration of this reactor. It will give a more detailed report at a later date.

Sincerely yours,

C. ROGERS McCULLOUGH

Chairman, Advisory Committee on Reactor Safeguards

REFERENCES:

(1) GEAP-1044—Preliminary Hazards Summary Report for the Dresden Nuclear Power Station, 3 September, 1957.
(2) Amendment No. 1 to Preliminary Hazards Summary Report for the Dresden Nuclear Power Station, 26 May, 1958.
(3) Amendment No. 2 to Preliminary Hazards Summary Report for the Dresden Nuclear Power Station, 25 August, 1958.
(4) Report to ACRS by Division of Licensing and Regulation, 3 October, 1958.

21 October, 1958

Honorable JOHN A. McCONE,
Chairman,
U.S. Atomic Energy Commission,
Washington D.C.

SUBJECT: *Yankee Atomic Electric Co.*

Dear Mr. McCone:

At its 10th meeting on 15 October, 1958, the Advisory Committee on Reactor Safeguards reviewed amendments No. 7 and No. 8 to the application of the Yankee Atomic Electric Co. for a license to operate the nuclear powerplant the company is constructing at Rowe, Mass. The Committee had available to it the material referenced on the last page.

Amendment No. 7 proposed an alternative to the measures recommended by the Committee in its letter of 16 September, 1957, for determining the effects of plutonium buildup on the nuclear characteristics and stability of this reactor. Amendment No. 8 described the waste disposal facilities planned for the Rowe plant.

Amendment No. 7

In Amendment No. 7, Yankee states that determination of the effect of plutonium buildup by measurements on synthetic fuel elements made up of uranium and plutonium in a part-core critical facility, as suggested in the Committee's earlier letter, would be difficult to interpret because of the impossibility of duplicating the temperatures and neutron spectrum of the actual Yankee reactor in this facility. As an alternative and more dependable procedure for determining the effect of plutonium, Yankee proposes an experimental program to measure temperature coefficients, prompt and overall, in the actual power reactor at start-up, after 2000 hr of operation, and at intervals while plutonium is growing into the core.

To establish that the reactor can be operated safely while plutonium is building up in the core during the interval between experimental measurements, Yankee described calculations made by Westinghouse and by Nuclear Development Corp. of America which show that the buildup of plutonium during even the entire anticipated fuel lifetime of 10,000 hr will have only a minor effect on the overall temperature coefficient of reactivity.

The Committee concurs with Yankee's proposal to determine these temperature coefficients in the actual reactor rather than in the part-core critical facility and agrees with Yankee's judgment that the effect of plutonium buildup on these coefficients will be small enough to permit these measurements to be made with safety in the actual reactor. In this, the Committee concurs with the determination of the Hazards Evaluation Branch.

The Committee recommends that Yankee be asked to provide a description of the specific experiments which will be made to determine the effects of plutonium buildup on prompt and overall temperature coefficients. It suggests that controlled transient experiments, with known sinusoidal or step changes in reactivity, may be a convenient means of measuring these coefficients.

Amendment No. 8

The committee heard a detailed description of the facilities planned by Yankee for disposal of gaseous, liquid and combustible solid wastes. It believes that the design of these facilities is conservative and concurs with the conclusion of the Hazard Evaluation Branch that the proposed facilities will permit the disposal of wastes without undue hazard to onsite or offsite personnel.

Sincerely yours,

C. ROGERS MCCULLOUGH

Chairman, Advisory Committee on Reactor Safeguards

REFERENCES:

(1) Amendment No. 3 to "Preliminary Hazards Summary Report" by Yankee Atomic Electric Co., dated April, 1957.

(2) Amendment No. 4 to "Preliminary Hazards Summary Report" by Yankee Atomic Electric Co., dated 15 July, 1957.

(3) Amendment No. 6 to "Preliminary Hazards Summary Report" by Yankee Atomic Electric Co., dated 5 March, 1958.

(4) Amendment No. 7 to "Preliminary Hazards Summary Report" by Yankee Atomic Electric Co., dated 21 July, 1958.

(5) Amendment No. 8 to "Preliminary Hazards Summary Report" by Yankee Atomic Electric Co., dated 28 July, 1958.

(6) Report to ACRS by Division of Licensing and Regulation, dated 29 September, 1958.

(7) Memorandum from R. C. Dalzell to H. L. Price, subject: "Yankee, Atomic Electric Co. License Application Amendments No. 7 and No. 8," dated 2 September, 1958.

21 October, 1958

Mr. Harold L. Price,
Director,
Division of Licensing and Regulation,
U.S. Atomic Energy Commission,
Washington, D.C.

Subject: *Yankee Atomic Electric Co.*

Dear Mr. Price:

The Committee finds that there are some matters relating to the Yankee reactor on which it should have more information before it can give its final opinion on the overall safety of this reactor. These are:

(1) The means that will be used to estimate the distribution of neutron flux and heat flux in the reactor so that the margin of operation below burnout conditions can be determined.

(2) The results of experiments on the extent of precipitation of solids from water containing the amounts of boric acid and lithium hydroxide expected in the operating reactor, under the conditions of temperature, pressure, and radiation intensity which will be experienced in the reactor.

(3) The principles and procedures to be used in operating this reactor.

Would you please arrange to have this information developed for us?

Sincerely yours,

C. Rogers McCullough

Chairman, Advisory Committee on Reactor Safeguards

8 March, 1958

Honorable Lewis L. Strauss,
Chairman,
U.S. Atomic Energy Commission,
Washington, D.C.

Subject: *General Electric Boiling Water Reactor at Vallecitos*

Dear Mr. Strauss:

The problems of the Vallecitos boiling water reactor were considered by the Advisory Committee on Reactor Safeguards with representatives of the licensee, General Electric Co., and the Hazards Evaluation Branch. The pertinent documents are listed below.

General Electric has asked for three revisions to their license, construction permit CPPR-3 and license Nos CX-2 and DPR-1 as amended, for this reactor Amendment No. 14 would increase the number of rod-type elements in a loading, from 1 to 14 for thermal powers up to 20 megawatts. Amendment No. 15 would increase allowable thermal power with mixed flat plate and 14 rod-type elements, from 20 to 30 megawatts. Amendment No. 16 would allow heating of the reactor from cold to operating temperature by nuclear heat at a rate not exceeding 1 megawatt instead of by heat supplied to the coolant externally.

In view of the reported performance so far obtained in operations of this reactor, the Committee sees no significant increase in hazard in following these proposals provided that rod-type elements be so dispersed in the core as to be as uniformly surrounded by plate type elements as possible.

There seems to be some possibility of a "cold-water" type accident in operation of the system and its controls as now constituted, i.e., sudden introduction of cold coolant into reactor causing excessive increase in activity and power. This possibility should receive further study and if modifications of control systems or operating procedures are necessary to ensure against such an accident they should be made before resuming operation.

Sincerely yours,

C. ROGERS MCCULLOUGH

Chairman, Advisory Committee on Reactor Safeguards

REFERENCES:

(1) GE-BWR Report SG-VAL, 1.
(2) GE-BWR Report SG-VAL 2.
(3) Amendments Nos. 13 through 16.
(4) HEB Staff Report of 25 February, 1958.

5 August, 1958

Honorable JOHN A. MCCONE,
Chairman,
U.S. Atomic Energy Commission,
Washington, D.C.

SUBJECT: *General Electric Vallecitos Boiling Water Reactor (GE-VBWR)*

Dear Mr. McCone:

The Advisory Committee on Reactor Safeguards at its ninth meeting on 4 August, 1958, considered amendment No. 24 to the license application of the General Electric Co. for the operation of the Vallecitos boiling water reactor. This amendment is designed to provide the operator with greater latitude in the choice of fuel elements and operating limits. In reviewing the amendment application the Committee considered the supporting material submitted in the reports referenced below and held a meeting with representatives of the applicant and members of the Hazards Evaluation Branch.

The Committee believes that the technical specifications set out in section 1 of amendment No. 24 define a scope of operations within which it is possible to operate without undue hazard to the public. This belief is based on the assumption that the specifications outlined do not affect the magniture of the postulated maximum credible accident, which the applicant has shown does not result in the release of dangerous amounts of radioactivity beyond the site boundary provided the container maintains its specified leak tightness. While, as stated, it is the Committee's belief that the proposed technical limitations have no influence on the magnitude of the maximum credible accident, this has not in fact been clearly demonstrated by the applicant, and we believe this point should be documented more definitely.

On the other hand, the Committee would like to emphasize that the technical specifications alone do not guarantee the safety operation of the reactor, especially from the standpoint of hazards within the boundaries of the site.

The Committee is concerned with the mounting number of amendments to the licensee's application on the VBWR operation which apparently stems from attempts to cover a multiplicity of specific situations pertinent to the safe operation of the reactor. Rather than improving the safety it seems possible that the real issues of safe operation may get beclouded by changing one set of circumstances for another.

The Advisory Committee on Reactor Safeguards suggests that the applicant be permitted within the scope of amendment No. 24 to assume technical responsibility concerning hazards in connection with his experimental program.

The latitude contemplated in the proposed amendment imposes a special responsibility on the applicant to review each proposed change in operating conditions in the light of its effect on the probability of an accident and on the possible severity of the accident if it occurs.

Sincerely yours,

C. ROGERS MCCULLOUGH

Chairman, Advisory Committee on Reactor Safeguards

REFERENCES:

(1) Amendment No. 24 to License Application for Vallecitos Boiling Water Reactor, 14 May, 1958.
(2) SG-VAL-2, Second Edition, Final Hazards Summary Report, 8 May, 1958.
(3) Key to Second Edition of SG-VAL-2, 8 May, 1958.
(4) Report to ACRS by Division of Licensing and Regulation on GE-VBWR, 1 August, 1958.

12 July, 1958

DR. WILLARD F. LIBBY,
Acting Chairman,
U.S. Atomic Energy Commission,
Washington, D.C.

SUBJECT: *General Electric Test Reactor* (*GETR*)

Dear Dr. Libby:

The Advisory Committee on Reactor Safeguards reviewed at its eighth meeting, 10–12 July 1958, the proposal of the General Electric Co. to operate the General Electric test reactor. The Committee had previously offered advice on this reactor at its second meeting, 1–3 November, 1957, in connection with the General Electric request for a construction permit.

In its current review, the Committee had access to the reports referenced below and discussed the proposal with representatives from both the General Electric Co. and the Hazards Evaluation Branch.

The GETR is a pressurized water reactor operating at 33 megawatts (thermal) located at the General Electric Vallecitos Atomic Laboratory, Pleasanton, Calif. A large body of information and experience exists on the nuclear, hydraulic, and mechanical behavior of the components of pressurized water reactor systems. The primary area of uncertainty, with regard to reactor safety, now concerns the transient response of this type of reactor to rapid additions of excess reactivity. Pertinent information is now being obtained as part of the SPERT program. However, in this interim period, considerable guidance as to reactor dynamics can be obtained from existing Borax and SPERT data. The ACRS believes that operation of this reactor, considered separately from the intended experimental program, presents no greater hazard than many other reactors now approved for operation.

A judgment as to the continuous safe operation of this reactor including its testing function presents an additional problem because of the inability to define precisely the specific characteristics of the future experimental program. Relatively more dependence must be placed upon the sound judgment of the operators of test reactors than upon that of operators of reactors for which less flexibility is required. While it is hoped that in the future more flexible definitions of the areas for the independent action on the part of operators for testing reactors will be developed, the ACRS believes that the General Electric Co. has proposed reasonably acceptable limitations within which the GETR staff may take action independent of prior AEC approval.

The Advisory Committee on Reactor Safeguards thus advises that the GETR may be operated as a testing reactor as proposed by the General Electric Co. without undue hazard to the public.

Sincerely yours,

C. Rogers McCullough

Chairman, Advisory Committee on Reactor Safeguards

References:

Amendment No. 3 to License Application for GETR, 26 February, 1958.

Amendment No. 4 to License Application for GETR, for Experimental Facilities 15 May, 1958.

Amendment No. 5 to License Application for GETR, 18 June, 1958.

HEB Staff Report on GETR, 27 June, 1958.

13 January, 1958

Honorable LEWIS L. STRAUSS,
Chairman,
U.S. Atomic Energy Commission,
Washington, D.C.

SUBJECT: *The MIT Reactor, Docket No.* 50-20

Dear Mr. Strauss:

The Massachusetts Institute of Technology has applied for license to operate its nuclear reactor located in Cambridge, Mass., construction of which is now nearing completion. This letter is in reply to a request by the Atomic Energy Commission for the advice of the Advisory Committee on Reactor Safeguards with respect to the safety of the proposed operation of this reactor.

The Committee's advice is based upon information contained in the application and amendments thereto. The former Advisory Committee on Reactor Safeguards reviewed the proposed design of the reactor prior to the issuance of a construction permit and submitted a report, dated 5 March, 1956, on this matter to the General Manager.

This is a research reactor designated for 1 megawatt (thermal) utilizing enriched alloy plate-type fuel elements and D_2O cooling and moderation. It incorporates many of the design principles of the CP-5 and MTR reactors. There are no novel features requiring demonstration and no important changes in design have been made since the review for a construction permit.

The reactor is located in a densely populated area close to public activities. Therefore, it is essential that effective administrative controls and effective operating and emergency procedures be established and maintained. The applicant has indicated provision for such controls and procedures, which appears to the Committee to be adequate.

While the Committee believes that any serious release of fission products is highly improbable, it is important that containment be maintained because of the location of the facility. The containment proposed is generally adequate. However, there is one point of weakness, namely, complete dependence on the reliability of the automatic valve closure mechanism in the ventilation system. The Committee recommends that provision be made for some effective auxiliary means of closing the inlet and outlet lines of the ventilation system. Such a requirement could be met by provision for manual operation of the present valves in addition to the automatic operation already installed or by some other effective means. However, such a requirement is not considered necessary prior to the commencement of the research program.

In the opinion of this Committee, this reactor can be operated with an acceptable degree of risk to the health and safety of the public.

Sincerely yours,

C. ROGERS MCCULLOUGH

Chairman, Advisory Committee on Reactor Safeguards

5 November, 1957

Honorable Lewis L. Strauss,
Chairman,
U.S. Atomic Energy Commission,
Washington, D.C.

Subject: *National Advisory Committee for Aeronautics (NACA)—Docket No.* 50-30

Dear Mr. Strauss:

This letter constitutes the report of the Advisory Committee on Reactor Safeguards on the application for a construction permit by the NACA, docket No. 50-30, in accordance with section 182 of the Atomic Energy Act of 1954, as amended.

The application is for a test reactor designed to operate at power levels up to 60 megawatts of heat. It is to be located 3 miles south of Sandusky, Ohio.

One purpose of the reactor is to test nuclear fuel bearing components to destruction or near destruction. This aspect of the experimental program leads the Committee to be especially concerned with the operation of this reactor at a site so close to a densely populated area.

The Committee is of the opinion that with the proposed container and at the selected site it is possible so to restrict the experimental program that the operation of the reactor will not result in appreciable hazard to the public. However, the necessary restrictions may add materially to the cost of the program and may impose serious time delays. Further, some experiments which fall within the general type of experimental program proposed by NACA may not be permissible at this location.

In view of the above, the Committee believes that the facility proposed would be more useful for the program proposed if it were located at a site less close to a center of population.

It is the opinion of the Committee that NACA is providing reasonable precautions to avoid the escape of radioactivity which is likely to be damaging to the health and safety of the public. Among these precautions are three important items:

(1) NACA proposes to place the reactor within a pressure vessel which has as its design criterion a maximum leakage rate of 115 cubic feet per day. Furthermore, the applicant has proposed a variety of measures to check the leak tightness of this container during operations. It is difficult to prove and maintain a leakage rate this low but if such a rate actually can be demonstrated and maintained the Committee believes that it would provide adequate protection to the health and safety of the public.

(2) NACA is proposing to enclose each test loop within a secondary tank or container which is designed to contain the possible releases of fission products and other radioactive materials in case of breakdown of the fuel elements and other components being tested. The committee believes that this would be a valuable additional safeguard but is not convinced that this secondary container can be depended upon under all circumstances.

(3) The proposed design includes means to prevent the release directly to the atmosphere of effluents from the operation of the reactor or from the experimental loops. Again, the Committee agrees that this is an important safeguard but does not believe that accidental release to the atmosphere can be entirely precluded.

The applicant proposes to establish a procedure for reviewing planned experiments in order to minimize the possibility of any failure which would release radioactivity even through the secondary enclosure.

The Committee believes that testing of fuel elements under conditions well within limits of possible failure does not offer a significant potential hazard provided that the experiments are properly designed and operated. However, testing of fuel elements in such a way that they are likely to be destroyed may not be permissible. Since NACA has not defined any specific experiments, the Committee is unable to state a more precise opinion than the above.

The Committee also believes that the operation of a test reactor at a site of this nature requires extensive area monitoring both on and off site so that any release of radioactivity to the environment may be detected as soon as possible and necessary protective or warning measures for the public carried on.

The Committee is aware of the risk that pressure may be brought to bear to permit a loosening of restrictions. This could come about as a result of a false sense of security which might develop from a period of successful operation and as a result of the importance of proposed experimental programs to the national defense. This problem would not be as serious if the proposed reactor were located at a less populated site.

The following are additional remarks by Dr. Abel Wolman:

"While I agree with all that the Committee has stated, I feel that I must add some remarks for purposes of clarifying my own position. In view of the prospect of future continuing debates as to the safety of conducting essential experiments at this site, I would recommend against the site on the information presently available. I believe that the applicant should be required to consider the availability of other sites at which operation of the reactor would be feasible and which would afford a higher degree of protection to the health and safety of the public.

"It is unrealistic to permit operation at this site if experiments of importance to the national defense are likely to have to be curtailed because of the site. The realities of human behavior are such that operation of experiments, the hazards of which may be uncertain, are likely to be permitted if they are important to the national defense.

"I do not believe that we should freeze on a site in a situation like this merely because an applicant has chosen it."

Sincerely yours,

C. ROGERS MCCULLOUGH

Chairman, Advisory Committee on Reactor Safeguards

16 March, 1959

Honorable JOHN A. MCCONE,
Chairman,
U.S. Atomic Energy Commission,
Washington 25, D.C.

SUBJECT: *Nuclear Merchant Ship Reactor Project* (*NS Savannah*)

Dear Mr. McCone:

At its 14th meeting (12–14 March, 1959) the Advisory Committee on Reactor Safeguards continued its review of the nuclear merchant ship (NS *Savannah*). Representatives were present from the Division of Reactor Development, Babcock & Wilcox Co., New York Shipbuilding Corp., Geo. G. Sharpe Co., States Marine Corp., and the U.S. Coast Guard. The ACRS has also had the benefits of thorough review of the ship by the Hazards Evaluation Branch and by Oak Ridge personnel. The pertinent documents are listed at the end of this letter.

Inasmuch as a final hazards review for the NS *Savannah* has not yet been submitted, and considerable information is still outstanding, the ACRS is not in a position to make a final recommendation to the Commission concerning the overall safety of this nuclear ship and the possible restrictions which may be required for the adequate protection of the public.

However, the committee has been asked to make an interim report. The ACRS has focused its attention primarily upon the nuclear propulsion system for which a large part of the information is available.

Pressurizer.—In the NS *Savannah*, the pressurizer not only maintains the primary system pressure, but also provides a heat source and sink for reducing pressure transients which occur during changes in load demand. The ACRS believes that the pressurizer can be made to work without jeopardizing the safety of the ship. The ACRS is somewhat uncertain as to the adequacy of the detailed design of the pressurizer inasmuch as the detailed information of transient response of the pressurizer has not been submitted to us. Under any circumstances, the actual performance of the system in addition to analog simulation, must be available before complete confidence in the adequacy of the system can be assured.

Rod control and scram system.—The design of this combined electrical rod drive and hydraulic scram system is new. The adequacy of this system should be demonstrated by extensive testing of prototypes and selected production units under all credible conditions of life, presence of solids in the water, misalignment, angle of roll and tilt, etc.

Containment.—The design of the containment vessel seems adequate provided the numerous penetrations do not themselves provide a channel through which fission products escape. The ACRS does not yet have sufficient information to decide whether the valving on these penetrations is adequate.

Interlocks on loop pumps.—A cold water accident initiated by the starting of an idle, cold-loop pump might create a serious nuclear excursion even if the scrams work. Therefore, it is essential that reliable and multiple interlocks be used to prevent this possible accident.

Miscellaneous comments.—The considerations of the shock loading under collisions appear to be adequate. However, ACRS has not fully completed its review in this area.

In view of the new design and unproven features associated with the reactor used in the NS *Savannah*, the Committee is of the opinion that the extensive shakedown and testing required should not be carried out over the full power range at the dockside location. The Committee advises that severe limits must be placed on the operations during the dockside testing and understands that the necessity for restrictions has been recognized.

The meteorological problems at this site (and also in general for rivers, estuaries, ports and at sea) have not been adequately resolved.

The required maneuverability of the ship results in a large amount of thermal cycling of the UO_2 fuel elements. This places an increased burden on the testing required to prove adequacy of fuel elements. It is obvious that a steam bypass around the turbine would reduce the problems associated with the terminal cycling.

The Committee is also not yet convinced that the operator will have a sufficiently informed staff to execute its overall responsibility for the safety of the nuclear ship. It is aware that crew members are in training. However, it urges that States Marine Corporation quickly acquire individuals who can understand and partake in the hazards analyses which are now underway.

Sincerely yours,

C. ROGERS MCCULLOUGH

Chairman, Advisory Committee on Reactor Safeguards

REFERENCES:

(1) Preliminary Safeguards Report—Babcock & Wilcox Co., BAW-1117, Volume I, revised 22 December, 1958. Volume II, revised 3 November, 1958.

(2) DLR Comments to ACRS on Nuclear Merchant Ship Reactor Project (NS *Savannah*), 4 November, 1958.

(3) DLR Report to ACRS on NS *Savannah* Control System, 5 January, 1959.

(4) DLR Report to ACRS on (NS *Savannah*) Nuclear Merchant Ship Reactor, 24 February, 1959.

16 March, 1959

Honorable John A. McCone,
Chairman,
U.S. Atomic Energy Commission,
Washington, D.C.

Subject: *S1C—Submarine Reactor*

Dear Mr. McCone:

At its 14th meeting, 12–14 March, 1959, the Advisory Committee on Reactor Safeguards reviewed the SlC—Combustion Engineering light water cooled reactor—which is a prototype submarine reactor installed within sections of a submarine hull modified to serve as a suitable containment vessel. This has been described in the listed documents. The installation is under construction at the Combustion Engineering project at Windsor, Corn. The reactor design deviates little from several other naval nuclear reactors reviewed previously and upon which operating experience is well known.

Based upon the data submitted, oral descriptions and discussions, the committee considers the SlC can be operated at this site without undue hazard to the health and safety of the public.

Sincerely yours,

C. Rogers McCullough

Chairman, Advisory Committee on Reactor Safeguards

References:

CEND-S1C-151, S1C Hazards Summary Reactor Report, 6 February, 1959.
DLR Report to ACRS on the S1C Reactor, 25 February, 1959.

12 January, 1959

Honorable John A. McCone,
Chairman,
U.S. Atomic Energy Commission,
Washington, D.C.

Subject: *Proposed Nuclear Power Reactor for the City of Piqua, Ohio*

Dear Mr. McCone:

At the 12th meeting of the Advisory Committee on Reactor Safeguards on 11–13 December, 1958, representatives of Atomics International, the city of Piqua, and the Division of Reactor Development described a revised plan for construction of an organic moderated nuclear powerplant in Piqua, Ohio. Earlier plans for this plant were reviewed by the Committee at its 10th and 11th meetings, and were the subject of letters to you dated 5 August and 12 November, 1958. The revised plan presented at the 12th meeting is described in NAA-SR-MEMO 3405 entitled, "Supplement III to Preliminary Safeguards Report for the Piqua Organic Moderated Reactor (NAA-SR-3100)."

A subcommittee reviewed supplement II prior to the 13th ACRS meeting. At the 13th meeting, a report by the Hazards Evaluation Branch was reviewed, and oral discussion by representatives of the Division of Reactor Development indicated some changes in containment had been proposed.

The site now proposed appears more suitable than the location originally selected. However, the Committee does not consider the installation at this site of a nuclear powerplant of this capacity of a relatively untried type to be without undue public hazard until the present proposed unconventional type of containment is replaced by a more substantial and dependable system.

Chairman C. Rogers McCullough did not participate in these reviews and discussions.

Sincerely yours,

W. P. CONNER, JR.

Acting Chairman, Advisory Committee on Reactor Safeguards

12 November, 1958

Honorable JOHN A. MCCONE,
Chairman,
U.S. Atomic Energy Commission,
Washington, D.C.

SUBJECT: *Proposed Nuclear Power Reactor for the City of Piqua, Ohio*

Dear Mr. McCone:

At its 10th meeting on 16 October, 1958, the Advisory Committee on Reactor Safeguards reviewed the organic moderated reactor proposed for installation at the Piqua municipal powerplant as a nuclear steam generator. Discussions of the proposal were held with the Division of Licensing and Regulation, representatives of the City of Piqua Municipal Power Commission and Atomics International. In addition, the Committee had available for reference purposes the preliminary safeguards report on the project, NAA-52-3100, and the report of the Hazards Evaluation Branch dated 14 October, 1958.

Although the Committee is favorably impressed with the organic moderated reactor concept, and is aware of the generally favorable results of the OMRE experience to date, it wishes to reaffirm its opinion that the Piqua site proposed on 16 October for installation of a nuclear powerplant based on this concept is unsuitable. That site is in an urban area and makes no provision for an exclusion zone. Both the meteorological and the hydrological conditions are unfavorable to the safe dispersal of radioactive by-products. The organic moderator presents a local fire hazard which is increased by the inclusion of a decay heat removal system employing a xylene boiler.

Representatives of Atomics International and the city of Piqua appeared before the Committee at its 11th meeting on 6 November, 1958, with oral proposals for a new location for the reactor farther removed from the populated area, and for better containment of the facility. Because of the preliminary nature of the information presented, the Committee has no basis for arriving at any firm conclusion with respect to these new proposals. However, it can be said that the proposed new location with its approximately quarter mile removal from immediately populated areas is an improvement over the site previously proposed; and, with adequate containment provisions, may prove to be acceptable for a reactor of the general type proposed.

Chairman C. Rogers McCullough did not participate in these reviews or discussions.

Sincerely yours,

R. C. STRATTON

Acting Chairman, Advisory Committee on Reactor Safeguards

5 August, 1958

Honorable JOHN A. MCCONE,
Chairman,
U.S. Atomic Energy Commission,
Washington, D.C.

SUBJECT: *Proposed Nuclear Power Reactor for the City of Piqua, Ohio*

Dear Mr. McCone:

At its ninth meeting, 4 August, 1958, the Advisory Committee on Reactor Safeguards was given an oral presentation by the Hazards Evaluation Branch of the general characteristics and site of the proposed nuclear power reactor for the city of Piqua, Ohio.

The matter has not been formally submitted to the Committee as yet and no other information has been made available.

The tentative view of the Advisory Committee on Reactor Safeguards is that the site is not a suitable one.

Sincerely yours,

C. ROGERS MCCULLOUGH

Chairman, Advisory Committee on Reactor Safeguards

15 December, 1958

Honorable JOHN A. MCCONE,
Chairman,
U.S. Atomic Energy Commission,
Washington, D.C.

SUBJECT: *Comparison of safety features of: G.E.-Vallecitos boiling water reactor (VBWR); sodium graphite reactor experiment (SRE); Shippingport pressurized water reactor (PWR); Rural Cooperative Power Association reactor, Elk River, Minn.*

Dear Mr. McCone:

The following is in reply to your request for comparison of the standards applied in evaluating the Elk River site with those that were applied to the VBWR, SRE, and PWR sites. Pertinent data tabulations are attached. These were furnished by the Hazards Evaluation Branch.

It need scarcely be emphasized that the question of site evaluation is complex. A large number of variable factors, many not strictly comparable from site to site, must be considered. Exact, completely objective, numerical site criteria are difficult to formulate, however convenient and desirable these might be. But the committee attempts to bring a consistent philosophy to the reactor hazards problem and to provide a common basis for site judgments.

Three distinct types of reactor are involved in the group in question. These are of the sodium graphite, pressurized water, and boiling water types.

SRE, a low power (20Mw thermal) reactor of the sodium graphite type, operates at atmospheric pressure in an underground location. The primary coolant is contained in a stainless steel shell which is in turn contained in a sealed concrete structure. Secondary coolant from primary heat exchangers located within the containment structures gives up its heat in external steam boilers. A rupture of the primary system will not cause melting of fuel or release of fission products therefrom. For these reasons, and because the SRE is located in a relatively large exclusion area (1.4 miles minimum radius), immediately surrounded by a sparsely populated district no containment vessel of the type used for pressurized reactors is employed.

The PWR reactor of the pressurized water type is provided with an exclusion distance of approximately 0.5 miles. It is fully contained and provided with biological shielding of the containment structures. It is designed to contain the vapor and energy released in the event of a rupture of the primary water system and one steam generator. In addition, the interconnected containment vessels are designed to contain the energy resulting from significant metal–water reactions.

The VBWR and Elk River reactors are of the boiling water type. In the VBWR the coolant is vaporized and is used for the direct drive of turbogenerators. In the Elk River reactor, radioactive steam is taken to a heat exchanger, providing a barrier. Both are provided with containers designed to prevent release of vapors resulting from a break in the cooling system. The VBWR has been designed to contain the results of a metal water reaction. The Elk River reactor containment vessel is provided with significant missile and biological shielding.

In attempting to decide for a particular reactor whether a given exclusion distance provides adequate protection for public safety, the Committee evaluates design features such as containment vessels, missile shields, biological shields, hydrology, meteorology, and geology, all of which affect reactor safety, particularly when a reactor is located near a populous area. Thus it was felt that the Elk River site would provide an acceptable degree of protection to the public in view of the isolated primary system and the vapor containment provided. Like considerations were applied in the case of the PWR reactor. The SRE has somewhat less containment, but has a greater exclusion radius than the others mentioned.

Population density is of concern to the Committee. Consequently, the relatively low population density nearby the Elk River site was considered to be a generally favorable element. On the other hand potential growth of the community needs also to be taken into account. The extent of this growth at Elk River is problematical; but the Committee felt it appropriate to express its consideration for the eventuality and to recognize dependence on engineering features for minimizing risk.

The fact that a highway and a railroad run comparatively near the Elk River site is not considered to increase the risk significantly, and conversely acceptance of these features does not imply a reduction in standards of population protection. Since highway and railroad occupancy is transient and intermittent, both the probability and the intensity of the risk are greatly reduced over those applying to stationary, permanent populations at the same distance. Moreover, access to a highway or railroad can be restricted in the event of an accident.

Sincerely yours,

C. ROGERS MCCULLOUGH

Chairman, Advisory Committee on Reactor Safeguards

5 August, 1958

Honorable JOHN A. MCCONE,
Chairman,
U.S. Atomic Energy Commission,
Washington 25, D.C.

SUBJECT: *Rural Cooperative Power Association, Elk River, Minn.*

Dear Mr. McCone:

At its ninth meeting in Washington, 4 August, 1958, the Advisory Committee on Reactor Safeguards considered the submission of the Rural Cooperative Power Association covering a proposed power reactor to be built under invitation as part of the Atomic Energy Commission power reactor demonstration program. Representatives of the contractor and the Hazards Evaluation Branch participated in the conference.

The proposal covers a natural circulation boiling-water-type reactor having 58.2 megawatts (thermal) capacity complete with a supplemental coal-fired superheater of 14.8 megawatts. The design is such that, if proven, the installation may have an ultimate capacity above 100 megawatts. The reactor, which will be provided with containment, is to have fuel elements using a combination of thoria and enriched uranium oxide in stainless steel cladding. The characteristics of the type of reactor are generally well known.

The site selected for this reactor upon property already owned by the cooperative does not now provide a sufficient exclusion distance but this situation is to be improved through the purchase of adjacent land which is State owned. Mutual agreement to this purchase already exists. Even with the increased exclusion distance in the event of a major accident a few persons might be exposed to higher radiation dosage than is considered acceptable.

As a matter of policy the Advisory Committee on Reactor Safeguards considers it not desirable to locate a nuclear reactor of this power level so close to a growing community. However, assuming the containment vessel will provide an adequate factor of safety and

Site Comparison of Elk River and Other Reactors

	Elk River	VBWR	SRE	PWR
Type	BWR	BWR	Sodium graphite	PWR
Power	116 TMw	50 TMw	20 TMw	230 TMw
Containment	Yes	Yes	No	Yes
Pressure	21 pounds per square inch gage	45 pounds per square inch gage	Conventional industrial building	52.8 pounds per square inch gage
Leak rate	0.1 per cent per day	1 per cent per day		0.15 per cent per day at 72 pounds per square inch gage
Criteria	Primary system rupture	Primary system rupture and metal–water reaction		Rupture of primary system and 1 steam generator
Hydrology	Mississippi River, 600 feet	No significant factor	Semi-arid	Ohio River
Wind	25 to 30 per cent toward Elk River	5 to 9 per cent toward Livermore	NW.–W	NSW.–WNW
		3 to 6 per cent toward Pleasanton	SSE.S	Inversion: SSE.–SE (valley forms trap)
Distance, site boundary	0.2 mile (excluding highway and railroad)	0.34 mile	1.4 mile	0.3 mile

Distance to highway	300 feet	0.34 mile	Approximately 3 miles	0.4 mile
Distance to railroad	250 feet	2 miles	Beyond 2 miles	Less than 0.1 mile
Nearest residence	Slightly over 0.25 mile	0.34 mile	2 houses within 1.7 mile	0.5 to 0.625 mile
Nearest community	Elk River (1,600), 0.5 mile	Pleasanton (2,800), 4 miles	Susana Knolls (750), 2.5 miles	Midland (6,500), 1 mile
Other towns or cities	Anoka (10,000), 11 miles; Minneapolis 25 miles	Livermore (10,000) 6 miles	Chatsworth, 6 miles; Santa Susana, 3 miles; Los Angeles, 30 miles	East Liverpool (25,000), 7.5 miles; Pittsburgh, 25 miles
Nearby facilities	Light industry, Elk River 1 mile	Veterans hospital 4 miles east	2 small firms under 10 miles, training school 1.8 miles	1 mile
Future population	Some beyond 0.25 mile	Agriculture and waste	None under about 2 miles	Some beyond 1 mile
Population versus distance:				
0.25 mile	0	0	0	0
0.5 mile	60	(1)	(1)	0
1.0 mile	700	(1)	(1)	(1)
5.0 miles	7,500		(1)	20,000
10.0 miles	(1)	22,000	(1)	125,000
(1) Data not readily available				

will meet the specified leakage rate and that the State-owned land will be acquired, the Advisory Committee on Reactor Safeguards believes that a power reactor of this general type and size may be operated at this site without undue hazard to the public and its property.

This represents some relaxation of previous practice in regard to exclusion distance. In this particular case it appears tolerable because of the present low population density adjacent to the proposed exclusion area and because of the low risk of damage.

Sincerely yours,

C. ROGERS MCCULLOUGH

Chairman, Advisory Committee on Reactor Safeguards

REFERENCES:

(1) Report No. 58–08813, 31 January, 1958—A Proposal for a Nuclear Power Steam Generating Plant for the Rural Cooperative Power Association, Elk River, Minn.

(2) Supplement A to Preliminary Hazards Evaluation of the RCPA—Elk River Site, 25 July, 1958.

(3) Supplement B to Preliminary Hazards Evaluation of the RCPA—Elk River Site, 1 August, 1958.

(4) Letter from Edward E. Wolter, RCPA, to Milton Klein, Chicago Operations Office, 9 July, 1958, with following attachments: Preliminary Hazards Evaluation of the RCPA—Elk River Site, 9 July, 1958; Meteorological Estimates for Elk River, Minn., by Special Projects Section, Office of Meteorological Research, June 1958; Hydro-Geological Aspects of the Elk River, Minn., Reactor Site, by E. S. Simpson, geologist (USGS); Excerpt from American Machine & Foundry Co. proposal, '*Technical Description of AMF Closed Cycle Boiling Water Reactor Power Plant for RCPA, Elk River, Minnesota,*' volume I, dated 22 March, 1957.

(5) Letter of 7 July, 1958, from Dr. Wexler to Dr. Lieberman, and hazard calculations for Elk River, Minn., reactor.

(6) Report to ACRS by Division of Licensing and Regulation on Elk River site evaluation. 31 July, 1958.

15 July, 1958

MR. HAROLD L. PRICE,
Director,
Division of Licensing and Regulation,
U.S. Atomic Energy Commission,
Washington, D.C.

SUBJECT: *Elk River Nuclear Powerplant*

Dear Mr. Price:

At its eighth meeting in Washington, 10–12 July, 1958, the Advisory Committee on Reactor Safeguards was furnished preliminary information relative to the evaluation of the site for the boiling water reactor to be constructed at Elk River, Minn., by ACF Industries. This reactor is to be operated by the Rural Cooperative Power Association.

The Committee was of the opinion that the information submitted was too meager to serve as a basis for a judgment. For example, the assumptions used in performing the radiation dose calculations are not clear. We all realize that the information was submitted to the Commission so late as to preclude adequate study prior to the Committee meeting.

It is my understanding that an early judgment on the safety aspects of this reactor project at this construction permit stage is desired. I will be happy to arrange further consideration either by subcommittee action or by including this project as an agenda item for the August meeting.

Sincerely yours,

C. ROGERS MCCULLOUGH

Chairman, Advisory Committee on Reactor Safeguards

15 December, 1958

Dear Mr. McCone:

At its 11th meeting (6–8 November, 1958), the Advisory Committee on Reactor Safeguards reviewed the design (at its present state of development) of the plutonium recycle test reactor. In addition to the reports referenced below, descriptions of the proposed designs were presented by members of the Hazards Evaluation Branch and representatives of Hanford. Additional clarifying information has been received since that meeting. The views of the Committee are summarized below:

(1) The proposed site appears to be suitable for the proposed facility in view of the power level, the inherent nuclear stability of the active lattice, and the intended containment features.

(2) The proposed scram mechanism is unusual but appears to be adequate.

(3) An untried method for fine control is proposed, but we do not feel that this feature increases the public risk appreciably.

The Committee understands that the design is proceeding toward the following objectives:

(4) The shim rods are to be designed so that the rods will move only when the drive motor is operating. Only one shim rod motor may be run at a time.

(5) Emergency light water cooling for the fuel elements will be applied automatically less than 30 sec after severe loss of pressure in the coolant system.

(6) To facilitate testing or maintenance, the designers propose that instruments which initiate a scram signal may be bypassed. A bypass which is to be used only when the pile is shut down must be interlocked so as to make operation or startup impossible until the bypass has been removed. In order to bypass such an instrument during operation, duplication of instrumentation must be provided so that the reactor will not operate without proper protective devices.

Sincerely yours,

C. ROGERS MCCULLOUGH

Chairman, Advisory Committee on Reactor Safeguards

REFERENCES:

HW-46461—Plutonium Recycle Program Demonstration Reactor Site Study, 7 November, 1956.

HW-48800—Plutonium Recycle Program Reactor Preliminary Safeguards Analysis, 12 July, 1957.

HW-48800—(Revised) Plutonium Recycle Test Reactor Preliminary Safeguards Analysis, 5 June, 1958.

Hazards Evaluation Branch Report to the Advisory Committee on Reactor Safeguards, 8 October, 1958.

U.S. Weather Bureau Comments, 16 September, 1958.

12 November, 1958

Honorable JOHN A. MCCONE,
Chairman,
U.S. Atomic Energy Commission,
Washington, D.C.

SUBJECT: *Heavy Water Components Testing Reactor*

Dear Mr. McCone:

At its 11th meeting on 6 November, 1958, the Advisory Committee on Reactor Safeguards reviewed the site selection for the heavy water components testing reactor. It is proposed to locate a 60 thermal megawatt, pressurized, heavy water cooled and moderated, testing reactor at the Savannah River plant. Containment is to be provided.

Data concerning the site were obtained from DPST 58–409 *et al.*, Hazards Evaluation Branch summary report, and through oral presentations by representatives of the contractor and by the Hazards Evaluation Branch.

The Committee considers the site proposed, under the conditions of design tentatively presented, including containment, to be acceptable from the standpoint of health and safety of the public.

It is understood that the detailed design features of the reactor when available will be submitted to the Committee for further review.

Sincerely yours,

C. ROGERS MCCULLOUGH

Chairman, Advisory Committee on Reactor Safeguards

21 October, 1958

Honorable JOHN A. MCCONE,
Chairman,
U.S. Atomic Energy Commission,
Washington, D.C.

SUBJECT: *SPERT III Reactor*

Dear Mr. McCone:

The operation of the SPERT-III reactor was reviewed by the Advisory Committee on Reactor Safeguards on 15 October, 1958, at the request of the Division of Licensing and Regulation. The Committee considered both the report of the Division of Licensing and Regulation and the report from the Phillips Petroleum Co., IDO-16425.

SPERT-III is a light water moderated and cooled experimental reactor designed to operate intermittently at pressures up to 2,000 pounds per sq in. and temperatures up to 670°F. It is designed for transient studies of pressurized water reactors operating under full power conditions. The SPERT-III program is an extension of the reactor kinetic studies which have been carried out successfully under the SPERT-I program.

The nature of the SPERT-III experimental program is such that this reactor will be operated closer to failure conditions than normal power reactors. The safe performance of these experiments rests largely on the competence of the operating staff. The Advisory Committee on Reactor Safeguards believes that the SPERT staff by performance has shown that it can handle this type of assignment. The Committee commends, and agrees with, the SPERT proposal to avoid running the more hazardous experiments during times of adverse meteorological conditions.

The Advisory Committee on Reactor Safeguards concurs with the Division of Licensing and Regulation that SPERT-III can be operated without undue hazard to the health and safety of the public.

The Advisory Committee on Reactor Safeguards wishes to comment again that the SPERT program is contributing extremely worthwhile information which is basic to many of the safety problems of the entire atomic reactor industry.

Dr. Richard L. Doan excused himself from participation in the discussion and recommendation in this case.

Sincerely yours,

C. ROGERS MCCULLOUGH

Chairman, Advisory Committee on Reactor Safeguards

21 October, 1958

Honorable JOHN A. MCCONE,
Chairman,
U.S. Atomic Energy Commission,
Washington, D.C.

SUBJECT: U^{233} *Loading of the Materials Testing Reactor* (*MTR*)

Dear Mr. McCone:

The Advisory Committee on Reactor Safeguards reviewed on 15 October, 1958, the proposed operation of the materials testing reactor (MTR) with a U^{233} core loading at the request of the Division of Licensing and Regulation.

The Committee had access to the report from the Phillips Petroleum Co., PTR-312, and was briefed by the staff of the Division of Licensing and Regulation.

The Committee agrees with the Hazards Evaluation Branch that the proposed operation of the MTR on a U^{233} core will not endanger the health and safety of the public. Also the Committee is of the opinion that such proposed operation will not expose the adjacent reactor operations to an unacceptable hazard.

In the Phillips proposal, it was recognized that the U^{233} loading will reduce the portion of delayed neutrons. To counteract the effect on the reactor kinetics of this decrease, the worth of the control rod and its rate of insertion will be correspondingly reduced. Also the maximum permissible worth of individual experiments will be reduced to a level compatible with the reduced portion of delayed neutrons. In addition, adequate means for the control of environmental hazards are available. The modifications of the MTR due to the U^{233} loading are very similar to those required for the operation of a plutonium core which has already been successfully carried out.

Dr. Richard L. Doan excused himself from participation in the discussion and recommendation in this case.

Sincerely yours,

C. ROGERS MCCULLOUGH

Chairman, Advisory Committee on Reactor Safeguards

8 March, 1958

Honorable LEWIS L. STRAUSS,
Chairman,
U.S. Atomic Energy Commission,
Washington, D.C.

SUBJECT: *The Operation of the MTR with Plutonium*-239 *Loading*

Dear Mr. Strauss:

On 7 March, 1958, the Advisory Committee on Reactor Safeguards reviewed with MTR personnel and the Hazards Evaluation Branch the proposed special plutonium core for the MTR. In this review the Committee had the benefit of a report by the Hazards Evaluation Branch and report PTR 224 prepared by MTR operating personnel.

The Committee agrees with the conclusion of the MTR personnel that, with the modifications proposed in the control system, the reactor kinetics for the plutonium loading is not significantly different from that of the current uranium-235.

It is understood that the plutonium fuel elements will be subject to the same inspection procedure and specifications as the present uranium-235 element. It is also understood that extensive reactor physics measurements at low power will be carried out in advance of full power operation.

The use of plutonium in the core does not add significantly to radiological problems associated with the fission product burden at the end of the normal operating cycle.

The Committee therefore believes that there are no reasons involving matters of safety which indicate that the plutonium loading should not proceed as proposed.

Sincerely yours,

C. ROGERS McCULLOUGH

Chairman, Advisory Committee on Reactor Safeguards

12 July, 1958

Dr. WILLARD F. LIBBY,
Acting Chairman,
U.S. Atomic Energy Commission,
Washington, D.C.

SUBJECT: *A1W Reactor*

Dear Dr. Libby:

At its seventh meeting, 8 June 1958, the Advisory Committee on Reactor Safeguards visited the A1W installation at the National Reactor Testing Station. At the eighth meeting, 11 July, 1958, the Committee reviewed the data submitted, as referenced below, during a conference with the Hazards Evaluation Branch and representatives of the Westinghouse Atomic Power Division and the Naval Reactors Branch.

The characteristics of the A1W installation differ little from other plants of the same type previously reviewed. The chief hazard is from loss of coolant. Discussion concerning such possible accidents developed the fact that adequate protection was provided by the fill and charging pumps or by auxilliary coolant water supply through available emergency connections, as indicated in the hazard summary report.

The Advisory Commmittee on Reactor Safeguards concludes that the A1W component of the NRF may be operated without undue hazard to public health and safety.

While the safety philosophy being developed for the A1W facility at NRTS appears to be adequate for shipboard installation, the fact that there will be several reactors in close operating relationship will require more further review as the project progresses and approaches completion.

Sincerely yours,

C. ROGERS McCULLOUGH

Chairman, Advisory Committee on Reactor Safeguards

REFERENCES:

WAPD-SC-600: A1W Hazards Summary Report, May 1958.
WAPD-187: A1W Plant Description, May 1958.
WAPD-A1W(S)-140: Self-Shutdown Characteristics of the A1W Cores, May 1958.
WAPD-A1W(S)-141: Analysis of Loss-of-Flow Accidents in the A1W Plant, May 1958.
WAPD-A1W(S)-142: A1W Reactor Protection Analysis, May 1958.
WAPD-A1W(S)-143: Analysis of Loss-of-Coolant Accidents in the A1W Plant, May 1958.
WAPD-A1W(S)-159: A1W Project—Meteorology of the A1W Site, 24 May, 1958.
WAPD-A1W(S)-146: Analysis of Primary Coolant Activity in the A1W Plant, May 1958.
WAPD-A1W(S)-147: Calculation of Doses Delivered from Release of Radioactive Material to the Atmosphere at the A1W Site, May 1958.

12 May, 1958

Honorable LEWIS L. STRAUSS,
Chairman,
U.S. Atomic Energy Commission,
Washington, D.C.

SUBJECT: *Brookhaven National Laboratory Research Reactor (BNL) Graphite Annealing*

Dear Mr. Strauss:

The Advisory Committee on Reactor Safeguards at its sixth meeting, 9 May, 1958, reviewed the present status and planned annealing of the Brookhaven National Laboratory research reactor at the request of the Commission.

The Committee had available and examined reports referenced below and had the benefit of discussion with the Hazards Evaluation Branch and a representative of the staff of Brookhaven National Laboratory. In addition the Committee reviewed the report of its subcommittee which considered the problem in detail with representatives of Brookhaven National Laboratory, Division of Reactor Development, Division of Licensing and Regulation, Brookhaven area office, and Oak Ridge National Laboratory, in New York, on 16 April, 1958. The Committee had previously been briefed by members of the U.S. team which had visited England and investigated the Windscale incident of October 1957. As supplementary information, members of the subcommittee and Hazards Evaluation Branch reported on the examination of the recent Harwell reports (see reference (3) and (4) covering the March 1958 annealing of BEPO).

The Committee is pleased that the Brookhaven National Laboratory staff has very carefully reviewed the problem of release of stored energy in the graphite moderator over a period of several years and have more recently, following the Windscale incident, correlated results of their studies with data available from the British. The Committee concurs with the conclusion of the Brookhaven National Laboratory staff that the next scheduled annealing of the Brookhaven Reactor will present no serious hazard to the reactor or to personnel.

Among the reasons which contribute to this conclusion are:

(a) Satisfactory annealing of BNL reactor graphite on seven prior occasions.
(b) Adequate controls to avoid excessive temperatures and an emergency action program to handle the unexpected.
(c) The recent refueling of the Brookhaven National Laboratory research reactor with uranium alloy fuel more resistant to oxidation than the natural uranium fuel previously used.

The Committee commends the Brookhaven National Laboratory staff on their continuing thorough study of problems peculiar to graphite reactors and encourages their future activities particularly along the lines of investigation of irradiated graphite oxidation phenomena.

Sincerely yours,

C. ROGERS MCCULLOUGH

Chairman, Advisory Committee on Reactor Safeguards

REFERENCES:

1. Final Report, Stored Energy, Growth, and Annealing Status of Graphite Moderator in the BNL Research Reactor. (Received by ACRS 11 April, 1958.)

2. Report on Graphite Annealing of the BNL Research Reactor by Division of Licensing and Regulation, 11 April, 1958.

3. Manual for the Second Wigner Energy Release, by J. L. Dickson B.E.P.O., memo 89, Engineering Division, A.E.R.E., Harwell (United Kingdom), March 1958.

4. BEPO Wigner Energy Release, Preliminary Report, by J. L. Dickson, P.O.C. memo 93, Reactor Services Group, Engineering Division, A.E.R.E., Harwell (United Kingdom), 28 March, 1958.

12 May, 1958

Honorable LEWIS L. STRAUSS,
Chairman, U.S. Atomic Energy Commission,
Washington, D.C.

SUBJECT: *Oak Ridge National Laboratory Research Reactor* (*ORR*)

Dear Mr. Strauss:

The Oak Ridge National Laboratory research reactor was reviewed at the sixth meeting of the Advisory Committee on Reactor Safeguards, 9 May, 1958, at the request of the Commission. The Committee, before it received statutory status, had previously submitted its recommendations on location and design of this reactor (reference 1) to the Commission.

For the present review, the Committee had access to information referenced below (2) to (9) inclusive. In addition, representatives from Oak Ridge presented amplifying comments orally.

The physics and engineering of reactors of the Oak Ridge research type are well understood. Considerable operating experience has been accumulated. Moreover, the Oak Ridge staff has demonstrated its ability to operate research and testing reactors.

The Advisory Committee on Reactor Safeguards concludes, in agreement with its initial recommendation and with the Hazards Evaluation Branch, that there is reasonable assurance that the Oak Ridge National Laboratory research reactor can be operated without endangering the health and safety of the public.

Sincerely yours,

C. ROGERS MCCULLOUGH

Chairman, Advisory Committee on Reactor Safeguard

REFERENCES:

1. ACRS 5th meeting, 21–23 April, 1954; 8th meeting, 21–24 October, 1954; 10th meeting, 3–4 February 1955; 11th meeting, 3 March, 1955. See letter C. Rogers McCullough to C. K. Beck, 29 January, 1958.

2. *The Oak Ridge National Laboratory Research Reactor Safeguard Report*, by F. T. Binford, T. E. Cole, and J. P. Gill, 7 October, 1954; Volume I, ORNL-1794; Volume II, TID-10083.

3. *A Method for Disposal of Volatile Fission Products From an Accident in the Oak Ridge Research Reactor*, by F. T. Binford and T. H. J. Burnett, 2 August, 1956, ORNL-2086.

4. The Oak Ridge National Laboratory Research Reactor (ORR), a General Description, by T. E. Cole, J. P. Gill, 17 January, 1958, ORNL-2240.

5. Letter J. A. Swartout to H. M. Roth, ORR test operation, 28 February, 1958 and attachments.

6. *Dispersion of Airborne Activity From a Cold Cloud Accident to the ORR*, by U.S. Weather Bureau Office, Oak Ridge, Tenn., 9 April, 1958.

7. Report to ACRS by Division of Licensing and Regulation on Oak Ridge Reactor (ORR), 15 April, 1958.

8. Meteorological Aspects of the Oak Ridge Research Reactor, by Special Projects Section, Office of Meteorological Research, Weather Bureau (Donald H. Pack), 22 April, 1958 (presented at the ACRS meeting).

9. Comments on ORR by John F. Newell, 25 April, 1958 (presented at the ACRS meeting).

12 May, 1958

Honorable LEWIS L. STRAUSS,
Chairman,
U.S. Atomic Energy Commission,
Washington, D.C.

SUBJECT: *Operation of S3G*

Dear Mr. Strauss:

The S3G General Electric light water-cooled reactor which is to be the prototype of a submarine reactor and which is being installed at West Milton, N.Y., in sections of a submarine hull modified to serve as a containment vessel, described in KAPL-ADM-1310, (reference (1)), has been referred to the Advisory Committee on Reactor Safeguards by the Commission with request for review prior to its operation.

At its 14th meeting in September 1955, this project, then described as SAR, Mark I hull containment, to be used at West Milton, was referred to the former Advisory Committee on Reactor Safeguards for consideration of containment only. At that time the Committee recommended "that the submarine hull, located at West Milton as described, as a container be permitted for the SAR reactor with the understanding that the final reactor and hull design will be such that the maximum credible accident can be contained." Since that time S3G designs for the project have been completed and construction is well advanced.

The Committee has reviewed the design and proposed operation of S3G as described in the references listed below (2) to (7), inclusive, and as amplified in discussion at the meeting with representatives of the General Electric Co., the Naval Reactors Branch, and the Hazards Evaluation Branch. The design incorporates certain experience obtained with several other reactors of this type. Questions raised during the review seem to be operational or items for developmental check and not of a nature to affect safety.

The Committee is of the opinion that there is reasonable assurance that the S3G can be operated at this site without endangering the health and safety of the public.

The Committee considers the careful selection and training of operating personnel by the Navy and the Naval Reactors Branch and the continued surveillance of this project important and a valuable contribution to safety.

Sincerely yours,

C. ROGERS McCULLOUGH

Chairman, Advisory Committee on Reactor Safeguards

REFERENCES:

1. "SAR-1 Hazards Evaluation Summary Powerplant Hull and Site," 1 July, 1955, KAPL-ADM-1310.
2. "S3G Hazards Summary Report," by KAPL staff, 3 March, 1958, KAPL-Internal-7.
3. "Details of Radiological Hazards of S3G," by KAPL staff, 3 March, 1958, KAPL-M-SSA-12.
4. "Report to ACRS on S3G Reactor," by staff of Division of Licensing and Regulation, 23 April, 1958.
5. "Addendum" to (4) by DLR staff (presented at the ACRS meeting).
6. Comments by the Division of Biology and Medicine, AEC, on S3G Hazards Summary Report, KAPL, 12 April, 1958 (presented at the ACRS meeting).
7. "Meteorological Aspects of S3G Hazards Summary Report", by Special Project Section, Office of Meteorological Research, Weather Bureau, 15 April, 1958 (presented at ACRS meeting).

8 March, 1958

Honorable LEWIS L. STRAUSS,
Chairman,
U.S. Atomic Energy Commission,
Washington, D.C.

SUBJECT: *Argonne Low-power Reactor* (*ALPR*)

Dear Mr. Strauss:

At the request of the Division of Licensing and Regulation, the Advisory Committee on Reactor Safeguards reviewed the design of the Argonne low-power reactor planned for construction and operation at the National Reactor Testing Station.

The Committee had available and examined reports referenced below and had the benefit of discussion with the Hazards Evaluation Branch and the staff of the Argonne National Laboratory.

In view of the low-power level proposed for this reactor, the remote location at which it is to be operated, and the broad experience of the Argonne National Laboratory with boiling water reactors of this type, the Committee agrees with the conclusion of the Hazards

Evaluation Branch report of 4 March, 1958, that this reactor can be constructed, and operated at power levels up to 3 megawatts, at the National Reactor Testing Station without undue risk to the health and safety of the public.

Sincerely yours,

C. ROGERS McCULLOUGH

Chairman, Advisory Committee on Reactor Safeguards

REFERENCES:

1. ANL-5744.
2. Letter from C. R. Baum to Martin Biles, 21 February, 1958.
3. HEB Staff Report dated 3 February, 1958.
4. HEB Staff Report dated 4 March, 1958.

13 January, 1958

Honorable LEWIS L. STRAUSS,
Chairman,
U.S. Atomic Energy Commission,
Washington, D.C.

SUBJECT: *Experimental Breeder Reactor II (EBR-II)*

Dear Mr. Strauss:

The Advisory Committee on Reactor Safeguards heard presentations of the EBR-II reactor by members of the staff of the Argonne National Laboratory on 19 December, 1957, during its third meeting. It again considered this reactor on 10 January, 1958, during its fourth meeting, and heard the comments of the Hazards Evaluation staff. The EBR-II is a fast, sodium-cooled, heterogeneous power reactor of 62.5 MW (thermal) designed by the Argonne National Laboratory as a step toward the achievement of an integrated fast breeder fuel cycle. The Committee considered only the enriched uranium loading of this reactor at the present time.

The Committee agrees with the conclusion of the Hazards Evaluation staff that a reactor of this type can be operated at the proposed location in the National Reactor Testing Station without undue hazard to the public.

Sincerely yours,

C. ROGERS McCULLOUGH

Chairman, Advisory Committee on Reactor Safeguards

REFERENCES:

1. Hazard Summary Report Experimental Breeder Reactor I (EBR-II) ANL-5719, May 1957.

2. Report to ACRS on Reactor Safeguards on Experimental Breeder Reactor II (EBR-II) by Hazard Evaluation Branch, Division of Licensing and Regulation, 31 December, 1957.

3. Addendum to Report of HEB (see reference 2) 3 January, 1958.

12 July, 1958

Dr. WILLARD F. LIBBY,
Acting Chairman,
U.S. Atomic Energy Commission,
Washington, D.C.

SUBJECT: *The Wahluke Slope*

Dear Dr. Libby:

At the request of the Atomic Energy Commission, the Advisory Committee on Reactor Safeguards has reviewed the current operational safety of the Hanford production reactors, with particular regard to whether or not these reactors continue to pose an undue risk to those choosing to live on the Wahluke Slope.

Pertinent to this question are the following facts:

(1) The Hanford reactors have been in successful operation for many years without experiencing any incident that created a significant hazard on the slope.

(2) There have been over the years continuing improvements in the design and operation of these reactors which have substantially reduced the probability of serious accidents

(3) Despite these favorable developments, the Hanford reactors continue to pose potential risks to the public that are greater than those of many other reactors, including the large power reactors now under construction at other locations. The reasons for this are associated partly with the early basic design of the Hanford reactors and partly with their role in national security.

(4) Recent studies have indicated the possibility of effecting a significant additional reduction in potential hazard to the public by improvements in the airtightness of the present reactor buildings and by the provision of suitable filters that will permit better confinement of any radioactive products that may be accidentally released from the reactors.

After careful consideration of all known factors affecting the overall safety of the Hanford operation, and to the things that have done and can still be accomplished to reduce the hazard to the public, the Committee has arrived at the following conclusions:

(A) While distance from the reactors offers no certain protection against the radioactivity that may be released in a reactor accident, it does provide an important factor of safety which should always be preserved at Hanford by the permanent retention of the exclusion area known as the primary control zone.

(B) The settlement of the Wahluke slope, to the extent that it attracts settlers from distant locations, will expose increasing numbers of people to the possible consequences of a reactor accident.

(C) The risks of living on the slope, while not negligible, are significantly less than they have been in the past and are presently not much greater than those existing at more distant locations.

(D) The positive values to be expected from various degrees of building confinement as described before, while promising, have not yet been fully evaluated. Until the advantage of partial confinement in reducing the hazards to the Wahluke slope and other areas has been adequately resolved, the Committee feels that the status of the secondary zone should remain unchanged.

(E) As soon as it is firmly established that confinement, which will offer adequate protection, can be achieved and a decision is made to proceed with such installation, the secondary zone could be released at once, since population growth in the area will be slow in the first few years.

Sincerely yours,

C. ROGERS MCCULLOUGH

Chairman, Advisory Committee on Reactor Safeguards

15 December, 1958

Honorable JOHN A. MCCONE,
Chairman,
U.S. Atomic Energy Commission,
Washington, D.C.

SUBJECT: *The Wahluke Slope*

Dear Mr. McCone:

At the request of the Atomic Energy Commission the Advisory Committee on Reactor Safeguards summarizes its views with respect to the proposal to remove restrictions which have heretofore limited the development of the so-called secondary zone of the Wahluke slope as an agricultural area.

The program, which the Commission has underway, increasing the degree of confinement of fission products in case of accidents, to the Hanford reactors will substantially decrease the hazard to the occupants of the Wahluke slope. After these changes have been completed, the risk from the reactor plant to the health and safety of occupants of this area should be low enough to allow normal use of the secondary zone. Since it is expected that the population growth in this area will be slow in the first few years and the changes will be completed in this time the secondary zone may be released now.

Pertinent to this question are the following facts:

(1) The Hanford reactors have been in successful operation for many years without experiencing any incident that created a significant hazard on the slope.

(2) There have been over the years continuing improvements in the design and operation of these reactors which have substantially reduced the probability of serious accidents.

(3) Despite these favorable developments, the Hanford reactors continue to pose potential risks to the public that are greater than those of many other reactors, including the large power reactors now under construction at other locations. The reasons for this are associated partly with the early basic design of the Hanford reactors and partly with their role in national security.

(4) Recent studies have indicated the possibility of effecting a significant additional reduction in potential hazard to the public by improvements in the airtightness of the present reactor buildings and by the provision of suitable filters that will permit better confinement of any radioactive products that may be accidentally released from the reactors.

After careful consideration of all known factors affecting the overall safety of the Hanford operation, and to the things that have been done and can still be accomplished to reduce the hazard to the public, the Committee has reasoned as follows:

(A) While distance from the reactors offers no certain protection against the radioactivity that may be released in a reactor accident, it does provide an important factor of safety which should always be preserved at Hanford by the permanent retention of the exclusion area known as the primary control zone.

(B) The settlement of the Wahluke slope, to the extent that it attracts settlers from distant locations, will expose increasing numbers of people to the possible consequences of a reactor accident.

(C) The risks of living on the slope, while not negligible, are significantly less than they have been in the past and with the proposed changes in confinement will not be much greater than those existing at more distant locations.

Sincerely yours,

C. ROGERS MCCULLOUGH

Chairman, Advisory Committee on Reactor Safeguards

INDEX